Advances of Artificial Intelligence in a Green Energy Environment

Advances of Artificial Intelligence in a Green Energy Environment

Edited by

Pandian Vasant
Research Associate at MERLIN Research Centre, TDTU, Vietnam

Joshua Thomas
Associate Professor, UOW Malaysia KDU Penang University College, Malaysia

Elias Munapo
Professor, North West University, South Africa

Gerhard-Wilhelm Weber
Professor, Poznan University of Technology, Poland

Academic Press is an imprint of Elsevier
125 London Wall, London EC2Y 5AS, United Kingdom
525 B Street, Suite 1650, San Diego, CA 92101, United States
50 Hampshire Street, 5th Floor, Cambridge, MA 02139, United States
The Boulevard, Langford Lane, Kidlington, Oxford OX5 1GB, United Kingdom

Copyright © 2022 Elsevier Inc. All rights reserved.

No part of this publication may be reproduced or transmitted in any form or by any means, electronic or mechanical, including photocopying, recording, or any information storage and retrieval system, without permission in writing from the publisher. Details on how to seek permission, further information about the Publisher's permissions policies and our arrangements with organizations such as the Copyright Clearance Center and the Copyright Licensing Agency, can be found at our website: www.elsevier.com/permissions.

This book and the individual contributions contained in it are protected under copyright by the Publisher (other than as may be noted herein).

Notices
Knowledge and best practice in this field are constantly changing. As new research and experience broaden our understanding, changes in research methods, professional practices, or medical treatment may become necessary.

Practitioners and researchers must always rely on their own experience and knowledge in evaluating and using any information, methods, compounds, or experiments described herein. In using such information or methods they should be mindful of their own safety and the safety of others, including parties for whom they have a professional responsibility.

To the fullest extent of the law, neither the Publisher nor the authors, contributors, or editors, assume any liability for any injury and/or damage to persons or property as a matter of products liability, negligence or otherwise, or from any use or operation of any methods, products, instructions, or ideas contained in the material herein.

ISBN: 978-0-323-89785-3

For information on all Academic Press publications visit our website at
https://www.elsevier.com/books-and-journals

Publisher: Charlotte Cockle
Acquisitions Editor: Lisa Reading
Editorial Project Manager: Chris Hockaday
Production Project Manager: Prem Kumar Kaliamoorthi
Cover Designer: Greg Harris

Working together
to grow libraries in
developing countries

www.elsevier.com • www.bookaid.org

Typeset by TNQ Technologies

Contents

Contributors	xiii
About the editors	xvii
Preface	xxi
Acknowledgments	xxvii

1. Application of some ways to intensify the process of anaerobic bioconversion of organic matter

Andrey A. Kovalev, Dmitriy A. Kovalev, Victor S. Grigoriev and Alexander Makarov

1.1	**Introduction**	1
1.2	**Methods for improving the process of methane digestion of manure and manure runoff and their analysis**	3
1.3	**Application results**	11
	1.3.1 Temperature regime	11
	1.3.2 Separation of the processing process into two or more stages	14
	1.3.3 Application of biological and physicochemical methods for pretreatment of organic waste	15
	1.3.4 Combining the processes of biological and thermo-chemical gasification of organic waste and intermediate products of their processing using the advantages of each process	17
	1.3.5 Carrying out the process under various influences on the processed substrate	19
	1.3.6 Application of heat energy recovery directly and using thermal transformers	22
	1.3.7 Complex application of biogas plant and additional sources of thermal and electric energy based on other RES	25
1.4	**Energy model**	25
1.5	**Conclusion**	27
	References	28

vi Contents

2. **Disasters impact assessment based on socioeconomic approach**
Igor Grebennik, Yevhen Hubarenko, Maryna Hubarenko and Sergiy Shekhovtsov

2.1	Introduction	35
2.2	Statistics and tendencies of natural and man-made disasters	37
2.3	Overview of institutions dealing with risk management	43
2.4	Promising approaches to decrease risks and losses caused by disasters	44
2.5	The issues of effective disaster countering organization	46
2.6	Socioeconomic approach to the estimations of risk dynamics	48
2.7	Conclusions	50
	Appendix A: An example of selecting the optimal list of prevention or mitigation measures	51
	References	55

3. **Uninterruptible power supply system of the consumer, reducing peak network loads**
V.A. Gusarov, L. Yu Yuferev and O.F. Gusarova

3.1	Introduction	57
3.2	Methodology of the work	58
3.3	Work results	62
3.4	Conclusions	63
	References	64

4. **Optimization of the anaerobic conversion of green biomass into volatile fatty acids for further production of high-calorie liquid fuel**
Marina A. Gladchenko and Sergey N. Gaydamaka

4.1	Introduction		67
4.2	Experimental part and methods		69
	4.2.1	Biocatalyst	69
	4.2.2	Substrate	71
	4.2.3	Methods	71
4.3	Results and discussion		72
	4.3.1	Specific acid-producing activity of the biocatalyst	72
	4.3.2	Biomass pretreatment	73
	4.3.3	Obtaining VFAs and ethyl alcohol during the conversion of pretreated green biomass with a biocatalyst in submesophilic and mesophilic modes	73
	4.3.4	Increase in the yields of volatile fatty acids and ethanol alcohol during the transformation of pretreated green mass with a biological catalyst modified by bacteria *Clostridium acetobutylicum* in submesophilic and mesophilic modes	74

Contents **vii**

4.3.5	Increase in the yields of volatile fatty acids and ethyl alcohol with adding a cosubstrate of glycerin to pretreated green biomass during fermentation with the selected biocatalyst in submesophilic and mesophilic modes
	76
4.3.6	The efficiency of conversion of substrates into volatile fatty acids and ethanol, depending on the temperature and time of their contact with the biocatalyst
	79

4.4 Conclusions   79
Abbreviations   81
References   81

5. Life cycle cost and life cycle assessment: an approximation to understand the real impacts of the Electricity Supply Industry

Joaquina Niembro-García, Patricia Alfaro-Martínez and Jose Antonio Marmolejo-Saucedo

5.1	Importance of Electricity Supply Industry	83
5.2	The economics of Electricity Supply Industry	87
5.3	The life cycle of Electricity Supply Industry	94
5.4	Importance of life cycle assessment of Electricity Supply Industry	98
5.5	Life cycle cost of Electricity Supply Industry	102
5.6	Mobilizing industry for a clean and circular economy	105
References		108

6. Comparison of open access multi-objective optimization software tools for standalone hybrid renewable energy systems

Mahesh Wagh, Purshottam Acharya and Vivek Kulkarni

6.1	Introduction		111
6.2	Literature review		112
6.3	Open access multi-objective optimization tools		113
	6.3.1	HOMER	113
	6.3.2	iHOGA	114
	6.3.3	Load dispatch strategies	115
	6.3.4	HYBRID-2	116
	6.3.5	Comparison of multiobjective optimization algorithm	119
6.4	Case study		119
	6.4.1	Input of load	119
	6.4.2	Solar radiation data input	122
	6.4.3	Wind speed data input	122
	6.4.4	Selecting other components data	124
	6.4.5	Selection of battery and inverter	124

viii Contents

	6.4.6 Calculation time	125
6.5	Conclusion	127
References		127

7. Optimization of the organic waste anaerobic digestion in biogas plants through the use of a vortex layer apparatus

Andrey A. Kovalev, Dmitriy A. Kovalev, Yuriy V. Litti, Inna V. Katraeva and Victor S. Grigoriev

7.1	Introduction	129
7.2	Anaerobic processing of organic waste: general characteristics of fermentation	130
	7.2.1 General characteristics of the methanogenesis	130
	7.2.2 General characteristics of the hydrolysis process	131
	7.2.3 The impact of external factors on the intensity of the gas formation process	133
7.3	Methods of preprocessing	135
7.4	Vortex layer apparatus	138
7.5	Application of the vortex layer apparatus in biogas plants	145
	7.5.1 Pretreatment of the substrate in the vortex layer apparatus together with biomass recycling	147
7.6	Conclusion	148
References		149

8. Search of regularities in data: optimality, validity, and interpretability

O.V. Senko, A.V. Kuznetsova, I.A. Matveev and I.S. Litvinchev

8.1	Introduction	151
8.2	Occam's razor principle for verification of parametric regression models	153
	8.2.1 Choice between complex and simple models	153
	8.2.2 Permutation test technique	154
	8.2.3 Choice of simple model	157
	8.2.4 Using Occam's razor principle to evaluate significance of piecewise linear models	159
8.3	Occam's razor principle for verification of regression models based on optimal partitioning	163
	8.3.1 Optimal partitioning	163
	8.3.2 Statistical significance	165
	8.3.3 Multiple testing	168
8.4	Conclusion	169
Acknowledgment		170
References		170

Contents **ix**

9. Artificial intelligence techniques for modeling of wind energy harvesting systems: a comparative analysis

Tigilu Mitiku Dinku and Mukhdeep Singh Manshahia

9.1	Introduction	173
9.2	Review of related works	173
9.3	Modeling of wind energy harvesting system	174
	9.3.1 Turbine model	174
	9.3.2 Modeling of PMSG	176
9.4	Maximum power point tracking system	177
9.5	Load side converter control	181
9.6	Results and discussion	182
	9.6.1 Case 1: step change in wind speed and fixed load	182
	9.6.2 Case 2: continuous change in wind speed	185
9.7	Conclusion	190
	Acknowledgments	190
	References	190
	Further reading	192

10. Human paradigm and reliability for aggregate production planning under uncertainty

Selma Gütmen, Gerhard-Wilhelm Weber, Alireza Goli and Erfan Babaee Tirkolaee

10.1	Introduction	193
	10.1.1 Human paradigm in APP	194
	10.1.2 Reliability in APP	196
	10.1.3 Uncertainty in APP	197
10.2	Literature review	197
10.3	Discussion and conclusion	199
	References	200

11. Artificial intelligence—based intelligent geospatial analysis in disaster management

R. Subhashini, J. Joshua Thomas, A. Sivasangari, P. Mohana, S. Vigneshwari and P. Asha

11.1	Introduction	203
11.2	Related work	204
11.3	Proposed work	208
	11.3.1 Preparation of various thematic maps	210
	11.3.2 Thematic maps	210
	11.3.3 Land use and land cover	210
	11.3.4 Digital elevation map	212
	11.3.5 Slope	212
	11.3.6 Slope gradient	212
	11.3.7 Lineaments	213

x Contents

		11.3.8	Drainage	214
		11.3.9	Landslide susceptibility zone map preparation	214
		11.3.10	Convolution neural networks	216
	11.4	**Performance analysis**		217
	11.5	**Conclusion**		220
	References			220

12. Optimizing the daily use of limited solar panels in closely located rural schools in Zimbabwe

Elias Munapo

12.1	**Introduction**	223
12.2	**Modeling the solar panel problem**	224
	12.2.1 The TSP model	224
12.3	**TSP network features**	226
	12.3.1 Network feature 1	226
	12.3.2 Network feature 2	226
	12.3.3 TSP network feature 3	227
	12.3.4 TSP network feature 4	227
	12.3.5 TSP network feature 5	230
12.4	**Dummies and their use in elimination of subtours**	232
	12.4.1 Dummy schools	232
	12.4.2 Dummy point	232
	12.4.3 Identifying dummy points	232
	12.4.4 Subtour eliminators	234
12.5	**Proposed algorithm for TSP**	234
	12.5.1 Proposed algorithm	234
	12.5.2 Numerical illustration	235
12.6	**Other applications of the traveling salesman**	240
	12.6.1 Wiring problem	240
	12.6.2 Hospital layout	241
	12.6.3 Dartboard design	241
	12.6.4 Designing a typewriter keyboard	241
	12.6.5 Production	241
	12.6.6 Scheduling	241
12.7	**Conclusions**	242
References		242
Further reading		243

13. Review on recent implementations of multiobjective and multilevel optimization in sustainable energy economics

Timothy Ganesan, Igor Litvinchev, Jose Antonio Marmolejo-Saucedo, J. Joshua Thomas and Pandian Vasant

13.1	**Introduction**	245
13.2	**Economic load/emission dispatch**	251
13.3	**Bioenergy and biofuel supply chains**	257
13.4	**Sustainable capacity planning and optimization**	263

Contents **xi**

13.5	Outlook	267
References		269

14. Hybrid optimization and artificial intelligence applied to energy systems: a review

Gilberto Pérez Lechuga, Karla N. Madrid Fernández and Ugo Fiore

14.1	Introduction		279
14.2	Stochastic programming		281
	14.2.1	The general model	282
	14.2.2	Software for stochastic programming instances	283
14.3	Optimization in energy systems		283
	14.3.1	Energy system models and their optimization processes	284
	14.3.2	A classification according to the applications	284
14.4	Conclusions		293
Abbreviations			293
Acknowledgments			294
References			294
Further reading			298

15. A brief literature review of quantitative models for sustainable supply chain management

Pablo Flores-Sigüenza, Jose Antonio Marmolejo-Saucedo and Roman Rodríguez-Aguilar

15.1	Introduction		301
15.2	Theoretical foundation and literature reviews		303
	15.2.1	Supply chain management	303
	15.2.2	Sustainable supply chain management	303
	15.2.3	Literature reviews	304
15.3	Methodology		305
	15.3.1	Material collection	305
	15.3.2	Descriptive analysis	306
	15.3.3	Category identification	306
	15.3.4	Material evaluation	306
15.4	Results		307
	15.4.1	Material collection	307
	15.4.2	Descriptive analysis	308
	15.4.3	SCM dimension	311
	15.4.4	Modeling dimension	311
	15.4.5	Sustainability dimension	314
	15.4.6	Research gaps and future research perspectives	320
15.5	Discussion		321
15.6	Conclusion		322
References			322

xii Contents

16. Optimized designing spherical void structures in 3D domains

Tatiana Romanova, Georgiy Yaskov, Igor Litvinchev, Igor Yanchevskyi, Yurii Stoian and Pandian Vasant

16.1 Introduction	331
16.2 Problem formulation	332
16.3 Mathematical model	333
16.4 Mathematical model with balancing conditions	334
16.5 Numerical experiments	336
16.6 Conclusions	343
Acknowledgments	344
References	344

17. Swarm-based intelligent strategies for charging plug-in hybrid electric vehicles

Pandian Vasant, Anirban Banik, J. Joshua Thomas, Jose Antonio Marmolejo-Saucedo, Timothy Ganesan, Elias Munapo and Mukhdeep Singh Manshahia

17.1 Introductions	347
17.1.1 Study objectives	348
17.2 Problem formulation	348
17.3 Swarm-based intelligence approaches	349
17.3.1 Particle swarm optimization	349
17.3.2 Accelerated particle swarm optimization	350
17.3.3 Gravitational search algorithm	352
17.3.4 Hybrid PSOGSA algorithm	354
17.4 Results and discussions	356
17.4.1 Particle optimization swarm findings	356
17.4.2 Accelerated PSO findings	360
17.4.3 Gravitational search algorithm findings	362
17.4.4 Hybrid PSO and GSA (PSOGSA) findings	364
17.4.5 Comparative analysis	366
17.5 Conclusions	372
17.5.1 Future research direction	372
References	372

Index	375

Contributors

Purshottam Acharya, Department of Technology, Shivaji University, Kolhapur, Maharashtra, India

Patricia Alfaro-Martínez, Universidad Panamericana, Facultad de Gobierno y Economía, Ciudad de México, México

P. Asha, School of Computing, Sathyabama Institute of Science and Technology, Chennai, Tamil Nadu, India

Anirban Banik, National Institute of Technology Agartala, India

Tigilu Mitiku Dinku, Department of Mathematics, Bule Hora University, Bule Hora, Ethiopia

Ugo Fiore, Parthenope University, Naples, Italy

Pablo Flores-Sigüenza, Facultad de Ingeniería, Universidad Anáhuac México, Naucalpan de Juárez, Estado de México, México

Timothy Ganesan, University of Calgary, Alberta, Canada; Member of American Mathematical Society, AB, Canada

Sergey N. Gaydamaka, Department of Chemical Enzymology, Chemistry Department, Lomonosov Moscow State University, Moscow, Russia

Marina A. Gladchenko, Department of Chemical Enzymology, Chemistry Department, Lomonosov Moscow State University, Moscow, Russia

Alireza Goli, Department of Industrial Engineering and Future Studies, Faculty of Engineering, University of Isfahan, Isfahan, Iran

Igor Grebennik, Kharkiv National University of Radio Electronics, Kharkiv, Ukraine

Victor S. Grigoriev, Federal Scientific Agroengineering Center VIM, Moscow, Russia

Selma Gütmen, Faculty of Engineering Management, Poznan University of Technology, Poznań, Poland

V.A. Gusarov, FGBNU FNATS VIM, Moscow, Russian Federation

O.F. Gusarova, FAEIHE "RUT" Moscow, Russian Federation

Yevhen Hubarenko, Kharkiv National University of Radio Electronics, Kharkiv, Ukraine

Maryna Hubarenko, Kharkiv National University of Radio Electronics, Kharkiv, Ukraine

Inna V. Katraeva, Nizhny Novgorod State University of Architecture and Civil Engineering, Nizhny Novgorod, Russia

xiv Contributors

Andrey A. Kovalev, Federal Scientific Agroengineering Center VIM, Moscow, Russia

Dmitriy A. Kovalev, Federal Scientific Agroengineering Center VIM, Moscow, Russia

Vivek Kulkarni, Sanjay Ghodawat Group of Institutions, Atigre, Kolhapur, Maharashtra, India

A.V. Kuznetsova, Institute of Biochemical Physics, Moscow, Russia

Gilberto Pérez Lechuga, Instituto de Ciencias Basicas e Ingenieria-AAIA, Universidad Autónoma del Estado de Hidalgo, Pachuca, Hidalgo, México

Yuriy V. Litti, Federal Research Centre "Fundamentals of Biotechnology" of the Russian Academy of Sciences, Moscow, Russia

Igor Litvinchev, Neuvo Leon State University (UANL), Neuvo Leon, Mexico; Faculty of Mechanical and Electrical Engineering (FIME), Nuevo Leon State University, Monterrey, Nuevo Leon, Mexico

I.S. Litvinchev, Federal Research Center "Computer Science and Control" of Russian Academy of Science, Moscow, Russia

Karla N. Madrid Fernández, Instituto de Ciencias Basicas e Ingenieria-AAIA, Universidad Autónoma del Estado de Hidalgo, Pachuca, Hidalgo, México

Alexander Makarov, Federal Scientific Agroengineering Center VIM, Moscow, Russia

Mukhdeep Singh Manshahia, Department of Mathematics, Punjabi University, Patiala, Punjab, India

Jose Antonio Marmolejo-Saucedo, Universidad Panamericana, Facultad de Ingeniería, Mexico City, Mexico

I.A. Matveev, Federal Research Center "Computer Science and Control" of Russian Academy of Science, Moscow, Russia

P. Mohana, Centre for Remote Sensing and Geo-informatics, Sathyabama Institute of Science and Technology, Chennai, Tamil Nadu, India

Elias Munapo, Business Statistics and Operations Research Department, Economic Sciences, FEMS, NWU, Mafikeng Campus; North West University, Mahikeng, South Africa

Joaquina Niembro-García, Universidad Panamericana, Facultad de Ingeniería, Mexico City, Mexico

Roman Rodríguez-Aguilar, Facultad de Ciencias Económicas y Empresariales, Universidad Panamericana, Ciudad de México, México

Tatiana Romanova, Department of Mathematical Modelling and Optimal Design, Institute for Mechanical Engineering Problems of the National Academy of Sciences of Ukraine (IPMach NASA), Kharkiv, Ukraine; Kharkiv National University of Radioelectronics, Kharkiv, Ukraine

O.V. Senko, Federal Research Center "Computer Science and Control" of Russian Academy of Science, Moscow, Russia

Sergiy Shekhovtsov, Kharkiv National University of Radio Electronics, Kharkiv, Ukraine; Kharkiv National University of Internal Affairs, Kharkiv, Ukraine

Contributors **xv**

A. Sivasangari, School of Computing, Sathyabama Institute of Science and Technology, Chennai, Tamil Nadu, India

Yurii Stoian, Department of Mathematical Modelling and Optimal Design, Institute for Mechanical Engineering Problems of the National Academy of Sciences of Ukraine (IPMach NASA), Kharkiv, Ukraine

R. Subhashini, School of Computing, Sathyabama Institute of Science and Technology, Chennai, Tamil Nadu, India

J. Joshua Thomas, UOW Malaysia KDU Penang University College, George Town, Pulau Pinang, Malaysia

Erfan Babaee Tirkolaee, Department of Industrial Engineering, Istinye University, Istanbul, Turkey

Pandian Vasant, MERLIN Research Centre, TDTU, Ho Chi Minh City, Vietnam; Faculty of Science and Information Technology, Universiti Teknologi Petronas, Teronoh, Seri Iskandar, Malaysia

S. Vigneshwari, School of Computing, Sathyabama Institute of Science and Technology, Chennai, Tamil Nadu, India

Mahesh Wagh, Department of Technology, Shivaji University, Kolhapur, Maharashtra, India

Gerhard-Wilhelm Weber, Faculty of Engineering Management, Poznan University of Technology, Poznań, Poland; Institute of Applied Mathematics, Middle East Technical University, Ankara, Turkey

Igor Yanchevskyi, National Technical University of Ukraine "Igor Sikorsky Kyiv Polytechnic Institute", Department of Dynamics and Strength of Machines and Strength of Materials, Kiev, Ukraine

Georgiy Yaskov, Department of Mathematical Modelling and Optimal Design, Institute for Mechanical Engineering Problems of the National Academy of Sciences of Ukraine (IPMach NASA), Kharkiv, Ukraine

L. Yu Yuferev, FGBNU FNATS VIM, Moscow, Russian Federation

About the editors

Pandian Vasant is an Editor-in-Chief of International Journal of Energy Optimization and Engineering (IJEOE) and a Research Associate at MERLIN Research Centre of Ton Duc Thang University. He holds PhD in Computational Intelligence (UNEM, Costa Rica), MSc (University Malaysia Sabah, Malaysia, Engineering Mathematics) and BSc (Hons, Second Class Upper) in Mathematics (University of Malaya, Malaysia). His research interests include soft computing, hybrid optimization, innovative computing, and applications. He has co-authored research articles in journals, conference proceedings, presentation, special issues guest editor, book chapters (300 publications indexed in Google Scholar), and General Chair of the EAI International Conference on Computer Science and Engineering in Penang, Malaysia (2016), and Bangkok, Thailand (2018). In the year 2009 and 2015, Dr. Pandian Vasant was awarded top reviewer and outstanding reviewer for the journal *Applied Soft Computing* (Elsevier), respectively. He has 31 years of working experiences at various universities. Currently, he is an Editor-in-Chief of *International Journal of Energy Optimization and Engineering*, a Member of AMS (USA), NAVY Research Group (TUO, Czech Republic), MERLIN Research Centre (TDTU, Vietnam), and a General Chair of the International Conference on Intelligent Computing and Optimization (https://www.icico. info/). H-Index Google Scholar = 34; i-10-index = 144.

http://www.igi-global.com/ijeoe

E-mail: pvasant@gmail.com

J. Joshua Thomas is an Associate Professor in Computer Science at UOW Malaysia KDU Penang University College. He obtained his PhD (Intelligent Systems Techniques) in 2015 from University Sains Malaysia, Penang, and master's degree in 1999 from Madurai Kamaraj University, India. He served as the head and deputy head of department computing between the years 2012 and 2017. From July to September 2005, he worked as a research assistant at the Artificial Intelligence Lab in University Sains Malaysia. From March 2008 to March 2010, he worked as a research associate at the same University. His work involves intelligent systems techniques in which he adopts computational algorithm implementation in interdiscipline field areas. His expertise is evident in working with international collaborators in publications, visiting

xviii About the editors

research fellow to share knowledge. He works in deep learning specially targeting on graph convolutional neural networks and graph recurrent neural networks, hypergraph attentions, end-to-end steering learning systems, design algorithms in drug discovery, and quantum machine learning. He is a principal investigator and a co-investigator in various grants funding, including internal, national, and international levels. He is an editorial board member for the *International Journal of Energy Optimization and Engineering*, a book author, and a guest editor for Applied Sciences, Computations, Mathematics Biosciences and Engineering, and Computer Modeling in Engineering and Sciences. He has authored and edited books. He has published more than 50 papers in leading international conference proceedings and peer-reviewed journals. He delivered Invited, Plenary, Keynote speaker and workshop presenter in IAIM2019, LCQAI2021, ICRITCC′21, and IAIM2022. He is a "Visiting Research Fellow" to Sathyabama Institute of Science and Technology.

Weblink: https://www.uowmkdu.edu.my/research/our-people/dr-joshua-thomas/

Email: joshua.j.thomas@gmail.com

Elias Munapo has a PhD obtained in 2010 from the National University of Science and Technology (Zimbabwe) and is a Professor of Operations Research at the North West University, Mafikeng Campus in South Africa. He is a Guest Editor of the *Applied Sciences* journal and has co-published two books. The first book is titled *Some Innovations in OR Methodology*: *Linear Optimization* and was published by Lambert Academic publishers in 2018. The second book is titled *Linear Integer Programming*: *Theory, Applications, and Recent Developments* and was published by De Gruyter publishers in 2021. Professor Munapo has co-edited a number of books, is currently a reviewer of a number of journals, and has published over 100 journal articles and book chapters. In addition, Professor Munapo is a recipient of the North West University Institutional Research Excellence award and is a member of the Operations Research Society of South Africa (ORSSA), EURO, and IFORS. He has presented at both local and international conferences and has supervised more than 10 doctoral students to completion. His research interests are in the broad area of operations research.

Weblink: http://commerce-nwu-ac-za.web.nwu.ac.za/business-statistics-and-operations-research/elias-munapo

Email: Elias.Munapo@nwu.ac.za

Gerhard-Wilhelm Weber is a Professor at Poznan University of Technology, Poznan, Poland, at Faculty of Engineering Management. His research is on OR, financial mathematics, optimization and control, neuroscience and bioscience, data mining, education, and development; he is involved in the organization of scientific life internationally. He received his Diploma and Doctorate in Mathematics and Economics/Business Administration at RWTH Aachen and his Habilitation at TU Darmstadt. He held Professorships by proxy at University of Cologne and TU Chemnitz, Germany. At IAM, METU, Ankara, Turkey, he was a Professor in the programs of Financial Mathematics and Scientific Computing and an Assistant to the Director, and he has been a member of further graduate schools, institutes, and departments of METU. Furthermore, he has affiliations at the universities of Siegen, Ballarat, Aveiro, North Sumatra, and Malaysia University of Technology; he is an "Advisor to EURO Conferences" and an "IFORS Fellow."

Weblink: https://www.researchgate.net/profile/Gerhard_Wilhelm_Weber

Email: gerhard-wilhelm.weber@put.poznan.pl

Preface **xxiii**

In Chapter 3 on *"Uninterruptible Power Supply System of the Consumer, Reducing Peak Network Loads,"* *V.A. Gusarov, L.Yu. Yuferev,* and *O.F. Gusarova* study on centralized power supply for large cities and small settlements as a complex engineering structure. These structures have a single structural diagram of electrical equipment. There are two major cost indicators: the one of capital costs and the one of electricity generated. The authors describe the developed system of power equipment which allows reducing, or avoiding, peak loads on the generating equipment of the networks. Efficiency is achieved through uninterruptible power supply units.

In Chapter 4 on *"Optimization of the Anaerobic Conversion of Green Biomass into Volatile Fatty Acids for Further Production of High-Calorie Liquid Fuel,"* *M.A. Gladchenko* and *S.N. Gaydamaka* study the possibility of increasing the concentration of soluble organic substances and reducing sugars due to the pretreatment of green biomass. Enhancement of the acidogenic biocatalyst by *Clostridium acetobutylicum* made it possible to increase the yield of target products. In the process of straw conversion, the maximum yield among the studied substrates was obtained. Glycerin increases the yield of butanoic and ethanic acids during the conversion of pretreated biomass.

In Chapter 5 on *"Life-Cycle Cost and Life Cycle Assessment—An Approximation to Understand the Real Impacts of the Electricity Supply Industry,"* *J. Niembro-García, P. Alfaro,* and *J.A. Marmolejo-Saucedo* state that the real cost that consumers should pay for electrical energy could be estimated using a life cycle approach (LCT). Life cycle cost (LCC) is a new thought form on which tools such as LCA and LCC are based. It considers jointly the analysis of all the costs and the environmental repercussions. The costs of a product throughout its life cycle can easily be made visible. The authors use different approaches to calculate supply and demand and show perspectives.

In Chapter 6 on *"Comparison of Open Access Multi-Objective Optimization Software Tools for Standalone Hybrid Renewable Energy Systems,"* *M. Wagh, P. Acharya,* and *V. Kulkarni* state that renewable energy systems, used along with some conventional resources or with other renewable energy sources, give maximum benefit, reliability, and lower cost than a single renewable energy system. Optimization is required for lowering the two main parameters which are the total levelized cost of energy and the net present value of system. A comparison of open access software tools has been carried out.

In Chapter 7 on *"Optimization of Biogas Plant Operation Through the Use of Vortex Layer Apparatus,"* *A.A. Kovalev, D.A. Kovalev, Y.V. Litti, I.V. Katraeva,* and *V.S. Grigoriev* optimize the processing of liquid organic waste of production and consumption by anaerobic bioconversion in biogas plants through the use of an electromagnetic mill. Main characteristics of the process, methods of intensifying, and the operation of an electromagnetic mill are given. The authors describe methods for increasing the efficiency of anaerobic conversion of organic waste due to additional processing of waste in a vortex layer apparatus.

xxiv Preface

In Chapter 8 on *"Search of Regularities in Data: Optimality, Validity, Interpretability,"* *O.V. Senko, A.V. Kuznetsova, I.A. Matveev,* and *I.S. Litvonchev* discuss a method for verifying regression models based on the Occam's razor principle. A variant of the principle is used to make choice between simple and complex regression models. Randomized permutation test is used to test null hypothesis that the simple model is sufficient to describe a relationship in the data. The authors discuss successful applications of the Occam's razor principle for the verification of piecewise linear regression models and models based on optimal partitions.

In Chapter 9 on *"Artificial Intelligence Techniques for Modeling of Wind Energy Harvesting Systems: A Comparative Analysis,"* *T.M. Dinku* and *M. Singh Manshahia* present a comparative analysis of different maximum power point tracking (MPPT) control techniques such as PI, FLC, MLFFNN, RBFNN, and ANFIS for medium-sized variable speed wind energy harvesting system. MPPT algorithms are analyzed and compared. Comparative analysis shows that ANFIS controller gives better efficiency especially in comparison with PI and FLC under variable wind speed.

In Chapter 10 on *"Human Paradigm and Reliability for Aggregate Production Planning Under Uncertainty,"* *S. Gütmen, G.W. Weber, A. Goli,* and *E. Babaee Tirkolaee* recall that aggregate production planning (APP) within a supply chain is one of the main activities in general planning of large and leading companies. They survey the importance of three main factors: (1) human paradigm, (2) reliability, and (3) uncertainty in the APP. To do so, these factors are investigated and the most significant challenges are discussed accordingly. Finally, the survey is concluded through discussing the main challenges, limitations, and recommendations for future research.

In Chapter 11 on *"Artificial Intelligence based Intelligent Analysis in Disaster Management,"* *R. Subhashini, J.J. Thomas, A. Sivasangari, P. Mohana, S. Vigneshwari,* and *P. Asha* use recent advanced AI and geographical information system (GIS) to demarcate the landslide susceptibility zones. Hilly zones of Sirumalai are taken as a study area. IRS P6 Satellite imagery is employed to prepare the layers. Integrating with the advanced AI, the multilayer feed forwarded CNN algorithm has been applied. Performance measures are evaluated for CNN and compared with RNN networks. The proposed classification outperforms in CNN architecture and provides better results.

In Chapter 12 on *"Optimizing the Daily Use of Limited Solar Panels in Closely Located Rural Schools in Zimbabwe,"* *Elias Munapo* presents and optimizes the daily use of limited solar panels in closely located rural schools in Zimbabwe, most of them with no electricity. Some donor agencies have donated solar panels. Solar panels are not enough for all the rural schools. Electrical power is necessary for computers and the science experiments. It is very necessary to move the limited solar panels from one school to the next school so that all rural schools can use the solar panels and the total distance traveled is minimal. This challenge is solved as a traveling salesman problem.

In Chapter 13 on "*Review on Recent Implementations of Multi-objective* and *Multi-level Optimization in Sustainable Energy Economics*," *T. Ganesan, I. Litvinchev, J.A. Marmolejo-Saucedo, J. Thomas, and P. Vasant* state that rapid progress in the sustainable energy industry concentrates on focus areas in the global energy ecosystem. This has generated avenues for the implementation of mathematical optimization and OR methodologies. The authors provide a concise review on recent implementations of MO and multilevel optimization on sustainable energy economic systems, featuring economic load/emission dispatch, bioenergy supply chains, and sustainable capacity planning.

In Chapter 14 on "*Hybrid Optimization and Artificial Intelligence Applied to Energy Systems: A Review*," *G. Pérez Lechuga, K.N. Madrid Fernández, and U. Fiore* explore and group into representative clusters the main application areas related to energy systems and the quantitative approaches used for them in design, creation, and scaling of emerging prototypes. They present some main methods and results obtained in the art of modeling and optimization of energy systems with emphasis on hybrid optimization and AI. They discuss differences between deterministic and stochastic AI and survey new benefits in nonpolluting energy systems.

In Chapter 15 on "*A Brief Literature Review of Quantitative Models for Sustainable Supply Chain Management*," *P. Flores-Sigüenza, Naucalpan de Juárez, J.A. Marmolejo-Saucedo, and R. Rodríguez-Aguilar* review 80 articles in peer-reviewed journals and apply a 4Ws analysis. Three dimensions are categorized: supply chain management, modeling, and sustainability. Among the results they evidenced a continuous growth in production of these articles, with a predominance of deterministic mathematical programming models from an environmental economic perspective, while identifying research gaps, highlighting the lack of life cycle analysis in the design of supply chain networks.

In Chapter 16 on "*Optimized Designing Spherical Void Structures in 3D Domains*," *T. Romanova, G. Yaskov, I. Litvinchev, I. Yanchevsky, Y. Stoian, and P. Vasant* optimize layout of variable sized spheres in a disconnected polyhedral domain, motivated by manufacturing. The spheres must be arranged in the container. The objective is to find coordinates of the centers and radii of the spheres maximizing the total volume of the spheres for two cases: with and without balancing conditions. Two nonlinear programming models are provided, and corresponding nonlinear optimization problems are formulated and solved. Numerical results are presented to illustrate the main constructions.

In Chapter 17 on "*Application of Natured-Inspired Optimization in Reservoir Storage Prediction: A Case Study*," *P. Vasant, A. Banik, J. J. Thomas, J.A. Marmolejo-Saucedo, T. Ganesan, E. Munapo, and M. Singh Manshahia* establish a control system for filling and promoting the penetration of upcoming plug-in hybrid power vehicles (PHEVs) on highways. Intelligent energy management needs statistical models via optimization strategy with computer intelligence. State-of-charge of PHEVs is optimized using particle swarm optimization (PSO), gravitational search algorithm (GSA), accelerated particle

xxvi Preface

swarm optimization, and a combination of PSO and GSA. Performance of the four techniques is defined by convergence speed, computation time, and best fitness.

As *editors*, we hope that the picked fields and chosen subjects reflect a core sample of international research facing emerging, challenging, complex, and even long-enduring problems of our *Green Energy Environment* and their fields in economics and finance, natural sciences and engineering, healthcare and medicine, city planning, and land use through the results and scientific tools of *AI* and *OR*. We are very grateful to the publishing house of *Elsevier* for the honor of hosting this first edition as a front running enterprise in intelligence and operations, science, and implementation. We extend particular thanks to the Directors, Editors, and Managers of *Elsevier* as well as to its Editorial Management and Team Members, for their continuous concern, interest and encouragement, advice, help, and guidance in every regard. We thank our dear authors for their hard work and readiness to share their youngest insights, findings, and results with our global community. Last but not least, we express our thankfulness for their diligence and wise recommendations to the anonymous reviewers of the chapters offered. Now we sincerely hope that our authors' research as gathered and arranged by us will be an inspiration of cooperation, advances, and friendship on a global stage and of a premium rank.

Acknowledgments

The editors would like to acknowledge the contributions of everyone involved in the development of this book and, in particular, would like to thank all authors for their contributions. Our sincere thanks go to the chapter authors who contributed their time and special thanks go to the people who, in addition to the preparations, actively participated in the evaluation process.

Our thanks go to the members of the editorial board who have done so much in making the book high quality. Especially, Professor Gerhard Wilhelm Weber, Professor Pandian Vasant, Professor Ugo Fiore for his excellent progress and his valuable help.

The editors are confident that readers will find this book useful because, among other things, it gives them an opportunity to learn about the results of the published research. It would be an indication of the enormous role Wiley has played in publishing this information.

Editors would like to thank all the editorial staff at Elsevier for helping the editors and authors prepare the manuscript of this book in a professional and extremely kind manner, and they went a long way to make an exceptional quality edition of the book.

I would like to thank our Lord and Savior Jesus Christ who gave life and strength to all who worked throughout the book process to achieve the book during the time of pandemic.

Pandian Vasant
MERLIN, Ton Duc Thang University, Ho Chi Minh City, Vietnam

J. Joshua Thomas
UOW Malaysia KDU Penang University College, Malaysia

Elias Munapo
North West University, Mahikeng Campus, South Africa

Gerhard Wilhelm Weber
Poznań University of Technology, Poland

Chapter 1

Application of some ways to intensify the process of anaerobic bioconversion of organic matter

Andrey A. Kovalev, Dmitriy A. Kovalev, Victor S. Grigoriev and Alexander Makarov
Federal Scientific Agroengineering Center VIM, Moscow, Russia

1.1 Introduction

The rise in prices for energy resources, emergence, and aggravation of environmental problems has led to a significant interest in the use of bioconversion technology for organic waste for energy production [1]. The negative impact of human activity on the environment is associated not only with the increasing consumption of natural resources but also, to a greater extent, with the formation of liquid and solid waste from agricultural and processing industries [2,3]. The fact that animals poorly assimilate the energy of plant feed and that more than half of this energy is used unproductively—goes into manure—allows us to consider the latter not only as a valuable raw material for organic fertilizers but also as a powerful renewable energy source.

After the Second World War, due to the energy crisis, in many European countries, to cover the lack of liquid fuel, they paid serious attention to the possibility of obtaining biogas from animal waste, in particular from farm animal manure. The operation of several dozen installations built at that time confirmed the possibility of processing also the excrement of farm animals using methanogenesis. The biogas produced was mainly compressed and used to drive tractors. However, in competition with cheap traditional fuels, the inefficient and difficult production of biogas and its utilization turned out to be economically disadvantageous.

In recent years, the situation in agriculture with energy raw materials has changed fundamentally. The acute energy deficit, accompanied by rapidly growing oil prices, as a permanent factor in the global economy, leads to the

---

Advances of Artificial Intelligence in a Green Energy Environment
https://doi.org/10.1016/B978-0-323-89785-3.00002-5
Copyright © 2022 Elsevier Inc. All rights reserved.

accelerated implementation of research programs aimed at discovering and practical use of additional local fuel resources. Under such circumstances, the problematic process of processing animal excrement into biogas again comes to the fore. It should be emphasized, however, that the reasons leading to the renewed interest in anaerobic fermentation go beyond the purely energetic rationale. The transition of the agro-industrial complex to an industrial basis leads to a sharp increase in the waste of agro-industrial enterprises, which must be disposed of without environmental pollution.

One of the methods for the rational use of organic waste from the agro-industrial complex is their methane fermentation, which turned out to be a good means of neutralizing liquid manure and preserving it as fertilizer while simultaneously obtaining a local source of energy—biogas. The experience of practical verification of methanogenesis in the field of agricultural use shows that in the hierarchy of the contribution of this method, its ecological characteristics occupy the first place, then the aspect of the production of organic fertilizers follows, and only then does the energy aspect follow. The interest in obtaining only biogas was replaced by the understanding of the significance of this process for the environment, as an energy-saving process of manure treatment and sewage treatment. In Russia, anaerobic treatment of manure and slurry is used on a limited scale, the scope of which is determined by several pilot plants. Little operating experience and the same state of affairs with the development and research of the biogas production process at such plants does not allow us to accurately judge its efficiency and the possibility of widespread use in manure utilization technologies.

In this regard, the main purpose of the work was aimed at developing a process for the production of biogas and determining the efficiency of its use on livestock farms [1].

As was shown by Namsaraev et al., the accumulation of organic waste is a significant environmental problem in Russia. Organic waste is a valuable source of raw materials, which can be used to produce biogas and bio-fertilizers. According to the estimates by Namsaraev et al., the total amount of organic waste in Russia suitable for the formation of biogas and biofertilizers is about 167.8 million tons per year. The main part of this waste is made up of livestock waste (90%), 6% is a fermentable fraction of the municipal solid waste, and 4% originate from the sewage sludge. The Central and Volga federal districts possess the greatest potential for organic waste (54 and 43 million tons, respectively). According to the estimates by Namsaraev et al., at present, the actual use of organic wastes potentially capable of serving as raw materials for the production of biogas is 2-3 orders of magnitude lower than the available potential of the organic waste [4,5].

The modernization of existing biogas plants, as well as the introduction of new ones in the agro-industrial complex, is the basis for solving a number of energy and environmental problems: reduction of anthropogenic load and emission of greenhouse gases due to the disposal of readily decomposable organic matter (OM); processing organic waste with the simultaneous production of organic fertilizers; providing agricultural enterprises with uninterrupted power supply. Therefore, the development of biogas production technologies is an urgent problem, the solution of which will allow the energy potential of manure to be involved in the national economy of the country, while reducing its negative impact on the environment. The main link of the biogas plant is a reactor for fermentation of manure, the required volume of which is determined by the daily output of manure from the livestock farm, temperature, and duration of processing. So for the most common type of cattle farm in Russia for 400 heads, the total volume of waste supplied for processing can be from 20 to 30 m^3 per day. For the anaerobic processing of such a quantity of waste, it will be necessary to build a reactor with a volume of 300—600 m^3, depending on the residence time of the substrate in the reactor, which in turn depends on the temperature regime of processing and is 10 days under thermophilic conditions and 21 days under mesophilic conditions [1].

1.2 Methods for improving the process of methane digestion of manure and manure runoff and their analysis

Interest in methane digestion of organic waste from agricultural production has now significantly increased in almost all countries, due to the further increase in world prices for oil and other fuels. Hence it follows that the priority tasks are the creation of technological lines operating in an intensive mode with maximum energy efficiency and the development of a universal project for a biogas plant, which, with minor changes and additions, could be carried out in any region, for example, in the conditions of Russia, this obliges the presence of a wide range of climatic zones.

Currently, scientific research on the intensification of the process of anaerobic processing of manure runoff is carried out in the following main directions:

- study of the process of fermentation of highly concentrated organic waste with a concentration of solid particles of 30%—50%;
- multistage anaerobic digestion of organic waste, based on the activity of acidogenic and methane-forming bacteria;
- creation of highly active strains of microorganisms grown in special cultivators and introduced in the form of a starter culture into the digester;
- study of the process of methane fermentation with the participation of psychrophilic bacteria;
- recycling of the digested sludge and the use of anaerobic biofilters in the digestion chamber of the digester [6—9].

Improvement of the process of anaerobic digestion of manure

- This is, first of all, a decrease in the fermentation cycle or the residence time of organic biomass in the digester. It is known that the loading dose of the digester and the frequency of fermentation of organic raw materials depend on the following parameters:
- temperature of the technological process of fermentation and moisture content of fermented raw materials;
- volatile solids content in the fermented substrate;
- concentration of hydrogen ions (pH) and oxidation and reduction potential of the fermented substrate in the digester;
- technology of loading and mixing the fermentable substrate (continuous or periodic) [10–13].

In this regard, the methods of improving the process of anaerobic digestion of manure in digesters can be conditionally divided into the following types (Fig. 1.1):

- mechanical effect on the fermented substrate (grinding the initial mass before loading into the digester, mixing the fermented raw material with recirculating fermentation gases, and recirculating the sludge);
- biochemical methods of influencing the fermentable substrate (alkaline agents, enzymes, powdered activated carbon, surfactants, heat treatment, electromagnetic treatment);
- thermal and electromagnetic treatment of the fermented substrate;
- microbiological methods, i.e., immobilization of methane-forming microorganisms on various carriers, where bacteria are fixed on special inert bacteria carriers, as a result of which an increased concentration of microorganisms in the digester is achieved.

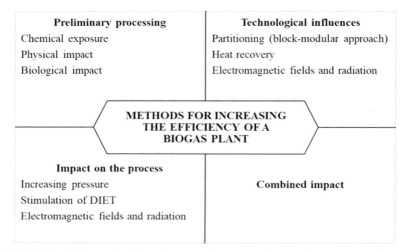

FIGURE 1.1 Existing methods for improving the process of anaerobic processing of organic waste.

The mechanical effect affecting the intensity of the methane fermentation process is the preliminary preparation of the feedstock for fermentation. Therefore, hard materials, especially of plant origin, must be prepared using cutting, tearing, or flattening devices in order to obtain smaller particles as a result of effective mechanical action, because grinding the initial mass before loading or the quality of raw material preparation, dosed and controlled feeding into the reactor affect the degree of decomposition of the OM of the fermented product and the rate of gas evolution. As a result of grinding, for example, manure, a homogeneous mass is obtained, the temperature of which at the outlet of the grinder is 6−8°C higher than the temperature of the feedstock. Further, the crushed mass is kept at a certain temperature and time, at which, due to the activity of microorganisms, oxygen is removed, the oxidation-reduction potential of the medium decreases, and carbohydrates, alcohols, and volatile fatty acids (VFAs) are formed. When the prepared raw material is fermented, the release of biogas begins within a few hours from the beginning of the experiment [14,15].

Further, it follows that the metabolic and reproductive capacity of methane-forming microorganisms is functionally dependent on temperature. Thus, temperature affects the amount of gas that can be obtained from a given amount of OM over a given time. If the mass is fermented without preheating, then heat exchangers and mixing devices have to be installed in the fermentation chamber. Heat exchangers do not provide a uniform temperature field, and mechanical mixing devices create high mixing rates of the layers of the fermented mass and increase heat losses to the environment. In addition, at the time of loading the initial mass into the fermentation chamber with a temperature significantly different from the temperature of the selected operating mode in the digester, the processed mass is cooled. In numerous earlier works, two temperature optima were adopted (about 33°C and 54°C), which correspond to the highest values of the metabolic activity of methane-forming microorganisms, i.e., from 25 to 55°C, the intensity of the methane formation process increases, with a further increase in temperature to 60°C, the process of decomposition of OM slows down, and in the range of 60−70°C, a significant slowdown of the fermentation process is noted. The microbiological activity of microorganisms almost ceases if the temperature drops to about 15°C. Microorganisms are very sensitive to temperature changes, especially to its sudden drops, and react to this by reducing them to reproduction. Based on this, the optimal temperature parameters for the mesophilic mode of fermentation are 32−35°C, and the thermophilic mode is 52−55°C. Hence it follows that opinions and conclusions regarding the temperature mode of anaerobic fermentation of manure are ambiguous and insufficiently studied [14,16−18].

To maintain the required temperature in the digester fermentation chamber, there are two options: preheating the initial mass before loading, where heat exchangers are located outside the digester fermentation chamber, and heating the fermentation product directly inside the vessel. It should be noted that the

6 Advances of Artificial Intelligence in a Green Energy Environment

location of the heat exchangers in the digestion chamber of the digester does not ensure uniform distribution of the temperature field throughout the entire volume, if there is no mixing device for the digested mass.

To equalize the temperature in the fermentation chamber, all kinds of devices are often installed for mixing the mass. There are practically no recommendations on the choice of mixing methods, and the effect of mixing on the process of methane fermentation of manure has not been studied enough. There is every reason to believe that mixing to some extent should help to increase the contact area between microorganisms and the substrate, i.e., multilevel mixing and forced degassing speeds up the fermentation process. With stirring, it is possible to achieve a uniform distribution of the loaded manure and microorganisms in the reactor, and with the help of forced degassing, prevent the accumulation of intermediate and final metabolic products. Currently, the following methods of mixing manure in a digester are known: continuous mixing; mixing only at a certain time, immediately after loading the original manure; intermittent stirring, for example, 10 min every hour, i.e., there is no unequivocal opinion in the application of mixing methods for methane fermentation of manure. Specialists of the All-Russian Scientific Research Institute for the Electrification of Agriculture (currently FSAC VIM) tested several mixing methods: injectors, propeller mixers, recirculation pumps, and gas recirculation, but they failed to find an optimal solution. For example, in France, a study was carried out on the production of biogas from liquid pig manure, where continuous anaerobic fermentation was used with constant stirring of the substrate or fermentation with "free cells." Experiments have shown that this method is not reliable, since fermentation does not always take place or takes a long time. Further, an improvement of the continuous mixing system is the "contact method," which consists in capturing active biomass in a settler at the outlet of the reactor in order to reintroduce it along with the original manure (installation of the Belgian company "Bio-processing") [14,15,19,20].

It should also be noted that the addition of acetate to the digester stimulates the methane formation process and at the same time is the only methano-genesis substrate that is found in them. Hence, there is a mismatch between the rates of acetate formation and consumption. This may be due to the lack of acetate-using microflora and, probably, it is necessary to achieve a certain concentration of acetate for it to start developing actively [21,22].

Numerous studies are devoted to the intensification of the fermentation process of manure with biostimulants, i.e., biochemical effect on the fermentable substrate. In order to increase the intensity of methane formation and the accumulation of methane microflora, scientists have conducted studies of the effect of adding methanol, acetate, and cellulose in the form of crushed filter paper. To accelerate the release of the digester to the operating mode, selected associations of microorganisms were used by introducing it in the form of inoculum simultaneously with the prepared manure, i.e., the introduction of exogenous additives reduces the time the digester reaches the operating mode to 3—5 days, and the use of a balanced syntrophic association of microorganisms can reduce this period to 2—3 days and start a continuous process of manure fermentation with a rather high daily loading dose of 30%—50%. However, in the first case, additional consumption of chemical reagents is required, and in the second, the introduction of intermediate containers for growing the necessary seed [10,16,17].

The concentration of methane-forming microorganisms and the intensity of biogas formation sharply decrease with an increase in the dose of manure loading into the digester. This is due to the removal of colonies of methane-forming microorganisms from the digester with a processed dose of the unloaded manure, which, in turn, leads to a reduction in the total volume of gas production [23].

The main direction of increasing the intensity of gas formation during the fermentation of manure runoff is the artificial increase of methane-forming microorganisms in the digester, in which the removal of active microflora and nonfermented substrate is reduced. During anaerobic digestion of manure runoff, the preservation of methanogenic microflora in the digester is of paramount importance, since it is possible to compensate for its loss from the digestion chamber only by increasing the processing time, because there is a sufficient volatile solids content in the fermented substrate for the development of new syntrophic and acetate-using bacteria. However, this option, i.e., an increase in the duration of manure processing, leads to an increase in energy consumption and the cost of the manufactured product [24,25].

A necessary condition for improving the process of fermentation of manure runoff is the retention of methanogenic microflora in the digestion chamber of the digester. This is the main technological principle of operation of all anaerobic filters, the implementation of which is possible on the basis of the ability of methane bacteria to create fixed and well-sedimented macrostructures.

Immobilization of methane-forming microorganisms in an anaerobic biofilter occurs under conditions of changing external influences caused by numerous and varied factors: temperature, humidity, dose and frequency of digestion tank loading, physical and mechanical properties of manure runoff, OM content in raw materials, etc. [26].

In the modern world, the need of society for the consumption of the benefits of civilization and technological progress is growing at an increasing pace, which leads to an increase in energy generation, anthropogenic load, and environmental pollution. The main source of such pollution is the return to nature of a huge mass of waste that is generated in the production and consumption of human society. The related total pollution of water, soil resources, and the atmosphere leads not only to the degradation or complete destruction of ecosystems but also to one of the serious environmental problems of our time—a constant and significant increase in the concentration of carbon dioxide (CO_2) and methane (CH_4) in the atmosphere, which is considered as a possible significant factor in global climate change (methane is the second most important greenhouse gas after carbon dioxide in the Kyoto Protocol, since it accumulates infrared energy 30 times more efficiently than carbon dioxide).

The environmental and energy challenges of the 21st century have given rise to the need to develop and use new and renewable energy sources, including the processing of large quantities of organic waste generated in various sectors of the economy. The processing of organic waste to reduce the anthropogenic load on the environment remains an urgent task, one of the solutions to which is the use of methods of anaerobic conversion of volatile solids to obtain biogas and organic fertilizers [27–29].

The processes of decomposition of organic waste with the production of combustible gas and its use in everyday life have been known for a long time: in China, their history goes back 5000 years, and in India, 2000 years. In most developed countries, the production of heat and electricity from biogas plants accounts for about 3%–4% of all energy consumed in European countries. In Finland, Sweden, and Austria, which promote the use of biomass energy at the national level, the share of biomass energy reaches 15%–20% of the total energy consumed. State support from the state and states is in the United States. Developing countries have extensive experience in using biogas. There are about 12 million farms in China that use biogas energy for heating, lighting, and cooking; in India about 10 million; and in Nepal 140,000 [30].

In Russia, biogas technology has not yet reached the modern world level; anaerobic processing of organic waste, which includes wastes from the agro-industrial complex, food and alcohol industries, urban wastewater, and organic fraction of solid waste, is used on a limited scale, the framework of which is determined by several pilot plants. Insignificant operating experience and the same state of affairs with the development and research of the biogas production process at such plants does not allow us to accurately judge its efficiency and the possibility of widespread use. Moreover, traditional systems for anaerobic treatment of organic waste have a number of disadvantages: the level of knowledge about modern waste management technologies is quite low, and the energy market and regulatory mechanisms for the introduction of new biotechnologies are not properly developed to integrate these technologies into a commercial system. At the same time, Russia has the potential to produce about 90 billion cubic meters of biogas per year. This is about 300 million tons of organic waste in dry terms, of which 250 million tons are agricultural waste and 50 million tons MSW [27–30].

Anaerobic processes of decomposition of organic compounds to obtain biogas and its use for domestic purposes have been known for a long time. Modern technologies make it possible to process any types of organic raw materials into biogas, with the exception of waxes and untreated lignin, however, the most effective use of biogas technologies for processing waste from livestock and poultry farms, agro-industrial enterprises, and wastewater, since they are characterized by a constant waste flow over time and their simplicity collection.

At the same time, methane fermentation should be considered not only as a means of protecting the environment but also as a method for obtaining gaseous fuel, valuable organic fertilizers, and feed additives. In its composition, biogas contains 60%–70% methane, 15%–45% carbon dioxide, 2%–3% nitrogen, 1%–2% hydrogen, and about 1% oxygen; there are traces of hydrogen sulfide and other gases. The heat of combustion of biogas is 20–27 MJ/m^3. He, like natural gas, is one of the most environmentally friendly types of fuel [31,32].

However, despite the positive effects of anaerobic waste treatment in digesters, a serious obstacle to implementation is relatively low energy efficiency in biogas production (up to 60% of the emitted biogas is used by the unit for its own needs); low concentration of methane in biogas; and relatively long payback periods; in the available scientific and technical literature, information on the industrial application of effective biogas technologies is presented in small quantities. Therefore, the development of methods for increasing the efficiency of biogas technology is an urgent task.

The main directions of increasing the efficiency of biogas technology for biogas production are the following:

1. Dividing the processing process into two or more stages with the provision of optimal conditions for the life of the microbial community for each stage.
2. Application of biological and physicochemical methods of pretreatment of organic waste (anaerobic or aerobic hydrolysis, cavitation treatment, mechanical grinding, thermohydrolysis, ultrasonic treatment, alkaline or acid hydrolysis, enzyme addition).
3. Additive to the processed waste of high-energy co-substrates (slaughterhouse waste, substandard food raw materials and food, etc.).
4. Combining the processes of biological and thermochemical gasification of organic waste and intermediate products of their processing using the advantages of each process.
5. Carrying out the process under various influences on the processed substrate (increased pressure in the reactor, exposure to constant and alternating magnetic fields, light irradiation with a certain wavelength, electrophysical effects).
6. Application of heat energy recovery directly and with the use of thermal transformers.
7. Complex application of biogas plant and additional sources of heat and electric energy based on other RES.

Biomass is composed of water and nonvolatile and volatile solids. In this case, biomass can be converted into energy carriers of various phase states (solid [granules, briquettes], liquid [bioethanol], or gaseous [hydrogen, methane]) [33–36].

The use of biotechnological methods makes it possible to obtain from biomass the waste products of microorganisms (bacteria, yeast, fungi) in the form of microbial biomass, ethanol, biogas, and stillage in the form of waste. Stillage, a waste of alcohol and brewing production, in turn, can also be used with high efficiency to obtain additional biogas using microbiological processes. Waste from all types of food processing is rich in residual nutrients, which are an excellent substrate for anaerobic conversion [37–39].

By anaerobic bioconversion of volatile solids contained in biomass, biogas can be obtained in the form of hydrogen and/or methane. Calculations show that the degree of bioconversion of volatile solids into hydrogen does not exceed 20%–30%, while about 80% of volatile solids are converted into methane [40,41]. Almost any organic substrate, with the exception of waxes and untreated lignin, can undergo anaerobic bioconversion to hydrogen and/or methane [42–44].

Liquid organic waste, most often used as a substrate for anaerobic processing, includes wastewater from the food industry, manure and liquid manure (manure), sludge, and excess activated sludge from municipal wastewater mechanobiological treatment plants.

Currently, scientific research on the intensification of the process of anaerobic processing of manure runoff is carried out in the following main directions:

- study of the process of fermentation of highly concentrated organic waste with a concentration of solid particles of 30%–50%;
- multistage anaerobic digestion of organic waste based on the activity of acidogenic and methanogenic microorganisms;
- creation of highly active strains of microorganisms grown in special cultivators and introduced in the form of a starter culture into the digester;
- study of the process of methane fermentation with the participation of psychrophilic bacteria;
- recirculation of the digested sludge and the use of anaerobic biofilters in the digestion chamber of the digester [6–9].

Improving the process of anaerobic digestion of manure is, first of all, reducing the fermentation cycle or the residence time of organic biomass in the digester. It is known that the loading dose of the digester and the frequency of fermentation of organic raw materials depend on the following parameters:

- temperature of the technological process of fermentation and moisture content of fermented raw materials;
- the concentration of OM in the fermented substrate;
- concentration of hydrogen ions (pH) and oxidation and reduction potential of the fermented substrate in the digester;

- technology of loading and mixing the fermentable substrate (continuous or batch) [10–13].

In this regard, the methods for improving the process of anaerobic digestion of manure in digesters can be conditionally divided into the following types:

- mechanical effect on the fermented substrate (grinding the initial mass before loading into the digester, mixing the fermented raw material with recirculating fermentation gases, and recirculating the sludge);
- biochemical methods of influence on the fermentable substrate (alkaline agents, enzymes, powdered activated carbon, surfactants);
- thermal and electromagnetic treatment of the fermented substrate;
- microbiological methods, i.e., immobilization of methane-forming microorganisms on various carriers, where bacteria are fixed on special inert bacteria carriers, which results in an increased concentration of microorganisms in the digester.

Organic waste from agricultural production (manure, droppings, etc.) contains both readily degradable compounds, from which biogas is formed during one-stage anaerobic digestion, and difficult-to-decompose compounds, which require a longer retention time in the reactor for their conversion into biogas.

1.3 Application results

1.3.1 Temperature regime

In order to reduce the volume of the reactor and, consequently, capital costs, increase the biogas yield, ensure sanitary standards for disinfecting manure, and reduce energy consumption for the plant's own needs, a thermophilic treatment regime (T = 55°C) should be adopted.

From the experience of using digesters in Russia, as well as from research results, it is known that mesophilic (anaerobic and aerobic) sediment stabilization methods do not provide the necessary reduction in pathogenic bacteria and helminth eggs. This is also evidenced by foreign data. Thus, studies carried out in Sweden have shown that Salmonella is present in 74% of raw sediment samples, in 70% of excess sludge samples, and in 20% of sediment fermented in mesophilic conditions (the number of samples of each sediment is 190). Thus, with mesophilic fermentation, the number of Salmonella is reduced by 70% [31].

With mesophilic fermentation, the number of pathogenic Enterobacteriaceae may even increase. According to Köser [45], after 4 weeks of fermentation, the sediment was contaminated with Salmonella twice as much as the raw one (they were found in 90% of the fermented sediment samples, while in the raw one, in 45% of the samples).

12 Advances of Artificial Intelligence in a Green Energy Environment

Thermophilic fermentation has a significantly higher sanitary and hygienic effect. This is evidenced by the practical experience of the Moscow aeration stations [31].

The specific heat consumption for the biogas plant's own needs is

$$Q_{pon} = Q_H + Q_K + Q_{bg} \qquad (1.1)$$

where Q_H is the specific heat energy consumption for preliminary heating of the substrate to the fermentation temperature, $kW \cdot h/(kg_{is}/day)$;

Q_K is the specific daily heat energy consumption to compensate heat losses from building envelopes and pipes, $kW \cdot h/(kg_{is}/day)$;
Q_{bg} is the specific amount of thermal energy leaving with generated biogas, $kW \cdot h/(kg_{is}/day)$.

The specific heat consumption for preheating the substrate [$kW \cdot h/(kg_{is}/day)$] is defined as

$$Q_H = \frac{C_H \cdot (T_H - T_1) \cdot hrt}{3,6} \qquad (1.2)$$

where C_H is the heat capacity of the substrate, $kJ/(kg_{is} \, ^\circ C)$;

T_H is the final substrate heating temperature—anaerobic digestion temperature, $^\circ C$;
T_1 is the initial substrate temperature, $^\circ C$;
hrt is the hydraulic retention time, day.

Average daily specific heat consumption [$kW \cdot h/(kg_{is}/day)$], necessary to compensate for heat losses through the enclosing surfaces of the bioreactor at an average annual outdoor temperature, is given by

$$Q_K = k \cdot F \cdot (T_H - T_O) \cdot 10^{-3} \cdot 24 \qquad (1.3)$$

where k is the heat transfer coefficient, $W/(m^2 \cdot K)$;

F is the area of the enclosing surfaces of the bioreactor, m^2;
T_H is the substrate temperature in the bioreactor, $^\circ C$;
T_O is the outdoor temperature, $^\circ C$.

The area of the bioreactor enclosing surfaces depends on the geometry of the reactor, as well as on the daily loading dose:

$$F = f(V_H) \qquad (1.4)$$

The specific amount of thermal energy leaving with the generated biogas [$kW \cdot h/(kg_{is}/day)$] is defined as

$$Q_{bg} = \frac{C_{bg} \cdot G_{bg} \cdot (T_H - T_O)}{24 * 3600} \qquad (1.5)$$

where C_{bg} is the heat capacity of generated biogas, $kJ/(kg \cdot ^\circ C)$ [27].
G_{bg} is the specific daily amount of generated biogas, kg/kg_{is}.

The initial data for the calculation:

- bioreactor—bioreactor of modular construction with a volume of 60 m^3;
- the area of the enclosing surfaces of the bioreactor F = 113 m^2;
- bioreactor thermal insulation—mineral wool 300 mm thick;
- heat transfer coefficient of the enclosing surfaces of the bioreactor k = 0.136 W/(m$^2 \cdot$K);
- heat capacity of the substrate C_H = 4.06 kJ/(kg$_{is} \cdot$K);
- substrate density ρ_H = 1020 kg/m^3,
- initial substrate temperature T$_1$ equal to 4°C at average annual outdoor temperatures below 5°C;
- at average annual outdoor temperatures above 5°C, the temperature of the initial substrate T$_1$ is taken equal to the temperature of the outdoor air.

Based on the collection of Rosstat "Agriculture, Hunting and Forestry in Russia. 2009," the range of average annual temperature at individual points in Russia is 36°C (from −23 to +13°C) [46].

Initial data for the calculation:

- bioreactor—block module with a volume of 60 m^3,
- the area of the enclosing surfaces of the block module F = 113 m^2;
- thermal insulation of the block module—mineral wool 300 mm thick;
- the heat transfer coefficient of the enclosing structures of the bioreactor k = 0.136 W/m^2 · K;
- heat capacity of the substrate C_H = 4.06 kJ/(kg · K);
- substrate density ρ_H = 1020 kg/m^3,
- the temperature of the initial substrate T$_1$ is taken equal to 4°C at average annual outdoor air temperatures below 5°C;
- at average annual outside air temperatures above 5°C, the temperature of the initial substrate T$_1$ is taken equal to the outside air temperature.

Fig. 1.2 shows the specific costs of thermal energy for the auxiliary needs of a biogas plant in the mesophilic and thermophilic modes depending on the ambient temperature.

The data shown in Fig. 1.2 were obtained by the calculation method based on formulas (1.1−1.5) and the initial data for the calculation.

As can be seen from Fig. 1.2, the specific heat consumption for the own needs of a biogas plant in the mesophilic mode is higher than in the thermophilic one in the entire range of average annual outdoor temperatures in the territory of the Russian Federation. At the same time, the mesophilic temperature mode of biogas plants becomes more efficient from an energy point of view only at average annual outdoor temperatures above 19°C.

Thus, the thermophilic mode is energetically more advantageous than the mesophilic one, that is, in the thermophilic mode, the specific energy consumption for the auxiliary needs of the installation is less at average annual ambient temperatures of all constituent entities of the Russian Federation.

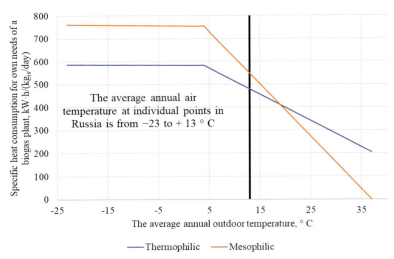

FIGURE 1.2 Specific heat consumption for own needs of a biogas plant.

In this case, the specific yield of biogas in the thermophilic regime is higher than in the mesophilic regime.

1.3.2 Separation of the processing process into two or more stages

1.3.2.1 Aerobic–anaerobic digestion

With aerobic decomposition of organic substances, such an amount of heat is released that, under favorable conditions, the substrate temperature can reach 65–70°C [47].

Since this thermal energy is generated by the same substances that emit biogas, a two-stage fermentation process, consisting of the first, aerobic phase, which has the goal of obtaining heat, and the second, anaerobic, which serves to produce gas, is always associated with a lower gas yield. In addition, one should take into account the fact that aerobic fermentation (composting) without additional energy consumption (apart from preparation) is possible only in the presence of solid and moist organic material, which has a porous structure favorable for gas exchange. Liquid substrates, on the other hand, require large amounts of energy to introduce air with simultaneous intensive mixing, which adversely affect the overall energy balance. Additional costs of funds in this case are also relatively high [29,48].

1.3.2.2 Biohydrogen production

Hydrogen (H_2) is one of the cleanest energy sources; its energy output is 122 kJ/g, which is 2.75 times more than that of fossil fuels. Biological methods to produce hydrogen in comparison with electrochemical or thermochemical methods have a number of advantages associated with their high environmental friendliness and low cost. Distinguish between light-dependent and dark biotechnological production of biohydrogen [49]. The dark process of obtaining hydrogen-containing biogas during the processing of organic waste under anaerobic conditions, which allows solving two problems: to obtain environmentally friendly energy and to utilize organic waste, is especially promising. With a comparable generation rate and cost of biohydrogen, the dark process, in contrast to the light-dependent one, is not so demanding on the composition and, most importantly, on the microbiological purity of the substrate used, which is very important when processing complex and far from sterile organic waste, for example, such as wastewater various industries, sewage sludge (WWS), etc. [50]. During the formation of biohydrogen in the dark process, only partial decomposition of OM occurs, VFAs, lactic acid, alcohols, and other products accumulate in the fermented mass. A promising technology is the organization of a two-stage anaerobic fermentation of organic substrates, in which, at the first stage of OM decomposition, conditions are created for hydrolysis and acidogenesis processes, as a result of which a large amount of hydrogen-containing biogas is formed [51]. At the second stage, the accumulated decomposition products of OM are transformed into methane-containing biogas. This organization of the anaerobic process will allow for a more complete decomposition of OM due to the creation of optimal conditions for both acidogenic and methanogenic microbial communities, as well as to increase the stability of the process as a whole [27]. Analysis of scientific literature shows that for the production of biohydrogen, substrates rich in carbohydrates are mainly used [52,53].

The use of biogas as an energy carrier is expedient as obtaining thermal energy, since the receipt of electrical energy is associated with a low 20%−40% efficiency of internal combustion engines (ICEs) operating in the drive of an electric generator. The production of biohydrogen makes it possible to use hydrogen in fuel cells (FCs) with conversion into electrical energy with an efficiency of 60%−80%. The use of FCs, in comparison with ICEs, has design advantages (minimal nomenclature of parts, absence of moving, coupled parts and mechanisms), which determine noiselessness and reliability in operation, as well as technological advantages—higher energy density and high sensitivity to load changes. The disadvantage is the need for a high degree of hydrogen purification, which leads to additional energy consumption and a relatively high cost of the fuel cells themselves. The use of fuel cells in the digesters complex is a promising direction due to active world research in the field of increasing efficiency, production technology, and reducing the cost of fuel cells.

1.3.3 Application of biological and physicochemical methods for pretreatment of organic waste

One of the methods to increase the efficiency of anaerobic conversion of organic waste to obtain biogas and organic fertilizers is to increase the solid retention time in the reactor [54].

For the efficient operation of a biogas plant, it is important to correctly correlate two parameters—the retention time of the initial substrate in the bioreactor and the degree of decomposition of volatile solids of initial substrate. The retention time is the period required to completely replace the substrate in the bioreactor. The degree of decomposition of volatile solids of initial substrate is the amount of decomposed volatile solids referred to the initial volatile solids. In the process of anaerobic conversion, volatile solids of waste are converted into biogas, i.e., the amount of total solids in the bioreactor is constantly decreasing. Biogas production rate is usually maximized at the start of the process (after loading new dose of initial substrate) and then the biogas production rate decreases. Often, the longer the substrate is held in the reactor, the more methane is recovered due to the increased contact time of the microorganisms with the nutrients. Batch systems generally have a higher decomposition rate than continuous systems. Theoretically, in batch systems, the degree of decomposition of volatile solids can reach 100%. However, in practice, complete (100%) decomposition of volatile solids and complete extraction of biogas is impossible. The degree of decomposition of volatile solids also depends on the type of initial substrate. Rapidly decomposing waste such as pomace from sugar beet can have a degradation rate of over 90%, while high-fiber forage crops degrade 60% in the same time [55].

Thus, based on the degree of decomposition of volatile solids, such a hydraulic retention time (HRT) of the substrate in the bioreactor is experimentally selected, which provides the maximum rate of biogas production with a relatively high degree of decomposition of volatile solid substrate. A constant volume of substrate is maintained in the bioreactor by feeding the initial substrate into the reactor while removing the effluent at regular intervals. The effluent removed from the reactor after the anaerobic conversion consists of water, including dissolved salts, inert materials, and non-decomposed volatile solids. The effluent also contains microbial biomass accumulated during anaerobic conversion in the bioreactor.

The HRT is the ratio of the volume of the bioreactor to the daily dose of loading, equal to the time during which the substrate in the form of a liquid remains in the bioreactor. The selection of the optimal HRT is important, since, on the one hand, long-term HRT promotes a deeper degree of decomposition by increasing the duration of contact between microorganisms and volatile solids. On the other hand, at low HRT, the growth rate of microorganisms will be less than the rate at which the effluent is removed from the bioreactor, which will lead to acidification and shutdown of the biogas plant. Typically, the doubling time of methanogenic microorganisms in bioreactors is more than 200 h. Therefore, HRT for classic digesters should be longer than the doubling time of microorganisms. Usually, for single-stage bioreactors, HRT ranges from 220 to 720 h; HRT for thermophilic reactors averages 66% of that for mesophilic ones.

One of the methods to increase the efficiency of the anaerobic conversion of volatile solids to obtain biogas and fertilizers is to increase solid retention time in the bioreactor.

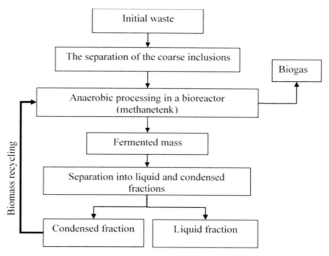

FIGURE 1.3 Block diagram of a method for increasing the efficiency of anaerobic bioconversion of organic waste to obtain a gaseous energy carrier and organic fertilizers based on recycling biomass.

By separating the effluent into a supernatant liquid and a solid digestate, it is possible to increase the residence time of the substrate without increasing HRT by recirculating the solid digestate.

The recirculated solid digestate consists not only of the biomass of microorganisms, but also of the incompletely decomposed volatile solid in the digestate. With this approach, SRT is increased by recirculation of volatile solids remaining in the solid phase, which require a longer decomposition time. The main advantage is the ability to increase SRT without increasing HRT while simultaneously recycling the biomass of microorganisms. Due to this, the required degree of decomposition can be achieved in bioreactors with reduced volume [54].

Fig. 1.3 shows a block diagram of a method for increasing the efficiency of anaerobic conversion of organic waste to obtain biogas and fertilizers based on biomass recirculation.

1.3.4 Combining the processes of biological and thermochemical gasification of organic waste and intermediate products of their processing using the advantages of each process

1.3.4.1 Thermochemical gasification

Mixing in digesters is carried out using mechanical stirrers, as well as by circulating the substrate and recirculating the gas. The latter method is less commonly used in practice, but it is more effective than the use of mechanical stirrers. There are several reasons for the beneficial effects of gas mixing on the fermentation process. When compressed gas is supplied, good mixing is created due to the intensive rise of gas bubbles. In this case, there is also a mechanical separation of small gas bubbles from methanogenic microorganisms, which facilitates their contact with the nutrient substrate. When compressed gas is supplied in the digester, the concentration of dissolved carbon dioxide increases, which, being an acceptor of hydrogen, reduces its partial pressure and thereby improves the living conditions of acetate-decomposing methanogens, as a result of which the yield of methane increases. With an increase in CO_2 concentration, the load on the digester can be increased. An increase in the concentration of carbon dioxide can be achieved by introducing flue gases, as well as increasing the pressure in the digester.

Therefore, the development of a power supply system for a biogas plant, through the use of thermochemical (gasification) processing of organic waste, is relevant. To increase the low energy efficiency of the bioenergy plant, a constructive scheme has been developed that combines the process of anaerobic digestion and pyrolysis, which allows to exclude the consumption of biogas for own needs due to the use of gas generator gas for heating the digester in winter [55].

1.3.4.2 Immobilization of methane-forming microorganisms on carriers

At present, along with the improvement of methods of aerobic biological treatment, the technology of anaerobic treatment is intensively developing, which is used mainly for highly concentrated wastewater. The rapid development of anaerobic technology is associated with the desire to create compact and efficient devices, distinguished by reliability and flexibility of operation, low capital, and operating and energy costs for wastewater treatment.

The development of anaerobic technology proceeded from primitive mixers, anaerobic contact apparatus to anaerobic filters, and modern UASB and EGSB reactors. In foreign practice, over the past 10−15 years, anaerobic reactors of the second generation have become the main structures for the treatment of highly concentrated wastewater, providing effective wastewater treatment in a very wide range of concentrations (BODtotal $= 0.3 \div 100$ g/L), with the residence time of water in devices from 30 min up to 48 h. The anaerobic method is used to treat a wide variety of wastewater and, first of all, for the pulp and paper industry (24%), sugar production (19%), and breweries (13%).

Conversion of volatile solids of highly concentrated wastewater under anaerobic conditions, in comparison with aerobic treatment, allows to reduce capital costs by almost 10 times, to place facilities on a much smaller area (approximately 10−20 times). At the same time, operating costs, in comparison with aerobic cleaning, are reduced by almost three times [56]. The disadvantages of the anaerobic method associated with the low growth rate of anaerobic bacteria; their high sensitivity to changes in pH, temperature and fluctuations in the concentration of pollutants in wastewater, and the lower rate of anaerobic processes, in comparison with aerobic ones, are successfully overcome by the undoubted advantages of modern anaerobic systems. A significant advantage of the anaerobic method is a decrease in the amount of excess sludge (3–10 times), its stability, and the possibility of obtaining additional energy due to the resulting biogas. The technology of anaerobic treatment can be implemented in reactors with suspended sedimentation and with attached biomass.

The main direction in the development and improvement of anaerobic reactors, just as in the case of aerobic bioreactors, is the search for structures that ensure the maintenance of a high dose of active biomass in devices. This is most successfully achieved in UASB and EGSB reactors using granular activated sludge. It is these devices that have received the greatest distribution in the world in recent years. Granular sludge has a high activity, rather than high strength of the granules and good sedimentation properties [57]. For this reason, the concentration of sludge in the core of the apparatus can reach 5080 kg/m^3, due to which it is possible to achieve high volumetric loads. However, granular sludge is not formed in all wastewater; in addition, for its formation and growth, conditions such as removal of suspended solids, a certain degree of acidic fermentation, restrictions related to the residence time of wastewater in the apparatus, etc., must be observed. Of the reactors with attached biomass, the possibility of increasing the dose is successfully implemented in reactors with a fluidized bed of loading [58].

1.3.5 Carrying out the process under various influences on the processed substrate

1.3.5.1 Increased pressure in the reactor

By raising the pressure in the digester to 0.15 MPa, it is possible to provide good thermophilic fermentation performance at higher (2–3 times) loads (13.5–18 instead of 6 kg/m^3 day). This technique can be effective when fermenting highly concentrated sediments [31].

When compressed gas is supplied in the digester, the concentration of dissolved carbon dioxide increases, which, being an acceptor of hydrogen, reduces its partial pressure and thereby improves the living conditions of acetate-decomposing methanogens, as a result of which the yield of methane increases. With an increase in CO_2 concentration, the load on the digester can be increased. An increase in the concentration of carbon dioxide can be achieved by introducing flue gases, as well as increasing the pressure in the digester [59].

Of particular note is the new technology of anaerobic high pressure digestion (AHPD) to obtain gas with a methane content of about 95% without interfering with the technology [60]. In the AHPD concept, methanogenic biomass creates pressure inside the reactor, since carbon dioxide has a higher solubility than methane; it almost completely dissolves in the liquid phase at high pressures. This method is simple and allows you to additionally dissolve other gas components, such as hydrogen sulfide and ammonia.

In the laboratory of bioenergy and supercritical technologies of the "Federal Scientific Agroengineering Center VIM," a number of experimental studies have been carried out to increase the organic load in the digester of a biogas plant by increasing the pressure in the reaction zone. A new method of increasing the pressure in the reactor due to the recirculation of biogas is proposed. An increase in pressure in the reaction zone of the digester from 2 to 150 kPa made it possible to double the load on the reactor for OM from 4 kg OM/m^3 · day to 7.95 kg OM/m^3 · day, while the process remains stable. An increase in pressure during anaerobic bioconversion of a synthetic substrate made it possible to increase the degree of decomposition of OM to 83.9%, i.e., by 14% [61].

1.3.5.2 Electrophysical impact

Studies of recent years show that in microbial electrolysis cells (MEC), exoelectrogenic, that is, capable of emitting and transferring electrons, bacteria (for example, representatives of the genus Geobacter), under the influence of low voltage electric current, can convert organic waste into different products such as hydrogen and methane [62,63]. It is shown that in a single-chamber MEC, for example, in an anaerobic reactor with a pair of electrodes placed in it, the processes of hydrolysis and methanogenesis are accelerated. The increase in methane production in MEC is mainly due to the cathodic reduction of carbon dioxide to methane using a biofilm grown on the cathode containing hydrogenotrophic methanogens; in this case, the biofilm is a kind of biocathode and presumably carries out the reaction $CO_2^0 + 8H^+ + 8e^- = CH_4 + 2H_2O$ ($E_0 \approx -0.44$ V vs. a standard hydrogen electrode) [62]. The electrons used for the cathodic reduction of CO_2 are formed as a result of the anodic oxidation of substrates due to the transfer of electrons from organic substances to the electrode with exoelectrogens.

In the laboratory of bioenergy and supercritical technologies of the Federal Scientific Agroengineering Center VIM, a number of experimental studies were carried out on the influence of electrophysical effects on the methane fermentation process, which resulted in an increase in the yield of biogas and methane content, the degree of removal of OM, and a decrease in the concentration of VFAs. In the work performed, a significant, but not fully understood, positive effect of attached microorganisms (biofilm), electrically conductive material, and electric current on the characteristics of the process of methanogenic fermentation of municipal organic waste was shown. The results obtained are of practical interest, since the approaches used in this work have a simple design and are quite simple to operate [64].

1.3.5.3 Application of conductive materials for direct interspecies electron transfer

Direct interspecies electron transfer (DIET) is an anaerobic syntrophic process in which microorganisms exchange electrons directly from cell to cell without the use of reduced molecules (hydrogen, formate) [65]. The electron transfer rate with DIET is significantly higher than with other syntrophic processes, which contributes to a faster and more complete decomposition of OM [64]. In nature, DIET is carried out by the aggregation of microorganisms, e-pili, cytochromes C, and other biological structures [65,66]. In anaerobic reactors, DIET can be stimulated by introducing conductive materials [65–69].

The most common are two types of such materials: carbon-based (granular activated carbon, carbon fabric, biochar, graphite, graphene) and iron-based (magnetite, hematite, stainless steel).

Conductivity is critical to DIET; carbon fabric has the greatest potential for biogas plants. In work [70] points to experimental studies of the process of methanogenesis with the use of poorly conductive cotton fabric, it was found that the electrically conductive characteristic of carbon fabric is the main reason for the acceleration of the process.

Microorganisms are able to attach to the surface of conductive materials and use them as conductors for electron exchange. In addition, the use of conductive materials contributes to the accumulation and retention of active biomass in the reactor—currently, it is the effect of carbonic tissue and granular active carbon. The addition of carbon fiber stimulates the process of thermophilic fermentation of rice straw, contributing to an increase in the rate of methanogenesis by 3.5 times, the methane content in biogas by 16%, and the degree of decomposition of OM by 32% [71]. Using carbon fabric to activate DIET during fermentation of lychets, it was possible to achieve optimal operation of the reactor at a load of 49.4 kg of COD m-3 per day, which is currently the highest load for anaerobic reactors processing lychets [72]. During the fermentation of concentrated food waste with the addition of granular activated carbon, it was possible to achieve an 18-fold increase in the methane yield and a stable operation of the reactor [73]. The introduction of granular activated carbon more effectively contributes to the restoration of methanogenesis with an excessive accumulation of VFAs than the introduction of other conductive materials [74]. Activated carbon stimulates the development of syntrophic bacteria and methanogenic archaea, due to which the concentration of VFAs is kept at a low level, which makes it possible to increase the organic load on the reactor [73]. The use of some conductive materials, in particular graphene and carbon nanotubes, for DIET in methane fermentation is currently not economically feasible due to the very high cost of starting materials.

22 Advances of Artificial Intelligence in a Green Energy Environment

The use of conductive materials allows

- toreduce the start-up period of the installation,
- to increase methane production rate,
- to increase degree of decomposition of volatile solids and reduce volume of the bioreactor, and
- to increase the adaptive ability of the microbial community to unfavorable conditions (for example, excessive accumulation of VFAs or H2, decreased pH) [75].

1.3.5.4 Application of increased pressure in the reactor space and conductive materials

In work [75], a technological scheme of a line for processing the organic fraction of solid municipal waste or bedding manure in anaerobic bioreactors with the use of process intensification means—increased pressure in the reactor and conductive (electrically conductive) materials as a carrier of biomass in a second generation bioreactor—is proposed. The possibility of using the recently discovered syntrophic process of DIET to increase the efficiency of anaerobic digestion of complex organic waste is considered.

1.3.5.5 Integrated preheating from a heat exchanger using waste heat of the effluent and heating using microwave radiation

Also, in the laboratory of bioenergy and supercritical technologies of the Federal Scientific Agroengineering Center VIM, the process was studied with a combined substrate preheating system, which includes heating from a heat exchanger using waste heat of the effluent, and heating using microwave radiation.

As can be seen from the experiment, the maximum heating rate was achieved with the simultaneous switching on of the heat exchanger and microwave radiation. The lowest energy consumption for heating the substrate will be when using only microwave radiation. However, the heating rate of the substrate using microwave radiation is comparable to the heating rate using a heat exchanger. In addition, the use of only microwave radiation is disadvantageous from the economic and energy points of view. High power magnetrons are expensive and consume a lot of electricity. Thus, the combined use of heat exchangers and microwave radiation in the system for preheating the substrate before anaerobic digestion will increase the energy and economic efficiency of the process when using waste heat of the effluent [76].

1.3.6 Application of heat energy recovery directly and using thermal transformers

1.3.6.1 Direct heat recovery

The heat contained in the effluent is an additional reserve of energy, which should, if possible, be used to heat the loaded substrate and compensate for heat losses in the reactor. The simplest solution is to install a recuperative heat exchanger (RHE) of the "influence-effluent" type on the effluent discharge line from the bioreactor. This scheme ensures the use of the heat of the fermented substrate for partial heating of the influent. Its use reduces the energy consumption for the fermentation of the substrate. The most effective scheme can be used in the thermophilic mode in a bioreactor [29].

As heat exchangers, spiral heat exchangers of the "influence-effluent" type are usually used. However, schemes in which the influence passes through the effluent accumulator have a simpler design solution, but in these cases, a relatively small part of the energy is reused due to losses in the sludge accumulator.

Organic waste from animal husbandry, as a rule, has high stickiness and viscosity and is very diverse in terms of dispersed composition. Therefore, the speed of movement of the substrate should be at least 3—5 m/s, due to which the heat of the effluent does not have time to be transferred to the substrate loaded into the digester.

1.3.6.2 Heat recovery using thermal transformers

An alternative to traditional heat supply methods based on fuel combustion or direct heat recovery is heat generation using a heat pump [29].

The substrate (effluent) treated in the anaerobic bioreactor is directed to the effluent accumulator and continuously pumped through the heat exchanger-evaporator of the heat pump. The heat energy from the effluent through the low-boiling coolant after increasing the temperature potential in the compressor is transferred through the heat exchanger-condenser to the initial substrate circulating according to the scheme "bioreactor—pump—heat exchanger-condenser—bioreactor." Thus, the thermal energy of the heated substrate removed from the bioreactor is useful for heating the influent. When the conversion factor of the heat pump is at the level of 4-5, for every 3—4 kW of heat power removed from the bioreactor with the effluent, 4—5 kW of heat power supplied to the initial substrate can be obtained. This consumes ∼1 kW of mechanical power on the compressor drive.

The main disadvantage of this technical solution is the formation of deposits on the heat exchange surfaces from the manure side, which leads to significant losses of thermal power or to the need for a significant increase in expensive heat exchange surfaces. Another disadvantage is the low intensity of the main processes that determine the productivity of the technological line "receiving tank—bioreactor—effluent settler."

Due to the lack of preliminary microbiological treatment of the original manure in order to increase the degree of dissolution of OM and obtain an initial substrate with an increased content of components that contribute to intensive methanogenesis, the specific productivity of the line for biogas and substrate decreases. In the effluent settler, due to residual gas evolution, the intensity of the process of separating the treated substrate into solid and liquid fractions is significantly reduced, which ultimately leads to an increase in the mass and size characteristics of the effluent settler [77].

1.3.6.3 Aerobic—anaerobic—aerobic treatment with heat recovery

The method of aerobic—anaerobic—aerobic treatment of manure, considered in Ref. [78], is based on the use of the bioenergetic potential of liquid organic waste in conjugated biological treatment processes. It includes three main processes: preliminary aerobic treatment of initial waste, anaerobic processing of thickened and liquid fractions, and aerobic additional purification of the liquid fraction. In the proposed method, several methods of increasing the efficiency of biogas technology are used in a complex at once: dividing the processing process into several stages, preliminary aerobic hydrolysis, heat energy recovery, using a biofilter.

The principle of the proposed method is as follows:

The original manure from the manure receptacle is pumped into the aerobic pretreatment apparatus (AAP), in which aerobic heating and hydrolysis of the original substrate occur. The AAP maintains optimal conditions for the pretreatment process using a motorized agitator and aeration device with a compressor. The prepared substrate is fed into a mechanical separation device with a drive, from which the thickened fraction is pumped into an anaerobic bioreactor (digester). In the digester, anaerobic digestion takes place to obtain biogas and fermented mass, which, after processing, enters a mechanical separation device with a drive. The digester maintains optimal conditions for the life of the microbial community: temperature using a built-in heat exchanger and mass transfer (mixing) using a pump. The liquid fraction obtained on mechanical separation devices is pumped into an anaerobic biofilter with feed material. Residual OM is removed from the liquid fraction passing through the feed material due to its assimilation by anaerobic microorganisms attached to the feed material. The processed liquid fraction from the biofilter enters the RHE and from it into the aeration tank equipped with an aeration device with a compressor and a driven agitator. From the aeration tank, a mixture of the treated liquid and excess activated sludge enters the settling tank. The supernatant liquid from the settling tank is used for irrigation of agricultural crops, and part of the excess activated sludge is pumped into the aerobic pretreatment apparatus to use its bioenergy potential. Thermal energy obtained from the processed liquid fraction from the biofilter in RHE is directed to the built-in heat exchanger of the anaerobic digester bioreactor to maintain the temperature of the anaerobic digestion process.

For calculations, it is assumed that, depending on the technical means used, the efficiency of converting biogas energy into thermal energy is 80%—95% (hot water boilers) and 45%—50% (cogeneration plants) and into electricity is 10%—45% [27]; according to experimental data, the heat of combustion of biogas is 23,550 kJ/kg.

For the initial manure, a substrate with a moisture content $W = 92\%$, ash content $A = 25\%$, and an organic matter content of $OM = 60$ g/L was taken.

In comparison with anaerobic treatment in traditional digesters [29], the proposed scheme, with a flow of initial manure of 65 m^3/day, with the properties indicated in the calculation allows to reduce the consumption of thermal energy for auxiliary needs from 3990 to 993 kW h heat/day, with an increase in electricity consumption from 80 to 317 kW h of electricity/day.

1.3.7 Complex application of biogas plant and additional sources of thermal and electric energy based on other RES

To compensate for the losses of heat and electricity from the BEU, as well as to increase the energy efficiency of an enterprise working in conjunction with the BEU, it is advisable to use other types of renewable energy sources, if they can be located and their own quality characteristics are manifested. The works [79] consider the possibility of using solar collectors and thermal photoelectric modules of various designs in the heat supply system of a cattle farm from a block modular biogas plant.

The use of a block modular design of a biogas plant will allow to intensify heat and mass transfer in reactors modules, which, in turn, will lead to an increase in productivity by 10%−20% in comparison with traditional digesters; promptly manage the most important indicators of optimal conditions for the life of the microbial community in the reactor space, which will lead to an increase in the system's resistance to external influences; reduce capital costs for construction; and ensure high quality of equipment manufacturing and installation.

The use of solar collectors in the heat supply system of the cattle farm from a biogas plant will allow to reduce the consumption of biogas for maintaining the temperature regime in the bioreactor; to increase the energy efficiency of the system for processing organic waste from animal husbandry; and to increase the energy efficiency of the heat supply system of the cattle farm from the biogas plant. The use of thermal photovoltaic modules of various designs will allow using not only the above advantages but also compensate for the needs for electricity. This will make the power supply system more versatile in operation, as well as increase the energy independence and autonomy of the biogas plant and the cattle farm [80−82].

1.4 Energy model

An energy model was developed for the system of anaerobic processing of organic waste to generate electricity to assess the energy efficiency of technical means for anaerobic bioconversion of organic waste into biogas with subsequent conversion into electrical energy.

Fig. 1.4 shows a block diagram of the technological line for anaerobic processing of organic waste from the agro-industrial complex into organic fertilizers and electricity, which includes the main elements necessary for the production of electricity with high efficiency.

For the energy assessment of existing and developed methods for intensifying the process of anaerobic bioconversion, as well as methods for

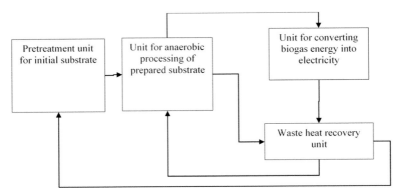

FIGURE 1.4 Block diagram of the line for anaerobic processing of organic waste from the agro-industrial complex into organic fertilizers and electricity.

converting a gaseous energy carrier obtained in the process of anaerobic bioconversion into electricity, an energy model of the system of anaerobic processing of organic waste to generate electricity was developed.

Thus, based on the works [1,78,83–87] and the developed energy model, the following types of energy appear in the energy balance of the line for anaerobic processing of organic waste from the agro-industrial complex into organic fertilizers and electricity:

- the energy of chemical bonds of the volatile solids of the substrate;
- thermal energy;
- electric energy;
- mechanical energy.

A block diagram of the energy model of a system for anaerobic processing of organic waste to generate electricity is shown in Fig. 1.5.

Thus, the indicators of an energy-efficient system for anaerobic processing of organic waste to generate electricity must correspond to the system of equations:

$$\begin{cases} \phi = \dfrac{OM_{is}}{OM_{eff}} \to max \\ \sum Q_\text{к} \to min \\ COP^Q = \dfrac{Q_{com}}{Q_{on}} \to max \\ COP^E = \dfrac{E_{com}}{E_{on}} \to max \\ hrt \to min \end{cases} \quad (1.6)$$

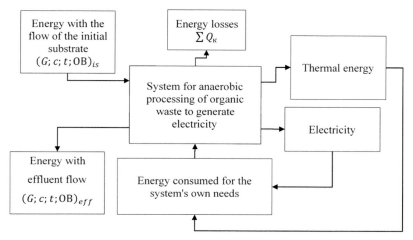

FIGURE 1.5 Block diagram of an energy model of a system for anaerobic processing of organic waste to generate electricity.

1.5 Conclusion

With the help of each technology for processing raw materials, you can get its own specific advantages: increased biogas yield; improving the qualitative composition of biogas; obtaining biohydrogen; increasing the degree of purification of raw materials; reduction in processing time; and getting better quality fertilizers.

The use of methods for compensating thermal and electric energy will increase the yield of commercial biogas and will give a positive economic effect not only by saving energy costs but will also allow solving the most important problem of external sources of power supply for the ECU and the enterprise in combination with ECU—to compensate for uneven loads, which will increase the operational reliability.

The use of methods to improve the efficiency of biogas technology will allow to reduce the size and cost of equipment, as well as the area for its placement; reduce the payback period of equipment; reduce the amount of fines for environmental violations; are more likely to participate in government subsidy programs; increase energy efficiency; and increase the level of power supply to local consumers in the absence of centralized power supply.

Each technology has its own advantages and disadvantages; therefore, it is advisable to comprehensively apply ways to improve the efficiency of biogas technology, using the advantages of each of them, while taking into account the initial composition of raw materials and the needs of the enterprise.

According to literature sources, some of the considered methods of increasing the efficiency of biogas technology have already found their

28 Advances of Artificial Intelligence in a Green Energy Environment

successful application in industry, but many of them are poorly studied even at the research level, which negatively affects widespread adoption.

The developed energy model of the system of anaerobic processing of organic waste to generate electricity allows one to evaluate both the energy efficiency of the technical means of individual blocks of the system and the energy efficiency of the system as a whole.

Based on the developed model, the assessment of energy efficiency should be carried out on the basis of the following conditions:

- an increase in the degree of transformation of volatile solids;
- minimization of energy losses;
- increase in the conversion factor of electricity;
- increase in the heat conversion coefficient.

In addition, the analysis of the above methods of intensification allows us to conclude that the maximum specific growth rate of microorganisms in the digester, on which the rate of biogas output depends, is a function of the energy supplied to both the substrate and the consortium of microorganisms, which includes the following types of energy:

- the energy of chemical bonds of substrate volatile solids;
- thermal energy;
- electric energy;
- mechanical energy.

Further optimization of the process of anaerobic bioconversion of OM should be aimed at identifying the amounts and types of energy that must be supplied to the substrate for an efficient and stable process of anaerobic bioconversion.

The main disadvantage of this approach is the need to experimentally determine the BOD of the initial waste, which can lead to additional costs at the design stage of an anaerobic bioconversion system.

The main advantage of this approach is the ability to simulate energy fluxes of various types using various methods for intensifying the process of anaerobic bioconversion. In this case, the energy efficiency of both the intensification methods themselves and the anaerobic bioconversion system as a whole will be taken into account under the influence of various external factors (initial substrate, external temperature, temperature regime, HRT, etc.). At the same time, the scientific literature contains averaged data on the BOD of various wastes, which makes it possible to assess the energy efficiency of using various methods of intensifying the process of anaerobic bioconversion with sufficient accuracy.

References

[1] A. Kovalev, Andreevich Technologies and Feasibility Study of Biogas Production in Manure Utilization Systems of Livestock Farms (theses of doctoral dis.), 1998 (in Russ).

Anaerobic bioconversion of organic matter **Chapter | 1 29**

[2] A.Y. Izmaylov, Y.P. Lobachevskiy, A.V. Fedotov, V.S. Grigoryev, Y.S. Tsench, Adsorption-oxidation technology of wastewater recycling in agroindustrial complex enterprises, Vestnik mordovskogo universiteta = Mordovia Univ. Bull. 28 (2) (2018) 207−221, https://doi.org/10.15507/0236-2910.028.201802.207-221.

[3] A.V. Artamonov, A.Y. Izmailov, Y.A. Kozhevni-kov, Y.Y. Kostyakova, Y.P. Lobachevsky, S.V. Pashkin, O.S. Marchenko, Effective puri-fication of concentrated organic wastewater from agro-industrial enterprises, prob-lems and methods of solution, AMA, Agric. Mech. Asia, Afr. Lat. Am. 49 (2018) 49−53.

[4] Z.B. Namsaraev, P.M. Gotovtsev, A.V. Komova, R.G. Vasilov, Current status and potential of bioenergy in the Russian Federation, Renew. Sustain. Energy Rev. 81 (2018) 625−634.

[5] Z.B. Namsaraev, Y.V. Litti, A.N. Nozhevnikova, Analysis of the raw material potential for biogas production in the Russian Federation, J. Phys. Conf. Ser. 1111 (012012) (2018) 1−5, https://doi.org/10.1088/1742-6596/1111/1/012012.

[6] Biogas - 85. Problems and Solutions: Materials of the Sov. - Fin. Symposium. February, 4−6 1985 - Moscow, Helsinki, 1985, 279 p. (in Russ).

[7] V.T. Taysaeva (Ed.), Renewable Energy Sources for Sustainable Development of the Baikal Region: Proceedings of the 2nd International Scientific Conference, Ulan-Ude, June 21−22, 2004, Publishing House of the Belarusian State Agricultural Academy, Ulan-Ude, 2005, p. 172 (in Russ).

[8] A.A. Kovalev, The use of animal waste to obtain biogas, Scientific. Works of VIESH, in: A.A. Kovalev, P.I. Gridnev (Eds.), Energy Saving Technologies in Agricultural Production, vol. 64, 1985, pp. 107−114 (in Russ).

[9] R.A. Melnik, Biogas technologies - ecology, energy, agrochemistry, profitability and efficiency, in: R.A. Melnik, I.I. Evdokimenko, V.I. Borodin, A.G. Puzankov (Eds.), Energy Saving in Agriculture. Proceedings of the 2nd International Scientific-Practical. Conference (3−5 October 2000, Moscow - VIESH). To the 70th Anniversary of VIESH, Part 2. - M: VIESH, 2000, 544 p. (in Russ).

[10] A.B. Babayants, E.S. Pantskhava, Y.A. Khanukhaev, Intensification of thermophilic methane fermentation in the production of feed concentrate of vitamin B-12, Prikl. Biochem. Microbe. 12 (1976) 250−264. A.S. 1599319 USSR (in Russ).

[11] Y.T. Badmaev, Method of calculating the optimal dose of loading a bioreactor during anaerobic digestion of manure, in: Y.T. Badmaev (Ed.), Bulletin of the Buryat University. Ser. 9. Physics and Technology. Issue 5, Publishing House of the Buryat State. University, Ulan-Ude, 2006, pp. 210−214 (in Russ).

[12] A.G. Vedenev, T.A. Vedenev, Guidance on Biogas Technologies, DEMI, Bishkek, 2011, 84 p. (in Russ).

[13] R.A. Melnik, Investigation of the chemical and technological foundations of the intensification of the methane fermentation process, in: R.A. Melnik, I.I. Evdokimenko, V.I. Borodin, A.G. Puzankov (Eds.), Abstracts of the Conference "Research, Design of Construction of Systems of Facilities for Methane Fermentation of Manure." - Tallinn, 1982, pp. 52−56 (in Russ).

[14] V.P. Druzhanova, E.N. Kobyakova, Bulletin of the Buryat State Agricultural Academy named after V.R. Filippov, Substantiation of the Parameters of the Mixing Device in a Biogas Plant for Small Livestock Farms, vol. 4 (49), 2014, pp. 13−19 (in Russ).

[15] A.A. Kovalev, V.P. Losyakov, Research results of an experimental biogas plant, Mech. Electr. Agric. 11 (1987) 60−62 (in Russ).

[16] Y.T. Badmaev, High-intensity Technology of Anaerobic Processing of Organic Animal Waste in the Republic of Buryatia: Scientific and Methodological Recommendations,

30 Advances of Artificial Intelligence in a Green Energy Environment

Publishing House of the Belarusian State Agricultural Academy V.R. Filippova, Ulan-Ude, 2014, 104 p. (in Russ).

[17] N.V. Gvozdev, Intensification of Digestion Tanks/Author's Abstract, Diss. for a Job. Uch. Step. Cand. Tech. Sciences (05.23.04), 1983, 20 p. (in Russ).

[18] P.I. Gridnev, Research of the Process and Substantiation of the Parameters of Technological Equipment for Processing Cattle Manure in Anaerobic Conditions: Author's Abstract, Diss. for a Job. Scholarly Degree. Cand. technical sciences, 1982, 15 p. VASKHNIL (in Russ).

[19] V. Ermolenko, N. Naryshkin, Linnik, I. Shkodkin, Technology of preparation of manure runoff for use, Pig Breeding 2 (1989) 33—35 (in Russ).

[20] S.S. Zatsepin, Influence of various methods of pretreatment of cattle manure on the intensity of its methane digestion, Riga, in: S.S. Zatsepin, V.I. Sklyar, S.V. Kalyuzhny, S.D. Varfolomeev, A.G. Puzankov (Eds.), Abstracts of the Meeting. "Biogas-87", 1987, p. 47 (in Russ).

[21] Y.T. Badmaev, Y.A. Sergeev, Bulletin of the Buryat State Agricultural Academy named after V.R. Filippov, Results of Laboratory Studies of the Immobilization of Methane-Forming Microorganisms on Solid Carriers, vol. 3 (43), 2017, pp. 70—77 (in Russ).

[22] A.N. Nozhevnikova, R.A. Melnik, T.G. Yagodina, Search for microbiological ways to intensify the process of methanogenesis on animal waste, Biol. Thermophilic Microorg. (1986) 244—248 (in Russ).

[23] Y.T. Badmaev, Analytical review of methods for processing organic waste in digesters with an anaerobic biofilter, in: Y.T. Badmaev, V.A. Taishin, A.A. Kovalev (Eds.), Energy-saving and Environmental Technologies (Meeting at Lake Baikal): Materials of the II International Scientific-Practical Conference on August 20—25, 2003, Publishing House of VSSTU, Ulan-Ude, 2003, pp. 362—366 (in Russ).

[24] Y.T. Badmaev, Investigation of the process of immobilization of methane-forming microorganisms on micro - and macrocarriers, in: Y.T. Badmaev, V.R. Kryukov, A.A. Kovalev, V.A. Taishin (Eds.), Ecological Safety, Preservation of the Environment and Sustainable Development of the Regions of Siberia and Transbaikalia: Materials of the All-Russian Scientific-Practical Conference, VSGTU Publishing House, Ulan-Ude, 2002, pp. 220—223 (in Russ).

[25] N.S. Egorov, A.V. Oleskin, V.D. Samuilov, Biotechnology: textbook. Manual for universities in 8 v, in: N.S. Egorova, V.D. Samuilov (Eds.), Book. 1: Problems and Prospects, Nauk. dumka, 1989, 152 p., ill. (in Russ).

[26] B.Y. Tsyrendorzhievich, Improvement of Anaerobic Processing Technology of Pig Breeding Manufactures in the Conditions of the Republic of Buryatia (theses of candidate dis.), 2018 (in Russ).

[27] A.N. Nozhevnikova (Ed.), Biotechnology and Microbiology of Anaerobic Processing of Organic Municipal Waste: Collective Monograph/under Total, University Book, 2016, 320 p. (in Russ).

[28] V.V. Snakin, A.V. Doronin, G. Freibergs, I. Shcherbitskis, I.V. Vlasova, I.V. Chudovskaya, Methane in the atmosphere: dynamics and sources, Life Earth 39 (4) (2017) 365—380 (in Russ).

[29] A.A. Kovalev, Increasing the Energy Efficiency of Biogas Plants, Dis. Candidate of Technical Sciences: 05.20.01, 2014, 119 p. (in Russ).

[30] A.V. Vinogradov, B.V. Leonov, Potential Assessment and Experiment on the Use of Biogas Plants for Processing Waste from Pig-Breeding Enterprises in the Oryol Region: Monograph, Publishing House of the Federal State Budgetary Educational Institution of Higher Education Oryol GAU, Oryol, 2016, 136 p. (in Russ).

[31] L.I. Gunther, L.L. Goldfarb, M. Digesters, Stroyizdat, 1991. 128 p. (in Russ).
[32] V.S. Shevelukha, E.A. Kalashnikova, E.Z. Kochieva, et al., Revised. and add. - 710 p. (in Russ), in: B.C. Shevelukhi (Ed.), Agricultural Biotechnology, third ed., Higher School, 2008.
[33] G. Fuasak, P. Plion, V. Fiche, F. Tabe, Briquette on the Basis of a Compressed Lignocellulose Body Impregnated with Liquid Fuel: Pat. 2507241 Rus. Federation 2012108893/04; declared 03/07/2012; publ. 02/20/2014, Bul. No. 5. 11 p., 2014 (in Russ).
[34] V.V.Starshikh, E.A. Maksimov, Method of Briquetting Waste Products of Animals and Poultry and a Device for its Implementation: US Pat. 2507242 Rus. Federation 2012146319/04; declared 10/30/2012; publ. 02/20/2014 Bul. No. 5. 6 p., 2014 (in Russ).
[35] P. Sommer, T. Georgieva, B.K. Ahring, Biochem. Soc. Trans. 32 (2) (2004) 283.
[36] O.V. Senko, M.A. Gladchenko, I.V. Lyagin, et al., Altern. Energy Ecol. 3 (107) (2012) 89 (in Russ).
[37] S.V. Kalyuzhnyi, M.A. Gladchenko, V.I. Sklyar, et al., Environ. Technol. 21 (2000) 919.
[38] V.I. Sklyar, A.N. Epov, M.A. Gladchenko, et al., Appl. Biochem. Biotechnol. 109 (1−3) (2003) 253.
[39] S.V. Kalyuzhny, M.A. Gladchenko, E.A. Starostina, et al., Production of Alcohol and Alcoholic Beverages, vol. 3, 2004, p. 10 (in Russ).
[40] S.D. Varfolomeev, S.V. Kalyuzhny, D.Y. Medman, Adv. Chem. 57 (1988) 1201 (in Russ).
[41] S.D. Varfolomeev, E.N. Efremenko, L.P. Krylova, Adv. Chem. 79 (2010) 544 (in Russ).
[42] M.G. Khamidov, S.A. Streltsov, D.A. Danilovich, Communal Complex of Russia, vol. 2, 2009, p. 56 (in Russ).
[43] E.A. Tsavkelova, A.I. Netrusov, Appl. Biochem. Microbiol. 48 (5) (2012) 469 (in Russ).
[44] S.V. Khitrin, D.S. Meteleva, O.A. Shmakova, et al., Vseros. Scientific. Conf. Theoretical and Experimental Chemistry through the Eyes of Young People, ISU Publishing House, Irkutsk, 2013, p. 151 (in Russ).
[45] I.S. Turovsky, Sewage Sludge Treatment, Rev. and additional - M: Stroyizdat, third ed., 1988, 256 p. (in Russ).
[46] K.E. Laikam, Agriculture, Hunting and Hunting, Forestry in Russia (Sel'skoye khozyaystvo, okhota i okhotnich'ye khozyaystvo, lesovodstvo v Rossii), Rosstat, Moscow, 2009 (in Russ.).
[47] O.M. Osmonov, in: O.M. Osmonov (Ed.), Fundamentals of Engineering Calculation of Solar Power Plants: Scientific, Publishing and Analytical Center "Energia", 2011, 176 p. (in Russ).
[48] A.A. Kovalev, D.A. Kovalev, Possible ways of increasing the energy efficiency of a biogas plant, Bull. VNIIMZh 4 (8) (2012) 36−41 (in Russ).
[49] R. Kothari, D. Buddhi, R.L. Sawhney, Comparison of environmental and economic aspects of various hydrogen production methods, Renew. Sustain. Energy Rev. 12 (2008) 553−563.
[50] A. Marone, O.R. Ayala-Campos, E. Trably, A.A. Carmona-Martínez, R. Moscoviz, E. Latrille, J.-P. Steyer, V. Alcaraz-Gonzalez, N. Bernet, Coupling dark fermentation and microbial electrolysis to enhance bio-hydrogen production from agro-industrial wastewaters and by-products in a bio-refinery framework, Int. J. Hydrogen Energy 42 (3) (2017) 1609−1621.
[51] M.A. Khan, H.H. Ngo, W. Guo, Y. Liu, X. Zhang, J. Guo, S.W. Chang, D.D. Nguyen, J. Wang, Biohydrogen production from anaerobic digestion and its potential as renewable energy, Renew. Energy 129 (2018) 754−768.
[52] J. Wang, W. Wan, Factors influencing fermentative hydrogen production: a review, Int. J. Hydrogen Energy 34 (2) (2009) 799−811.

32 Advances of Artificial Intelligence in a Green Energy Environment

[53] Disposal of Liquid Organic Waste to Obtain Biohydrogen, I.V. Katraeva, E.R. Mikheeva, D.L. Vorozhtsov, E.A. Moralova, D.A. Kovalev, In the collection: problems of nature management and the ecological situation in European Russia and adjacent territories, in: M.A. Poland (Ed.), Proceedings of the VIII International Scientific Conference, 2019, pp. 289–292 (in Russ).

[54] M.V. Kevbrina, Y.A. Nikolaev, A.G. Dorofeev, A.Y. Vanyushina, A.M. Agarev, Highly efficient technology for methane digestion of sewage sludge with biomass recycling, Water Supply Sanit. Equip. 10 (2012) 61 (in Russ).

[55] A. Schnurer, A. Jarvis, Microbiological Handbook for Biogas Plants, Swedish Gas Centre Report 207, 2010, pp. 13–138.

[56] S.V. Kalyuzhny, High-intensity anaerobic biotechnologies for industrial wastewater treatment, Catal. Ind. 6 (2004) 42–50 (in Russ).

[57] S.V. Kalyuzhny, D.A. Danilovich, A.N. Nozhevnikova, Results of Science and Technology. Series "Biotechnology", Anaerobic Biological Wastewater Treatment, vol. 29, 1991, p. 155 (in Russ).

[58] I.V. Katraeva, Modern anaerobic devices for purification of concentrated wastewater, Izvestiya KazGASU 2 (16) (2011) 179–184 (in Russ).

[59] D.A. Kovalev, Determination of Optimal Parameters of the Reactor-Module of Biogas Plants of Block-Module Type, VNIIMZh Bulletin No. 4 (12), 2013, pp. 173–177 (in Russ).

[60] R.E.F. Lindeboom, Autogenerative High Pressure Digestion: Biogas Production and Upgrading in a Single Step (Ph.D. thesis. Netherlands, Wageningen), 2014. S. Achinas, V. Achinas, G.J.W. Euverink. A technological overview of biogas production from biowaste. Engineering. 3 (2017) 299307.

[61] D.A. Kovalev, A.A. Kovalev, Y.V. Karaeva, Preliminary studies of anaerobic bioconversion of organic waste in a reactor at elevated pressure, Trans. Academenergo 2 (2018) 98–105 (in Russ).

[62] S. Cheng, D. Xing, D.F. Call, B.E. Logan, Direct biological conversion of electrical current into methane by electromethanogenesis, Environ. Sci. Technol. 43 (2009) 3953–3958.

[63] B.E. Logan, D. Call, S. Cheng, H.V.M. Hamelers, T.H.J.A. Sleutels, A.W. Jeremiasse, R.A. Rozendal, Microbial electrolysis cells for high yield hydrogen gas production from organic matter, Environ. Sci. Technol. 42 (2008) 8630–8640.

[64] Y.V. Litty, D.A. Kovalev, A.A. Kovalev, Y.I. Russkova, A.N. Nozhevnikova, Investigation of the process of processing municipal organic waste in an anaerobic bioreactor with an electrophysical effect on the methanogenic microbial community, Water Mag. 6 (94) (2015) (in Russ).

[65] C.D. Dubé, S.R. Guiot, Direct interspecies electron transfer in anaerobic digestion: a review, Adv. Biochem. Eng. Biotechnol. 151 (2015) 101–115.

[66] Z. Zhao, et al., Enhancing syntrophic metabolism in up-flow anaerobic sludge blanket reactors with conductive carbon materials, Bioresour. Technol. 191 (2015) 140–145.

[67] S. Barua, B.R. Dhar, Advances towards understanding and engineering direct interspecies electron transfer in anaerobic digestion, Bioresour. Technol. 244 (2017) 698–707.

[68] D.R. Lovley, Syntrophy goes electric: direct interspecies electron transfer, Annu. Rev. Microbiol. 8 (71) (2017) 643–664.

[69] Q. Cheng, D.F. Call, Hardwiring microbes via direct interspecies electron transfer: mechanisms and applications, Environ. Sci. Process. Impacts 18 (8) (2016) 968–980.

[70] G. Baek, et al., Role and potential of direct interspecies electron transfer in anaerobic digestion, Energies 11 (2018) 1–107.

[71] K. Sasaki, et al., Efficient degradation of rice straw in the reactors packed by carbon fiber textiles, Appl. Microbiol. Biotechnol. 87 (2010) 1579−1586.

[72] Y. Dang, et al., Stimulation of the anaerobic digestion of the dry organic fraction of municipal solid waste (OFMSW) with carbon-based conductive materials, Bioresour. Technol. 238 (2017) 30−38.

[73] Y. Lei, et al., Stimulation of methanogenesis in anaerobic digesters treating leachate from a municipal solid waste incineration plant with carbon cloth, Bioresour. Technol. 222 (2016) 270−276.

[74] Y. Dang, et al., Enhancing anaerobic digestion of complex organic waste with carbonbased conductive materials, Bioresour. Technol. 220 (2016) 516−522.

[75] A.A. Kovalev, A.A. Nikitina, Y.V. Litty, A.N. Nozhevnikova, Y.V. Karaeva, Application of increased pressure in reactor space and conductive materials for intensification of anaerobic digestion, Trans. Academenergo 3 (2018) 73−81 (in Russ).

[76] D.A. Kovalev, A.A. Kovalev, Y.A. Sobchenko, Increasing the energy efficiency of the process of preheating liquid organic waste before anaerobic treatment, Bull. VNIIMZh 4 (32) (2018) 92−95 (in Russ).

[77] A.A. Kovalev, D.A. Kovalev, O.M. Osmonov, Methods for increasing the yield of commercial biogas during anaerobic conversion of organic waste in bioenergy installations, Bull. FGOU VPO MGAU 2 (2012) 64−67 (in Russ).

[78] D.A. Kovalev, A.A. Kovalev, Y.V. Karaeva, N.A. Kolesnikova, Energy analysis of the system of aerobic-anaerobic-aerobic manure treatment, Bull. VIESH 1 (26) (2017) 129−135 (in Russ).

[79] O.M. Osmonov, Scientific and Technical Basis for the Creation of Autonomous Bioenergy Installations for Peasant Farms in the Mountainous Regions of Kyrgyzstan (thesis … Doctor of Technical Sciences). Moscow, 2012 (in Russ).

[80] A.A. Kovalev, A.G. Makarov, The use of solar collectors to increase the energy efficiency of the heat supply system of the cattle farm from a biogas plant, Bull. VNIIMZh 4 (36) (2019) 20−23 (in Russ).

[81] V.A. Kovalev, V. Panchenko, V. Kharchenko, The use of solar thermal photovoltaic modules for power supply of a biogas plant with a heat pump, Innovations Agric. 5 (20) (2016) 233−240 (in Russ).

[82] A.A. Kovalev, V.A. Panchenko, The use of solar thermal photoelectric modules for power supply of a biogas plant, in: Scientific Bulletin of NUBiP of Ukraine. Series: Technology and Energy of the Agro-Industrial Complex, vol. 240, 2016, pp. 134−143 (in Russ).

[83] A.A. Kovalev, D.A. Kovalev, V.S. Grigoriev, Energy efficiency of pretreatment of the digester's synthetic substrate in the vortex layer apparatus, Eng. Technol. Syst. 30 (1) (2020) 92−110, https://doi.org/10.15507/2658-4123.030.202001.092-110 (in Russ).

[84] D.A. Kovalev, A.A. Kovalev, Y.V. Karaeva, I.A. Trakhunova, Analysis of energy efficiency of a biogas plant with effluent waste heat recovery, Int. Sci. J. Altern. Energy Ecol. 5 (169) (2015) 45−54 (in Russ).

[85] A.A. Kovalev, D.A. Kovalev, Recycling of low-potential waste heat of anaerobically treated organic substrates with electric power production, Innovations Agric. 1 (22) (2017) 138−141 (in Russ).

[86] A.A. Kovalev, D.A. Kovalev, Energy aspects of use of biogas units for heat supply, Innovations Agric. 3 (28) (2018) 191−196 (in Russ).

[87] A.A. Kovalev, Comparison of energy indicators of energy supply systems in anaerobic fermentation process in first generation bioreactors, Innovations Agric. 1 (26) (2018) 179−184 (in Russ).

Chapter 2

Disasters impact assessment based on socioeconomic approach

Igor Grebennik[1], Yevhen Hubarenko[1], Maryna Hubarenko[1] and Sergiy Shekhovtsov[1,2]
[1]*Kharkiv National University of Radio Electronics, Kharkiv, Ukraine;* [2]*Kharkiv National University of Internal Affairs, Kharkiv, Ukraine*

2.1 Introduction

Nowadays society is at the top of its development. An undeniable confirmation of this statement is the current world population. Technological progress and improved standards of living contribute to the growth of population, ensuring favorable conditions for the dynamic development of societies. Children mortality rates have dropped, whereas life expectancy has increased. Logistics, construction, and food production technologies make it possible to supply for a vast number of people. Nevertheless, despite such astonishing successes, humanity also faces problems. Resources are finite, our environment is degrading, and the amount of garbage keeps growing [1,2]. The process of globalization, relocation of the production, minerals extraction, and management facilities all lead to new hazards. Alongside natural disasters, which have always been causing economic losses and human casualties, there is an increasing number of man-made disasters, the scope of which is now comparable to the one caused by natural disasters. Biological, chemical, and radiological hazards have intensified. Some of the emerging hazards may be hardly classified within the usual framework, but the scope of the damage that they cause (allergic reactions, terrorist threats, etc.) is nearly the same.

There is a range of international organizations founded to diminish the risks of disaster outbreaks; they are currently engaged in various information activities. The latter list risk estimations, elaboration of measures and methods to lower the risks, analysis of the risk dynamics, and timely warnings of the population and politics about important tendencies. Thus, the majority of hazards have been addressed only following the signals from the international organizations.

Advances of Artificial Intelligence in a Green Energy Environment
https://doi.org/10.1016/B978-0-323-89785-3.00009-8
Copyright © 2022 Elsevier Inc. All rights reserved.

36 Advances of Artificial Intelligence in a Green Energy Environment

Some of the main problems of both general population and politics are negligence and an irresponsible character of decision-making. This is caused not only by the subjective processes (e.g., laziness or self-interest) but also by the objective ones (e.g., absence of similar precedents or low frequency of their occurrence, so that there is no constant threat to beware of). Therefore, it is of great importance to design the decision-making support systems capable of modeling the administrative processes and making efficient decisions based on the results of the modeling. These models must be constantly improved so that they reflect complex assessments including economic, social, and ecological factors within the society [3].

Various conflicts, including political, military, and economic, etc., also constitute one of the factors hindering the process of countering and/or minimizing the impact of disasters [4]. Unfortunately, a part of the international organizations is currently unable to act completely independent from the influence of the involved states, which affects the credibility of these organizations and consequently reduces the efficiency of measures against various disasters and hazards. Humane and unbiased perspective on the current situation may help to foster credibility and ensure sustainable development of the society [5].

We propose to use a formalized decision-making approach, considering economic, social, and environmental complex factors, the so-called socio-economic approach, when choosing measures to combat the consequences of natural disasters. For this purpose, we have adapted and improved the means of multicriteria assessment of a set of measures to prevent damage or reduce the consequences of natural disasters and man-made disasters.

The novelty of the approach is as follows:

A method for assessing a set of measures to prevent damage or reduce the risks of consequences of natural disasters and man-made disasters using the mathematical apparatus of multicriteria assessment is proposed.

A method of administrative management of the probability of occurrence of disasters and reducing of consequences is elaborated.

To increase the efficiency of assessing measures to prevent damage or reduce the risks of consequences of natural disasters and man-made disasters, the means of multicriteria assessment were modified, and the method of comparative identification was used to reduce the subjective impact in the formation of estimates.

Research results: The proposed approach makes it possible to evaluate measures to prevent damage or reduce the risks of consequences of natural disasters and man-made disasters, due to the complex consideration of management factors and consequences.

As a result of reducing the risks of disasters, the readiness of the socio-economic system to counteract threats increases. This strengthens the foundation for the integrated management of social, economic, and environmental systems in a sustainable development environment.

The results obtained can be used in the implementation of management of socioeconomic systems at different levels: state, regional, and municipal.

The chapter has the following structure: Section 2.2 "Statistics and Tendencies of Natural and Man-made Disasters" describes general trends and statistical information about natural and man-made disasters. Attention is given to the scale of the events taking place and the importance of the problem of threat prevention and mitigation. Section 2.3 "Overview of Institutions Dealing With Risk Management" describes the main organizations and institutions dealing with the problems of natural and man-made disasters. These organizations' activities allow the formation of a base of measures for the prevention of threats and assessments of the effect of their implementation, as well as to build a system of partial criteria. An example of such a system of partial criteria is given in Section 2.6. Section 2.4 "Promising Approaches to Decrease Risks and Losses Caused by Disasters" describes examples of measures that can lead to threats reduction or reduce the negative consequences. Section 2.5 "The Issues of Effective Disaster Countering Organization" deals with the problems that reduce the effectiveness of measures aimed at overcoming the consequences of natural and man-made disasters. Section 2.6 "Socioeconomic Approach to the Estimations of Risk Dynamics" describes a method of forming a scalar generalized evaluation, which will make it possible to form a more effective list of measures for preventing or reducing the disaster consequences.

Appendix A describes an example and operation of a software tool that aims to select the optimal list of measures using various algorithms with a limited budget.

2.2 Statistics and tendencies of natural and man-made disasters

Natural disasters should be considered from two angels:

- as a phenomenon that threatens the life and health of the population—the main task in this case is to protect the human life and health, to develop technologies to anticipate natural disasters, population warning systems, mechanisms of population evacuation, and establishment of emergency first-aid services;
- as a phenomenon that causes damage to the infrastructure, production, resource extraction processes, and citizens' properties, which leads to reduced budget allocations, additional costs, and recovery and compensation payments, it all consequently resulting in a temporary or permanent reduction of living standards. Yearly damage caused by natural and man-made disasters around the world may reach milliards of US dollars and hundreds or even thousands of human lives.

A group of scientists from the Karlsruhe Institute of Technology analyzed media and archive data in more than 90 languages on floods, droughts, storms, volcanic eruptions, earthquakes, and forest fires [6]. In their databank they registered 35,000 disasters, which took more than 8 million human lives. The results of their research were presented within the framework of the European Geosciences Union Conference [7]. In the period between 1900 and 2015, about 40% of economic damage was caused by floods, 25% by earthquakes, 20% by storms, 12% by droughts, 2% by fires, and 1% by volcanic activity; see Fig. 2.1 (data from BBC News scientists have calculated the damage from the elements in the world for 115 years https://www.bbc.com/russian/science/2016/04/160419_natural_disasters_economic_losses).

In absolute terms, economic losses have been growing for the past 100 years (more details can be found in the source [8,9]), but their current value constitutes a smaller percentage of the overall infrastructure costs (buildings, roads, production facilities, etc.). Indirectly, this indicates that the introduced measures tend to be successful in preventing or reducing the impact of natural disasters.

The total amount of economic damage caused by natural disasters in the period between 1900 and 2015 ranged from 6.5 to 14 trillion US dollars, depending on exchange and inflation rates. Usually, inflation is adjusted via country-specific consumer price index and consideration of exchange rate fluctuations between local currency and US dollars.

As a rule, direct losses or restoration costs (insurance company payments) are calculated.

It is nearly impossible to calculate overall losses in such a way that all aspects are considered. Varying approaches and methods may lead to impact estimations greatly varying in value. Experts or groups of experts tend to opt for those approaches, which are believed to reflect most realistic values. Thus,

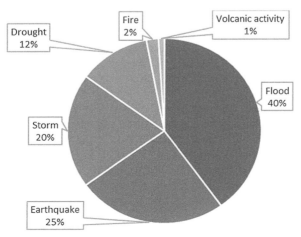

FIGURE 2.1 Share of disasters in the total scope of damage.

yearly improved, and more precise estimation methods as well as different states and sometimes different departments within the organization using varying approaches all make the calculations of overall losses a rather controversial issue.

As aforementioned, current approaches reflect those losses that may be calculated, averaged, and generalized. Natural disasters drastically alter the usual living conditions of population or usual functioning principles of infrastructure, production, or even complete industry sectors. Namely, droughts may provoke famine and increase of criminal activity (theft, robbery), which contribute to decrease of birth rates, increase of mortality rates (especially child mortality), and deteriorate the general well-being of population, etc.; consequently, all of it causes losses in retrospective. Examples of such cases are: the droughts in the USSR, which occurred in 1921–23 and 1932–33 and caused from 3 to 10 million deaths; in the China - 1939 (200,000 people), 1942–43 (about 3 million people); in the India - 1943 (about 1.5 million people); in the Kampuchea - 1975–78 (about 1 million people); in the Ethiopia - 1983–88 (about 1 million people); in the North Korea (about 2 million people). Damage caused by such disasters is difficult to calculate and estimate.

According to the UN [10] data on the number of deaths among all types of natural disasters, hydrometeorological disasters come first, geological second, and man-made disasters third.

We should also not forget about such phenomena as viruses and microorganisms, which may cause illnesses or poisoning and result in thousands of deaths yearly, constant, or temporary inability to work or pandemic of complete species of animals and plants. Biological hazards (unfortunately, unlike natural disasters, the term biological disasters are not particularly widespread, despite the similarities of their impacts and courses of development) list hazards of organic origin or those that can be transferred by biological carriers: toxins, pathogens, or biologically active substances. Apart from the aforementioned viruses, bacteria, and microorganisms (parasites), biological hazards also include poisonous insects, animals, and plants [11]. Threat to human life often gets most of the attention, whereas damage to flora and fauna is unnecessarily neglected. In their turn, harvest destroyed by illnesses, fields poisoned with pesticides or toxins, insect infestations, same as droughts and floods are almost sure to result in the reduction of consumption levels, famine, and need to address the issue of humanitarian help. According to Jason Burke in his Guardian News article, in February 2020, Eastern Africa (the whole region) found itself on the edge of famine due to the infestation of locusts. Locusts, being one of the most obvious examples of such disasters, are not the only ones threatening the crops.

According to the BBC, "The case of 2019-nCoV (Coronavirus) illustrates the scope of the impact of biological hazards, which may be similar and sometimes even bigger than the one caused by natural disasters. It started in China, where there had been several tens of thousands of infected and hundreds of deaths by the beginning of February 2020. As of 6 p.m. February 8, about 71.85 billion Yuan ($10 billion) was allocated on all levels to prevent and counter the epidemic," whereas actual costs have reached 31.55 billion Yuan ($4.5 billion). The virus has not been defeated and therefore the costs will continue to rise. It is already clear that it will have an enormous impact. Apart from human casualties, it is causing worldwide panic, cancellation of flights, and shutdown of railway and road connections. Delivery of goods and component parts has been stopped. Cultural exchange has also considerably reduced, if not stopped completely: businessmen have canceled their trips, scientists no longer conduct seminars, and all public events including international ones have been canceled or postponed.

Biological hazards and infectious diseases connected to them vary in scope and level of impact for public health. All people, regardless of their financial well-being, social role, or status, are put in danger. Every year millions of people fall ill, seriously affecting the production process and cultural exchange between states and regions. Given the current level of development of societies, no country is immune to the risk. Viruses can mutate and adapt; thus, new pathogens continue to emerge.

One should understand that infectious diseases travel easily across administrative boundaries; therefore, one of the biggest problems is localizing the outbreaks of the disease. The following documents were elaborated to address this issue: Sendai Framework, SDGs, and the Paris Agreement, which are complemented by the International Health Regulations (2005) (IHR) 64 and other relevant global, regional, national, and subnational strategies and agreements [12].

The current list of most dangerous infectious diseases includes cholera, plague, severe acute respiratory syndrome (SARS), Middle East respiratory syndrome (MERS), H1N1 pandemic influenza, Ebola, and AIDS (acquired immunodeficiency syndrome) [13]. Apparently, 2019-nCoV as an independent source of risk may also be added to the list, should no efficient way to counter it be found in the nearest future; now, however, researchers consider 2019-nCoV a mutation of SARS. The greatest danger of coronavirus is its antimicrobial resistance, which raises the issue of efficiency of infectious disease treatment [14].

There is also a certain difference between natural disasters (floods, earthquakes, storms, droughts, and other) and biological hazards. The causes (pathogenic agents) of biological hazards may exist and be constantly present on a certain territory without provoking any kind of inconvenience either to the local species of plants and animals or to the local human population as they might be constrained by certain factors, such as climate, immunity, or insensitivity. However, as soon as incidents disrupting the usual life or conduction patterns occur, they change the conditions in favor of diseases and deaths. The case becomes particularly intensive when combined with other disasters. Namely, droughts, floods, earthquakes, and conflicts arise, exacerbating the conditions favorable for disease transmission and causing population displacement [12].

Unfortunately, quite often societies underestimate the risks of biological hazards. Certain measures, such as food products inspections, work or public (including schools, shops, etc.) places checkups, tend to be taken negatively or indifferently by both owners and consumers. Disease risks can often be prevented or mitigated, and their harm reduced through vigilance coupled with a rapid response at all levels.

One of the ways to raise awareness with politicians and public about this issue is using the term biological risk, in other words, the risk of transformation of biological hazards into biological disasters. Risk is used as an indicator showing the efficiency of measures to counter biological hazards. Unfortunately, even at the lowest levels of biological hazards, there is still a chance of an epidemic outbreak. Still, its impact will be considerably lower in both social (fewer casualties and people temporarily or constantly unable to work) and economic terms.

Measures of risk management include the following:

- Methods to develop risk assessment, which should keep politicians and public informed and up-to-date about the current situation, tendencies, and efficiency of introduced measures. This includes measures to reduce impacts on the high-risk groups, preventing the spread and eliminating the source of biological hazards.
- Measures within the framework of civil prevention and medical and health care, such activities increase the resilience of the society in general and separate individuals to all kinds of emergencies; they improve the well-being, immunization, and nutrition of people and decrease their disease susceptibility.
- Ensuring primary health care in case of epidemics, disasters, and post-conflict situations, these measures aim to minimize as much as possible the impact of epidemic at its earliest stage, timely diagnose, and develop most efficient procedures and methods of disease treatments.
- Efficient planning of water supply systems, sanitation, and hygiene. Simple chlorination of water may prevent or decrease the risk of severe diarrheal diseases and other pathogens.
- Installation and usage of equipment to localize environmental hazards and to protect noncontaminated territory and population. This may include banal nets or microwave emitters to deter insects and bats as well as highly technological medical quarantine units to observe and isolate people during incubation period.
- Infection prevention and control.
- Methods of managing the behavior of people by means of improving awareness and informing the population about risks, efficient treatment, and/or planned or urgent vaccinations.
- Training the health care system workers and key personnel from other industries, such as logisticians, water systems, and sanitation engineers, as well as media workers.

42 Advances of Artificial Intelligence in a Green Energy Environment

Biological hazards may often be prevented, and their impact may often be reduced thanks to alertness combined with clear regulatory frameworks.

For instance, thanks to the IHR (2005) [15], it was possible to localize and prevent the spread of cholera, smallpox, and yellow fever. Border crossing protocols also belong to the framework. The potential of the international regulatory framework is not yet fully used. The spread of SARS in 2003, MERS in 2011, and 2019-nCoV in 2019 illustrates that IHR requires renewal, and the states will have to mind possible impacts of all biological hazards regardless of whether they happen naturally, by chance, or on purpose.

Apart from natural disasters, man-made disasters should also be considered; they happen often enough, but cause relatively smaller damage.

There are hundreds of disasters in the world. The ratio of natural catastrophes to man-made disasters and their dynamics is presented in Ref. [16].

One more phenomenon to be classified a disaster, or at least a hazard, are allergies and allergic reactions. A representative of the World Health Organization (WHO) called the new century "the age of allergy" and the disease itself "an epidemic": according to the WHO, at least 50% of the world population will be suffering from the chronic allergy by 2025. According to the European Academy of Allergy and Clinical Immunology (EAACI), about 20%–30% of the population in Europe currently suffer from chronic allergy.

It should be noted that allergy depends on the three following factors:

- Human body, given its varying genetic and physiological characteristics, reacts differently to different stimuli. It is rather difficult to define the general pattern and therefore to elaborate the general efficient treatment. Every separate case requires its unique approach.
- Concentrations—there is no person developing no allergic reaction to allergen, it all comes to the concentration of allergen, which the body can perceive without complications.
- Seasonal and territorial factor—about 90% of population in the Amur region suffer from pollinosis, which coincides with the period of wormwood bloom. People who do not suffer from any allergies may also have allergic reaction when moving in or traveling to the region.

The scope of the global market of allergy medication is now comparable to the market of beauty products and computer games. Every year hundreds of millions of people suffer from allergic reactions; often, additional costs and reduced ability to work make this state like a disease; however, the problem is that this phenomenon is nearly impossible (especially when it comes to adults) to cure, one can only temporarily remove the symptoms.

The number of people affected the decrease of work efficiency or complete inability to work as well as the general scope of this phenomenon all prove the importance of the issue and the necessity to mind the risks of allergic threats.

Many manufacturers take it seriously and invest into the production of allergy-free goods. In 2016, Nestle transferred $145 million for 15% of share capital to Aimmune Therapeutics, a Swiss company developing food allergy medicine. Employers strive to set the workspace in such a way that allergen exposure is minimized.

To date, the only way to treat allergies is immunotherapy. In its essence lies treating the patient with gradually increasing allergen doses. It is believed to teach the body to deal with stimuli.

2.3 Overview of institutions dealing with risk management

The UN Office for Disaster Risk Reduction (UNDRR) [17] was established in 1999 and serves as the focal point in the UN System for the coordination of disaster risk reduction. It supports the implementation of the Sendai Framework for Disaster Risk Reduction 2015—30. UNISDR and partners produce the biennial Global Assessment Report (GAR) on Disaster Risk Reduction. UNISDR also coordinates the Making Cities Resilient Campaign and Worldwide Initiative for Safe Schools and engages with governments in developing national disaster loss databases.

The Centre for Research on the Epidemiology of Disasters (CRED) is the world agency for the study of public health during mass emergencies, including the epidemiology of diseases. Based since 1973 in Belgium, CRED became in 1980 a WHO collaboration center.

CRED's Emergency Events Database (EM-DAT) contains the world's most comprehensive data on the occurrence and effects of more than 23,000 technological and natural disasters from 1900 to the present day, created with the support of the WHO and the Belgian government.

WHO [18] is a specialized agency of the United Nations that is concerned with world public health. It was established on April 7, 1948 and is headquartered in Geneva, Switzerland. Its current priorities include communicable diseases, in particular HIV/AIDS, Ebola, malaria, and tuberculosis; the mitigation of the effects of noncommunicable diseases such as sexual and reproductive health, development, and aging; nutrition, food security, and healthy eating; occupational health; substance abuse; and driving the development of reporting, publications, and networking.

The International Atomic Energy Agency (IAEA) is an international organization that seeks to promote the peaceful use of nuclear energy and to inhibit its use for any military purpose, including nuclear weapons. The IAEA was established as an autonomous organization on July 29, 1957. Though established independently of the United Nations through its own international treaty, the IAEA Statute, the IAEA reports to both the United Nations General Assembly and Security Council. The programs of the IAEA encourage the development of the peaceful applications of nuclear energy, science, and technology, provide international safeguards against misuse of nuclear technology and nuclear materials, and promote nuclear safety (including radiation protection) and nuclear security standards and their implementation.

The World Allergy Organization (WAO) is an international umbrella organization whose members consist of 89 regional and national allergology and clinical immunology societies from around the world. The organization was founded in 1951. WAO provides advice and active support to member societies with the mission of building a global alliance of allergy societies which will advance excellence in clinical care, research, education, and training in allergy.

There is also a wide range of other national and international organizations and statistical offices gathering, accumulating, and analyzing data in order to keep the population informed about the risks and development dynamics.

2.4 Promising approaches to decrease risks and losses caused by disasters

Modern societies must deal with the period of climate change, overall globalization, shifts of employment and management centers, and establishment of tighter connections with those regions that used to be isolated. This causes the increase of those risks that societies are already familiar with as well as those that are new and have no efficient countermeasures. Such situations create uncertainty for managing local processes (running a city, state, or region) and also for establishing global supranational alignments. Various international organizations play in such cases an important role; their task is to gather data on all possible kinds of accidents and elaborate strategies to reduce the risks. Risk management requires well-coordinated and urgent measures aimed at comprehensive risk reduction.

Countries adopted the Sendai Framework in 2015 to address a broader scope of hazards and risks. Thus, they defined the areas of further development and improvement of mechanisms to decrease the risks of disasters and mitigate their impacts. As aforementioned, not all hazards are taken seriously enough either by general population or by politicians. This is primarily explained by:

- lack of comprehensive, accessible, and timely data;
- complete or partial misunderstanding of disasters risks and impacts, which is sometimes caused by objective processes. For example, given the territorial and climate peculiarities, the frequency of natural anomalies in a certain area is quite low; consequently, the population and government of this area disregard the precaution measures, which may be normal for other states. According to the news agency "Sevodnia," starting from December 14, 2019, India suffered from abnormally cold temperatures of $7-11°C$, with the normal temperatures being around $17-22°$; it resulted in the death of at least 28 people in 2 months as well as relocation of part of the population to warmer regions, outflow of labor force, additional costs of facility heating, and other related economic losses for both the state and population;

- incorrect assumptions—this leads to incorrect risks perception and consequently to errors in the introduction of risk reduction and impact mitigation measures. There are numerous factors causing errors, including lack of information on the given process or phenomenon, incorrectly established patterns, wrong criteria, mistakes in model synthesis, or poor precision of calculations. Some errors are caused by the inability to gather enough data or lacking computing capacity to process and model these data. One of the best and most widespread examples is weather forecast, despite all the efforts to make precise forecasts, it still remains more of a joke than standard of reliability;
- lacking or limited funds to carry out risk reduction measures. Thus, countries with weak economies have highest death toll during disasters. That is why developed countries or private individuals by means of international organizations and charities send humanitarian help and support risk reduction measures;
- negligence—unfortunately, this factor reduces all effort and precaution measures to zero. Responsibility for taking the measures, meeting the safety requirements, reliable fulfillment of the set tasks, such as design of engineering protection facilities, development of software for modeling and forecasting, data collection, organization of workplaces, and simple fire evacuation training all of these and many other measures have to be taken seriously and in accordance with the given regulations; otherwise, all the effort made by the international organizations is a waste of time and resources.

These issues are solved partially by the aforementioned international organizations and by the mass media efforts. It is also worth mentioning the importance of educational institutions (schools, gymnasiums, lyceums, technical colleges) and family upbringing; their task is not only giving the knowledge but also training the sense of responsibility and sensible perception of risks.

The following technologies to reduce the number of victims and economic damage are the most promising ones:

- Early warning systems—it is necessary to implement a global system of hazard monitoring to mitigate the impact of natural and man-made disasters. It is important to understand that one country does not have the capacity of covering all means of monitoring, it is more meaningful to gather information from all around the world, identify the threats, and timely inform the population of all countries that are at risk by means of specialized software products and high-performance computer centers. Furthermore, a big problem is the irresponsible treatment of warning systems by the population itself, e.g., a 2018 Sulawesi earthquake and tsunami in Indonesia. The combined effects of the earthquake and tsunami led to the deaths of an estimated 4340 people. More than 70,000 houses are reported to be damaged [19].

46 Advances of Artificial Intelligence in a Green Energy Environment

- Construction technologies of protective facilities—protective facilities must be built everywhere, but not every country has the capacities for their efficient design, construction, and maintenance. Unfortunately, high costs, political situation, and sanctions placed on countries, construction companies, engineering offices, and even private individuals all hinder considerably the technological exchange and prevent the construction of protective facilities by those countries possessing the needed technologies and production capacities for such projects. One of the possible solutions would be to establish a register of companies and engineering offices under the jurisdiction and protection of international organizations with the following aims: construction of protective facilities; gathering experience and technologies; facility revision, maintenance, and repair; preparation of technical documentation, standards, and safety requirements; organization of seminars and conferences. However, creation of a generally accessible technology base would mean losing the advantage and creation of a closed base would mean protection only for the chosen ones, which contradicts the principles of modern society;
- Infrastructure construction technologies that would contribute to the risk reduction of natural and man-made disasters as well as impact mitigation;
- Forecasting technologies—mentioned above are the issues of model prognosis and creation, it is now worth mentioning the perspectives of the use of artificial intelligence and methods of data mining neural networks;
- Keeping the population informed, risk estimations, reflecting actual risk dynamics, accessibility of information on effective measures of hazard countering, and impact mitigation;
- Further development of international organizations offering humanitarian help. The world changes and climate and cultural changes contribute to the alteration of different hazards; therefore, international organizations cannot avoid transformations either, they have to be able to address the changing challenges and continue to productively counter the threats and reduce the risks.

2.5 The issues of effective disaster countering organization

The impact of natural and man-made disasters may be prevented or mitigated given the proper organization of risk reduction measures, use of protective facilities, implementation of preventive measures (trainings, repair, seminars, lectures, etc.), collection and distribution of actual and relevant data on both the threat itself and ways to counter it.

The main driving force of impact mitigation are state institutions and politicians. The role of international organizations is currently limited to information support, development of recommendations and protocols of conduct in case of different disasters and hazards, and raising public awareness of new hazards. Every state makes its own decision whether to accept or reject the assistance of international organizations.

There are three main problems hindering effective threat countering and risk reduction:

- lack of responsible approach from the side of politicians and negligence of citizens;
- lack of well-organized cooperation system between the states, conflicts, and sanctions;
- lack of balanced approach combining economic development, environmental issues, social development, and general safety of population.

Lack of well-organized cooperation system between the states, conflicts, and sanctions. Current trends see not only more frequent and powerful natural disasters but also an increase of political conflicts between different countries. Exact reasons for strained relations are difficult to single out, what can be objectively mentioned is that the period of unipolar world order has come to its end. It has become uncommon for one control center to manage the agenda and direct the scientific, public, and social development. Unfortunately, some organizations, including international ones, give in to the pressure coming from separate countries and adjust their position depending on temporary political interests of separate countries rather than objective tendencies. This leads to loss of credibility and poorer efficiency of management. International organizations and charities also lose credibility when special operations of various concerned parties are carried out under the cover of their activities. It makes no difference, whether such actions are mandated by the management of the organization or UN Council or whether only their emblems and symbols are used, organizations lose their credibility or permission to carry out legal activity. There have also been numerous cases of taking representatives of international humanitarian organizations as hostages or threatening them with death if the set requirements are not fulfilled. This chapter does not discuss the issue of terrorist threat, but disagreements between the countries contribute to intensification of the risks of terrorist threats. It is rather hard to find an effective way of solving this issue; one of the possible approaches is redefining the roles of countries in the modern society depending on their real political, financial, and military capacities. The key to success lies in objective assessment and humanitarian character of the aims to be set.

Lack of balanced approach combining economic development, environmental issues, social development, and general safety of population. It has become trendy to criticize the economic development of countries, or more precisely, the impacts of economic development: carbon footprint, environmental pollution, accumulation of waste, and inefficient use of nonrenewable resources. Sharp statements provoke in many people the negative perception of economic development in general. Undeniably, for the past decades the focus has mostly been on the economic development neglecting social and environmental aspects; however, it is important to understand that economic potential brings not only ecological problems but also resources that may help to solve many threats and challenges of humanity, namely, the issue of survival and colonization of space.

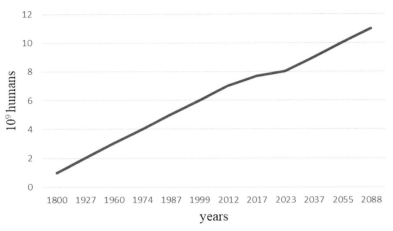

FIGURE 2.2 Population growth prospects.

The world population constitutes more than 7.7×10^9 people, and current dynamics show that it continues to grow (Fig. 2.2). Such high numbers result from the increase of the standards of living assured by the economic growth and the process of globalization rather than the natural factors of adaptability of human body or biological necessity [20]. Decrease of economic potential may lead to poorer standards of living and extinction of population. Research and modeling show that this process shall run uncontrolled and be followed by serious military conflicts. An effort to shift to Middle Age or earlier technologies may be compared in its scope and impact to the Biblical flood disaster. Modern technocratic society is strongly dependent on the levels of industrial development. Therefore, it is important to develop and increase the efficiency of new technologies to decrease the load on the environment instead of favoring emotional statements. The solution is to combine economic, social, and ecological factors as well as safety of population. Important is to implement the principles of sustainable development when assessing the risks and their dynamics. One should understand that modern society should learn to treat all decisions that have to be made with a certain level of responsibility and humanity. The decisions themselves must be balanced and reflect the variety of aspects.

2.6 Socioeconomic approach to the estimations of risk dynamics

The socioeconomic approach to the estimations of risk dynamics proposed in this research can be represented as the following sequence of steps:

- Selection of a set of criteria and indicators that will be used for decision-making. Such criteria are formed by international organizations and are

updated annually for various threats. Some examples of the criteria are response time to an ambulance call, average distance to a fire hydrant, access to a fire extinguisher indoors, average duration of providing victims with basic necessities, time to obtain personal protective equipment, and the time of placement in the protective structure. To implement the socioeconomic approach, it is necessary to consider the criteria in social, economic, and environmental spheres.
- Monitoring of the selected indicators. It is necessary to organize the timely collection of the necessary information, which will allow making adequate decisions.
- Formation of a generalized assessment of the level of expected losses and assessment of the level of risk of threats. For this it is proposed to use the theory of utility, formalization of particular criteria, and structural and parametric identification.
- Formation of a list of measures that are aimed at reducing losses or reducing the risk of a threat.
- Formation of assessments of various scenarios.
- Selection of a scenario with established restrictions.

Fig. 2.3 shows a schematic description of the method, where n is the number of possible options for alternative measures to reduce risks or mitigate consequences, and k is the number of lists that are composed of possible alternatives, taking into account the constraints (e.g., the budget size).

The development of methods and approaches to reduce the risks of natural disasters should be based on the systems analysis [21,22]; it should also comprehensively consider relevant factors. For more detailed information on the implementation of the marginal utility theory as well as the methods of systems analysis in the decision-making process aiming to reduce the risks of natural disasters, please refer to the paper. Additional information on systems theory and systems analysis can be found in the sources [23–27]. To implement specific problem structure associated with the organizational hierarchy and decomposition/aggregation techniques, the sources [28–32] can be used.

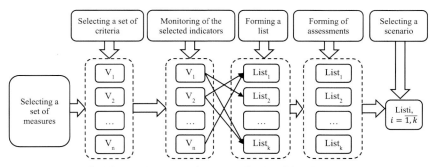

FIGURE 2.3 Conceptual framework for describing the method.

50 Advances of Artificial Intelligence in a Green Energy Environment

At present, literally the entire territory of our planet is at risk of natural or man-made disaster.

Monitoring and statistical analysis of data make it possible to identify the risks of disasters. They also help to quantitatively estimate the efficiency of various measures of disaster risk reduction or measures to mitigate the consequences of their occurrence.

In this context, efficiency of measures may encompass:

- reduced number of victims;
- reduction of material damage;
- reduced environmental impact.

In the process of setting the task of choosing various scenarios of minimizing the risks of natural disasters or their consequences, one may face the following:

- Scenarios to be chosen may be mutually exclusive, they may also decrease or increase the effect of each other's implementation, whereas the implementation effect of a few scenarios to reduce one certain threat may grow nonlinearly.
- Scenario selection process may take place in resource-limited settings.
- Implementation effect may depend on the quality of management, environmental, or cultural conditions, etc.

Currently, most measures to reduce the impact of natural disasters are limited to economic aspects; however, this approach cannot solve the issue.

One of the ways to reduce the consequences of natural disasters can be the reengineering of electric power systems [33]. This makes it possible to create a more flexible and stable system for delivering electricity to the consumer, which will allow building a system for countering threats at a higher quality level.

When organizing the process of monitoring, collecting statistical data, and obtaining expert assessments, it is important to ensure a prompt and nonintrusive authentication of users; for more details on various authentication methods, please refer to the paper [34].

2.7 Conclusions

One of the biggest challenges of modern society is to build resilience to natural disasters. As shown in the current study, there are two main ways to do so: reducing the risks of natural disasters and reducing the impact that natural disasters cause to the infrastructure and population. The authors conclude that to choose the most efficient solution to cope with natural disasters, it is necessary to list all possible measures, evaluate them by means of the marginal utility theory, decision theory, and elements of systems analysis, and consequently select those measures that would give the maximum effect with limited funding. Economic, social, and environmental factors should be considered when developing such a list.

Disasters impact assessment based on socioeconomic approach **Chapter | 2** **51**

The proposed methods make it possible to evaluate measures to prevent damage or reduce the consequences of natural disasters and man-made disasters, due to the complex consideration of management factors and consequences.

To evaluate alternatives, information provided by international organizations or experts is used. Unfortunately, expert assessments have a high level of subjectivity, and the assessments formed by international organizations often do not consider specific territorial features. This is due to the scales and averages that are objectively applied by international organizations. In the future, it is advisable to develop an approach that will allow the formation of more reliable assessments of the effectiveness of activities; for this purpose, the technology of neural networks and deep learning can be applied.

Appendix A: An example of selecting the optimal list of prevention or mitigation measures

We demonstrate the work of the method on the example of the developed software tool, which determines the action list to reduce the risks and losses caused by natural disasters under a limited budgetary resource.

The input data for the software tool are tables that contain the following information provided by experts:

— a list of disasters, the likelihood of their occurrence, and the losses they are likely to cause;
— a list of measures aimed at reducing the likelihood of occurrence and/or losses of a particular disaster;
— the size of the allocated budget.

The output data of the software tool are:

— list of measures that can be performed, under budget constraints;
— effect of the implemented actions.

An example of input data is represented in Tables 2.1 and 2.2. The tables contain data generated randomly. They cannot be used to characterize any region. The calculation aim is demonstration of the proposed method (Table 2.3).

The interface of the software tool is shown in Fig. 2.4. The tool allows to load data files and view their contents in the corresponding tabs: Disasters and Activities (Fig. 2.4).

The next step is selecting the algorithm for the list of measures, enter the budget size, and click "Calculate" (Fig. 2.5).

The calculation result is shown in Fig. 2.6 in the Results tab. This is a list of measures that will give the maximum reduction of losses and risks.

52 Advances of Artificial Intelligence in a Green Energy Environment

TABLE 2.1 Example of a list of natural disasters (part of the data is given).

Disaster code	Disaster	Likelihood of occurrence	Losses ($)
1	Earthquakes	0.62	42,916
2	Hurricane	0.80	37,948
3	Forest fires	0.88	46,951
4	Flood	0.92	48,806
5	Landslide	0.54	26,502
6	Volcanic eruptions	0.18	20,390
7	Tsunami	0.75	12,469
8	Typhoon	0.30	46,375
9	Dust storms	0.82	43,534
10	Epidemic	0.93	31,833
11	Drought	0.55	48,633
12	Extreme temperatures	0.86	23,838

TABLE 2.2 Example of a list of risk mitigation measures (part of the data is given).

Measure code	Disaster code	Reducing the likelihood of a natural disaster	Reducing disaster losses	Cost of implementation ($)
1	10	0.5	0.48	2335
2	6	0.66	0.35	9752
3	3	0.12	0.23	86,230
4	17	0.76	0.3	46,984
5	14	0.13	0.65	10,789
6	18	0.42	0	75,828
7	5	0.15	0.11	99,455
8	10	0.62	0.18	25,206
9	4	0.43	0.1	22,798
10	2	0	0.36	97,551

TABLE 2.3 Example of intermediate calculations, the definition of normalized values of the criteria in determining the generalized evaluation of the proposed measures.

Measure code	Disaster code	Reducing the likelihood of a natural disaster	Reducing disaster losses	Cost of implementation ($)	Normalized first criteria	Normalized second criteria	Summary evaluation
1	10	0,5	0,48	2335	0.657895	0.738462	0.722348
2	6	0,66	0,35	9752	0.868421	0.538462	0.604453
3	3	0,12	0,23	86,230	0.157895	0.353846	0.314656
4	17	0,76	0,3	46,984	1	0.461538	0.569231
5	14	0,13	0,65	10,789	0.171053	1	0.834211
6	18	0,42	0	75,828	0.552632	0	0.110526
7	5	0,15	0,11	99,455	0.197368	0.169231	0.174858
8	10	0,62	0,18	25,206	0.815789	0.276923	0.384696
9	4	0,43	0,1	22,798	0.565789	0.153846	0.236235
10	2	0	0,36	97,551	0	0.553846	0.443077
Weighting variant					0.2	0.8	

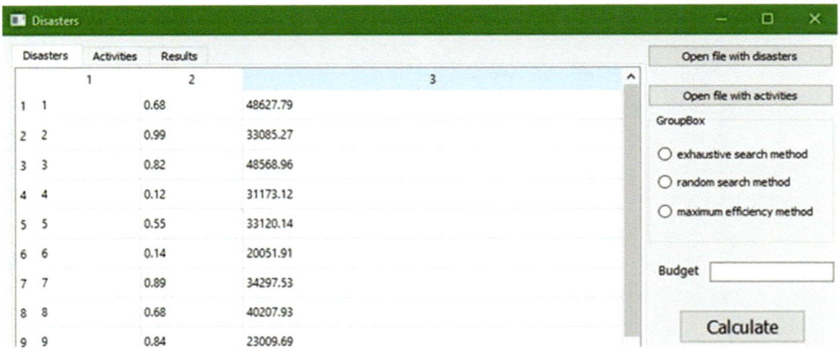

54 Advances of Artificial Intelligence in a Green Energy Environment

FIGURE 2.4 Example of loading and displaying input data. 1—disaster code; 2—likelihood of occurrence; 3—losses ($). The columns are the same as the columns in Table 2.1.

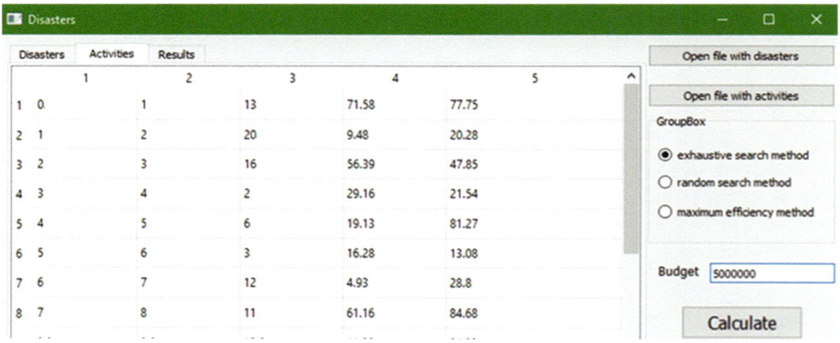

FIGURE 2.5 Algorithm selection and input the size of the allocated budget. 1—row index; 2—measure code; 3—disaster code; 4—reducing the likelihood of a natural disaster; 5—reducing disaster losses; 6—cost of implementation ($). (Not present in the screenshot, you have to scroll to the right in the software tool). The columns are the same as the columns in Table 2.2.

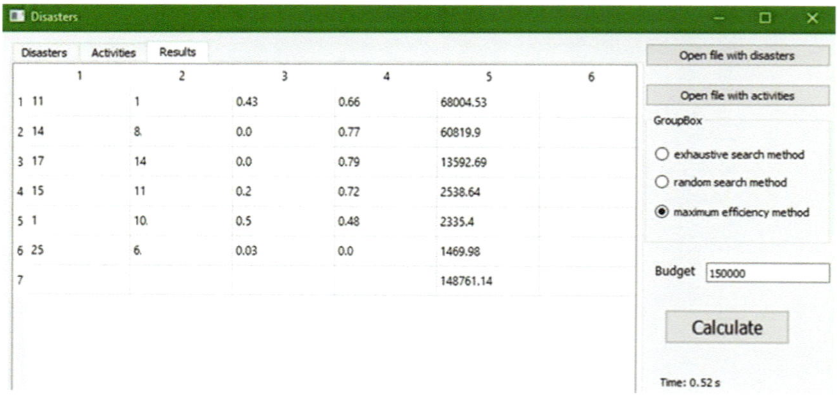

FIGURE 2.6 A list of measures that will give the maximum reduction of losses and risks.

References

[1] I. Grebennik, O. Khriapkin, A. Ovezgeldyyev, et al., The concept of a regional information-analytical system for emergency situations, in: Y. Murayama, D. Velev, P. Zlateva (Eds.), Information Technology in Disaster Risk Reduction. ITDRR 2017, vol. 516, 2019, pp. 25−35, https://doi.org/10.1007/978-3-030-18293-9_6.

[2] D. Debroasa, A. Monea, H.A. Ter, Plastics in the North Atlantic garbage patch: a boat-microbe for hitchhikers and plastic degraders, Sci. Total Environ. 599−600 (2017) 1222−1232.

[3] I. Grebennik, V. Semenets, Y. Hubarenko, Information technologies for assessing the impact of climate change and natural disasters in socio-economic systems, in: Y. Murayama, D. Velev, P. Zlateva (Eds.), Information Technology in Disaster Risk Reduction. ITDRR 2019, vol. 575, 2020, pp. 21−30, https://doi.org/10.1007/978-3-030-48939-7_3.

[4] J. Rossell O, S. Becken, M. Santana-Gallego, The effects of natural disasters on international tourism: a global analysis, Tourism Manag. 79 (2020) 104080, https://doi.org/10.1016/j.tourman.2020.104080.

[5] E. Esterwood, S.A. Saeed, Past epidemics, natural disasters, COVID19, and mental health: learning from history as we deal with the present and prepare for the future, Psychiatr. Q. 91 (2020) 1121−1133, https://doi.org/10.1007/s11126-020-09808-4.

[6] Karlsruhe Institute of Technology, Natural Disasters since 1900: Over 8 Million Deaths and 7 Trillion US Dollars Damage, Resource Document, Karlsruhe Institute of Technology, 2016, https://www.kit.edu/kit/english/pi_2016_058_natural-disasters-since-1900-over-8-million-deaths-and-7-trillion-us-dollars-damage.php. (Accessed 26 March 2020).

[7] J.D. Daniell, A.M. Schaefer, F. Wenzel, Losses associated with secondary effects in earthquakes, Front. Built Environ. 3 (2017) 30, https://doi.org/10.3389/fbuil.2017.00030.

[8] AON, Weather, Climate & Catastrophe Insight. 2019 Annual Report, Resource Document, AON, 2019, https://www.aon.com/weather-climate-catastrophe-insight-2019/index.jsp. (Accessed 26 March 2020).

[9] CRED, UNDRR, Economic Losses, Poverty & Disasters 1998-2017, Centre for Research on the Epidemiology of Disasters, UN Office for Disaster Risk Reduction, 2018. https://reliefweb.int/report/world/economic-losses-poverty-disasters-1998-2017. (Accessed 26 March 2020).

[10] S. Sebnem, S. Apurva, F. Alejandro, A. Bianca, Natural Hazards Unnatural Disasters, The World Bank, Washington DC, 2010, https://doi.org/10.1596/978-0-8213-8050-5.

[11] United Nations General Assembly, Conventional Arms Control at the Regional and Sub-regional Levels: Report of the Secretary-General, 2016. UN Doc. A/71/154.

[12] UNDRR, Global Assessment Report on Disaster Risk Reduction, United Nations Office for Disaster Risk Reduction (UNDRR), Geneva, Imprimerie Centrale, 2019. https://gar.undrr.org/sites/default/files/reports/2019-05/full_gar_report.pdf. (Accessed 26 March 2020).

[13] World Health Organization, World Health Statistics 2019: Monitoring Health for the SDGs, Sustainable Development Goals, WHO Press, Switzerland, 2019. https://apps.who.int/iris/bitstream/handle/10665/324835/9789241565707-eng.pdf?sequence=9&isAllowed=y. (Accessed 26 March 2020).

[14] World Health Organization, World Health Statistics 2015, WHO Press, Luxembourg, 2015. https://www.who.int/docs/default-source/gho-documents/world-health-statistic-reports/world-health-statistics-2015.pdf. (Accessed 26 March 2020).

56 Advances of Artificial Intelligence in a Green Energy Environment

[15] World Health Organization, World Health Statistics 2016: Monitoring Health for the SDGs, Sustainable Development Goals, WHO Press, France, 2016. http://apps.who.int/iris/bitstream/10665/206498/1/9789241565264_eng.pdf?ua=1. (Accessed 26 March 2020).

[16] L. Bevere, A. Ehrler, V. Kumar, R. Lechner, et al., Natural Catastrophes and Man-Made Disasters in 2018: "Secondary" Perils on the Frontline, Swiss Re Institute, Switzerland, 2019.

[17] United Nations Office for Disaster Risk Reduction, 2012. https://www.undrr.org. (accessed 20 March 2020).

[18] World Nuclear Association, Basic Documents Forty-Ninth Edition. Resource Document World Nuclear Association, 2019. https://apps.who.int/gb/bd/pdf_files/BD_49th-en.pdf#page=7. (Accessed 26 March 2020).

[19] S. Sassa, T. Takagawa, Liquefied gravity flow-induced tsunami: first evidence and comparison from the 2018 Indonesia Sulawesi earthquake and tsunami disasters, J. Landslides 16 (2019) 195−200, https://doi.org/10.1007/s10346-018-1114-x.

[20] United Nations Population Division, World Population Prospects the 2017 Revision Key Findings and Advance Tables, 2017. New York, United Nations, https://population.un.org/wpp/Publications/Files/WPP2017_KeyFindings.pdf. (Accessed 26 March 2020).

[21] P. Dudley (Ed.), Bogdanov's Tektology, Centre for Systems Studies, UK, Hull, 1996.

[22] L. Bertalanffy, Ludwig General System Theory, George Braziller, New York, 1969.

[23] N. Bourbaki, Théorie des ensembles, Diffusion, Paris, 1970.

[24] G.-H. Tzeng, J.J. Huang, Multiple Attribute Decision Making: Methods and Applications, CRC Press Taylor & Francis Group, US, 2011.

[25] J.-R. Wang, Y. Zhou, M. Medved, Existence and stability of fractional differential equations with hadamard derivative, J. Topol. Methods Nonlinear Anal. 41 (1) (2013) 113−133.

[26] T. Jech, Set Theory, Springer-Verlag, Berlin, 2006.

[27] L. Anastasakis, N. Mort, The Development of Self-Organization Techniques in Modelling: A Review of the Group Method of Data Handling (GMDH), The University of Sheffield, Sheffield, 2001.

[28] I. Litvinchev, Decomposition-aggregation method for convex programming problems, Optimization 22 (1) (1991) 47−56.

[29] I. Litvinchev, S. Rangel, Localization of the optimal solution and a posteriori bounds for aggregation, Comput. Oper. Res. 26 (10−11) (1999) 967−988.

[30] I. Litvinchev, M. Mata, S. Rangel, J. Saucedo, Lagrangian heuristic for a class of the generalized assignment problems, Comput. Math. Appl. 60 (4) (2010) 1115−1123.

[31] T.E. Romanova, P.I. Stetsyuk, A.M. Chugay, S.B. Shekhovtsov, Parallel computing technologies for solving optimization problems of geometric design, Cybern. Syst. Anal. 55 (2019) 894−904, https://doi.org/10.1007/s10559-019-00199-4.

[32] T. Romanova, Y. Stoyan, A. Pankratov, I. Litvinchev, J.A. Marmolejo, Decomposition algorithm for irregular placement problems, in: P. Vasant, I. Zelinka, G.W. Weber (Eds.), Advances in Intelligent Systems and Computing: Intelligent Computing and Optimization, vol. 1072, 2019, pp. 214−221, https://doi.org/10.1007/978-3-030-33585-4_21.

[33] I. Grebennik, A. Ovezgeldyyev, Y. Hubarenko, M. Hubarenko, Information technology reengineering of the electricity generation system in post-disaster recovery, in: Y. Murayama, D. Velev, P. Zlateva (Eds.), Information Technology in Disaster Risk Reduction. ITDRR 2019, vol. 575, 2020, pp. 9−20, https://doi.org/10.1007/978-3-030-48939-7_2.

[34] A. Nechiporenko, E. Gubarenko, M. Gubarenko, Authentication of users of mobile devices by their motor reactions, J. Telecommun. Radio Eng. 78 (11) (2019) 987−1003.

Chapter 3

Uninterruptible power supply system of the consumer, reducing peak network loads

V.A. Gusarov[1], L. Yu Yuferev[1] and O.F. Gusarova[2]
[1]*FGBNU FNATS VIM, Moscow, Russian Federation;* [2]*FAEIHE "RUT" Moscow, Russian Federation*

The purpose of this work is to develop a new method of highly efficient power supply to consumers, ensuring its high reliability, lower capital costs, and the cost of generated energy.

3.1 Introduction

Currently, many urban and rural areas are supplied with electricity from district thermal power station (DTPS), combined with a common power system or operating autonomously [1–3]. The calculation of the power of DTPS is carried out according to the peak daily power of the connected load, which is largely uneven [4–14]. In the daytime and in the evening, when industrial enterprises operate in cities, the load power is much higher than at night [15,16]. Fig. 3.1 shows an approximate graph of the daily load capacity of the city, where there are industrial enterprises.

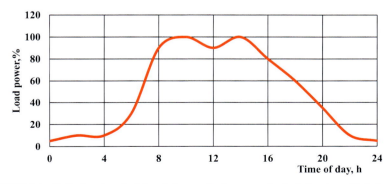

FIGURE 3.1 Daily hourly schedule of changes in the power load of the city with industrial enterprises.

58 Advances of Artificial Intelligence in a Green Energy Environment

In rural areas, where there are few industrial enterprises and where the private residential sector prevails, the maximum load is observed in the morning and evening.

For a rural settlement, consisting of 150 residential buildings with an average daily electricity consumption of 5 kWh per house, the total daily consumption of a rural settlement is 750 kWh [17−20].

Fig. 3.2 shows the daily load capacity graph of a small rural town.

DTPS contains boiler equipment, which can use natural gas, fuel oil, coal, diesel fuel, etc., as fuel, produces superheated steam, and supplies it to the turbine of the generator, the electricity of which, like the waste steam, is supplied through the conversion equipment to the electric and heat supply to consumers [21−23].

3.2 Methodology of the work

To maintain the standard frequency and voltage of the network, the DTPS steam generating device maintains a stable generator speed, and when the load power changes from 100% to 5%, in order to avoid dips in the generator speed and due to the energy content of the boiler equipment, fuel consumption is reduced by no more than 15%. For this reason, the generation of electricity at night is not very efficient and often unprofitable [24].

Fig. 3.3 shows the graphs of the load power and fuel consumption of an DTPS of a small rural town with 150 residential buildings per day.

Based on the graph shown in Fig. 3.3, the fuel consumption level of the DTPS at the minimum load power differs little from the flow rate with the maximum load power.

The dynamics of change in load power over time of DTPS is a set of graphs of electricity consumption by each consumer.

Electricity is generated by a 120 kW steam turbine unit with a specific fuel consumption of 0.275 kg/kW h.

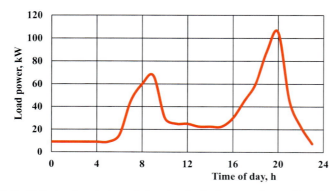

FIGURE 3.2 Daily hourly schedule of changing the load power of a small rural town with a private residential sector.

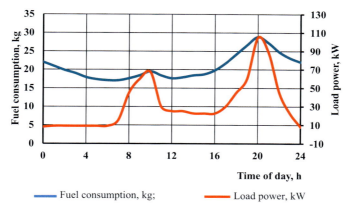

FIGURE 3.3 Hourly schedule of changes in the load power of a rural settlement and fuel consumption of DTPS.

In accordance with the consumption graph shown in Fig. 3.3, the daily fuel consumption of DTPS, $Q_{сут.}^{р.т}$, kWh, is

$$Q_{сут.}^{р.т} = \sum_{0}^{24} Q_{h}^{р.т} = 496 \text{ kg,} \qquad (3.1)$$

where $Q_{час}^{р.т.}$ is the average hourly fuel consumption.

The daily energy produced by the generator is equal to its daily consumption:

$$W_{сут.}^{пр.э} = W_{сут.}^{н.п.}, \qquad (3.2)$$

where $W_{сут.}^{н.п.}$ is the average daily electricity consumption by the settlement and $W_{сут.}^{пр.э}$ is the average daily power generation by DTPS.

To improve the efficiency of the power supply system, an uninterruptible power supply system (UPS) has been developed, which differs from the existing analogs, allows to reduce the peak power of the connected load, and is presented in Fig. 3.4 [25,26].

The UPS is located at the switchboard of each electricity consumer in the apartment of an apartment building, private residential building, office, etc. It

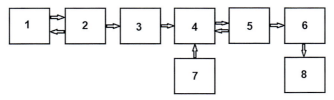

FIGURE 3.4 Block diagram of the uninterruptible power supply system. (1) DTPS; (2) metering device; (3) power limiter; (4) distribution block; (5) storage battery; (6) inverter; (7) renewable energy sources; and (8) load.

consists of an electronic conversion device and storage batteries (AB), the capacity of which is determined by the daily energy consumption. With an energy consumption of 5 kWh per day, it is recommended to install an AB with a consumption of 8–10 kWh; this is four ABs of 150–200 Ah each. In this case, the reserve of electricity ensures the operation of electrical equipment; in the event of a network outage, for 1 day, it will reduce the depth of discharge and increase the service life of the battery. UPS can operate without renewable energy sources.

Electricity of DTPS through network 1 goes to the consumer to the electric meter 2, after which to the power limiter (OM) 3, which regulates the required level of current supplied to the energy distribution unit (BR) 4. The BR converts the voltage and produces charges AB 5 from the network and from seven renewable energy sources connected to the device: solar PV modules, wind turbines, etc. An inverter (Inv) 6 is connected to the AB, which provides the load 8 with standard electricity [27,28]. The power of the inverter is selected depending on the maximum power of the electrical appliances only switched on in a certain period of time, but with a margin of 20%–30% of the power.

The hourly power graph of the connected electrical appliances of a rural residential building (Fig. 3.5) corresponds in shape to the graph shown in Fig. 3.2, where the maximum power in the period from 19^{00} to 20^{00} is 0.7 kW, and the total daily energy consumption is 5000 Wh.

The uniformity of energy consumption during the day by one rural house implies uniform consumption by the entire settlement, consisting of 150 houses.

The power limiter provides the battery with a set current level, which is calculated according to formula 3.3:

$$I = \frac{W^{с.д.}_{сут}}{24 \times U}, \qquad (3.3)$$

where $W^{с.д.}_{сут}$ is the daily consumption of electricity by a rural house, kWh;

U is the mains voltage, 220 V;

I is the battery charge current, A;

24—hours a day, h.

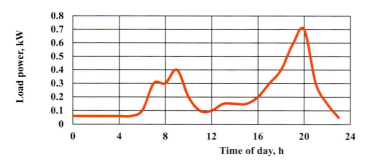

FIGURE 3.5 Hourly schedule of changes in the power of connected electrical appliances by the time of a rural residential building.

Uninterruptible power supply system of the consumer **Chapter | 3 61**

$$I = \frac{5000 \text{Вт} \cdot \text{ч}}{24 \times U} = 0,94 \text{ A} \approx 1,0 \text{ A}. \qquad (3.4)$$

Thus, the consumer, taking into account the battery charge, is uniformly supplied with electricity during the day with a voltage of 220 V and a current of 1.0 A. In the case of a full battery charge, the average daily energy received by the consumer is (Wh)

$$W_{\text{сут}}^{\text{с.д.}} = I \cdot U \cdot 24 = 220 \cdot 1 \cdot 24 \approx 5000 \text{ Wh}. \qquad (3.5)$$

The average daily consumption of electricity by a settlement is determined (kWh):

$$W_{\text{сут}}^{\text{н.п.}} = I \cdot U \cdot 24 \cdot 150, \qquad (3.6)$$

where $W_{\text{сут}}^{\text{н.п.}}$ is the daily consumption of electricity by the settlement, kWh; 150 is the number of houses in the settlement.

$$W_{\text{сут}}^{\text{н.п.}} = 1 \cdot 220 \cdot 24 \cdot 150 = 792 \text{ kWh}. \qquad (3.7)$$

Based on formula 3.2, the daily power generation of DTPS (kWh) is

$$W_{\text{сут.}}^{\text{пр.э}} = 792 \text{kWh}. \qquad (3.8)$$

Hence, the maximum current at the peak time of a rural house from 1900 to 2000, $I_{\text{н.п.}}$, A is

$$I_{\text{н.п.}} = P_{\text{мах}}^{\text{с.д.}} : U, \qquad (3.9)$$

where $P_{\text{мах}}^{\text{с.д.}}$ is the maximum power of working electrical appliances at the peak time of a rural house (W).

The maximum current of the consumer at peak time, $I_{\text{с.д.}}$, A, is

$$I_{\text{с.д.}} = 700 : 220 = 3,18 \text{ A}. \qquad (3.10)$$

The maximum power of the DTPS generator at peak time, $P_{\text{мач}}^{\text{Ген.}}$, KW, is

$$P_{\text{без СБП}}^{\text{Ген.}} = I_{\text{с.д.}} \cdot U \cdot 150, \qquad (3.11)$$

$$P_{\text{без.СБП}}^{\text{Ген.}} = 3,18 \cdot 220 \cdot 150 = 105 \text{ kW} \qquad (3.12)$$

Based on the peak load power, the DTPS electric generator corresponds to a power of 120 kW.

In power supply systems, where UPS is used with power redundancy and uniform consumption, the generator power, kW, is

$$P_{\text{СБП}}^{\text{Ген.}} = W_{\text{сут}}^{\text{н.п.}} : 24, \qquad (3.13)$$

$$P_{\text{СБП}}^{\text{Ген.}} = 792 : 24 = 33 \text{ kW} \qquad (3.14)$$

Based on the power of the load, acting evenly throughout the day of the settlement, an DTPS with an electric generator with a capacity of 40 kW is selected. The average daily fuel consumption of such an DTPS, $Q^{\text{р.т}}_{\text{сут. СБ}_\Pi}$, kg·h, is

$$Q^{\text{р.т}}_{\text{сут. СБ}_\Pi} = 33 \cdot 0,275 \cdot 24 = 217,8 \text{ kg}. \tag{3.15}$$

3.3 Work results

Fig. 3.6 shows a graph of the hourly fuel consumption of the DTPS and the consumption of electricity by a rural settlement with a UPS.

All peak load power in the morning and in the evening is fully covered by the accumulated energy of consumers. Comparative fuel consumption of a power supply system with a UPS system and a power supply system without a UPS system is shown in Fig. 3.7.

Fuel economy per day, $Q^{\text{эк.}}_{\text{сут.}}$, Kg, is

$$Q^{\text{эк.}}_{\text{сут.}} = Q^{\text{р.т}}_{\text{сут.}} - Q^{\text{р.т}}_{\text{сут.СБ}_\Pi}, \tag{3.16}$$

$$Q^{\text{эк.}}_{\text{сут.}} = 496 \text{ kg} - 217,8 \text{ kg} = 278,2 \text{ kg}. \tag{3.17}$$

The load current of one consumer at peak time without UPS (10) is 3.18 A; therefore, the entire settlement, $I^{\text{н.п.}}_{\text{без СБ}_\Pi}$, A, is

$$I^{\text{н.п.}}_{\text{без СБ}_\Pi} = 3,18 \cdot 150 = 477 \text{ A}. \tag{3.18}$$

The load current of one consumer with a UPS (4) is 1 A; the current strength of the entire settlement, $I^{\text{н.п.}}_{\text{СБ}_\Pi}$, A is

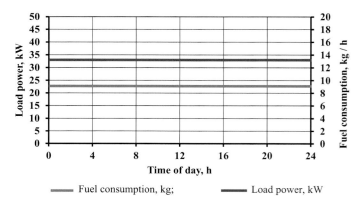

FIGURE 3.6 Schedule of DTPS hourly fuel consumption and electricity consumption of a rural settlement with uninterruptible power supply systems.

Uninterruptible power supply system of the consumer **Chapter | 3** 63

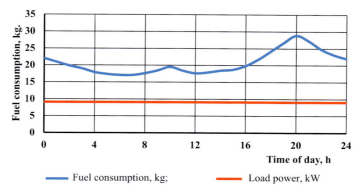

FIGURE 3.7 Comparative fuel consumption of a power supply system with an uninterruptible power supply system and a power supply system without an uninterruptible power supply system.

With a current density standard of 5 A/mm², the cross section of the wire providing power supply to the consumer without a UPS, $S_{\text{СБП}}^{\text{с.д..}}$, mm², is

$$S_{\text{СБП}}^{\text{с.д..}} = 1,0 \text{ A}: 5,0 \text{ mm}^2/\text{A} = 0,2 \text{ mm}^2. \quad (3.21)$$

The cross section of the wire providing power supply to the settlement without UPS, $S_{\text{без СБП}}^{\text{н.п.}}$, mm², is

$$S_{\text{без СБП}}^{\text{н.п.}} = 477 \text{ A}: 5 \text{ mm}^2/\text{A} = 95,4 \text{ mm}^2 \quad (3.22)$$

The cross section of the wire providing the power supply of the settlement with the UPS, $S_{\text{СБП}}^{\text{н.п.}}$, mm², is

$$S_{\text{СБП}}^{\text{н.п.}} = 150 \text{ A}: 5 \text{ mm}^2/\text{A} = 30 \text{ mm}^2. \quad (3.23)$$

Saving costs of cable products, relative to the traditional power supply system, is about 200%.

3.4 Conclusions

The use of a UPS unit with power redundancy allows the consumer to provide the necessary high-quality level of power supply, choosing the capacity of the battery station and installing its own additional autonomous power supply source of the required power. For capacity of storage batteries, the consumer, in an emergency, provides himself with a reserve of energy for several days. The developed power supply system can bring a tangible economic effect to power supply companies, stimulate the implementation of measures to reduce the cost of cable products and steam generating equipment, reduce harmful emissions, and thereby improve the environmental situation.

References

[1] D. Meha, A. Pfeifer, N. Duic, Increasing the integration of variable renewable energy in coal-based energy system using power to heat technologies: the case of Kosovo, Energy (2020). T. 212. Article number: 118762.

[2] O.V. Gazizova, R.M. Nigamatullin, Assessment of the influence of static load characteristics on the level of the network frequency when working separately with the power system, Bull. South Ural State Univ. Ser. Energy. 20 (4) (2020) 54−63.

[3] B.N. Abramovich, D.A. Ustinov, V.D. Abdalla, Substantiation of parameters of a DC-DC converter for a source of autonomous power supply, Bull. South Ural State Univ. Ser. Energy. 20 (4) (2020) 86−95.

[4] T. Deng, L. Tian, B. Hu, Dynamic availability of energy storage in district heating networks for automatic generation control of a CHP plant, Appl. Therm. Eng. (2021). T. 183. N. 1. Article number: 116198.

[5] S. Monie, A.M. Nilsson, J. Widen, A residential community-level virtual power plant to balance variable renewable power generation in Sweden, Energy Convers. Manag. (2021). T.228. Article number: 113597.

[6] M. Mahmoud, M. Ramadan, S. Naher, Recent advances in district energy systems: a review, Therm. Sci. Eng. Prog. (2020). T. 20. Article number: 100678.

[7] Y. Noorollahi, A. Golshanfard, S. Ansaripour, Solar energy for sustainable heating and cooling energy system planning in arid climates, Energy (2021). T.218. Article number: 119421.

[8] M. Doepfert, R. Castro, Techno-economic optimization of a 100% renewable energy system in 2050 for countries with high shares of hydropower: the case of Portugal, Renew. Energy (2021) 491−503. T. 165. N. 1.

[9] Y. Bayazit, R. Bakis, C. Koc, A study on transformation of multi-purpose dams into pumped storage hydroelectric power plants by using GIS model, Int. J. Green Energy 18 (3) (2021) 308−318.

[10] L. Al-Ghussain, A.M. Abubaker, A. Darwish, Superposition of renewable-energy supply from multiple sites maximizes demand-matching: towards 100% renewable grids in 2050, Appl. Energy (2021). T. 284. Article number: 116402.

[11] Y.M. Golembiovskiy, Y.B. Tomashevskiy, A.A. Shcherbakov, D.Y. Lukov, A.V. Starkov, Standalone single phase inverter with high quality output voltage, Bull. South Ural State Univ. Ser. Energy 18 (1) (2018) 75−81.

[12] I.G. Chernenko, On the need for a comprehensive modernization of heat supply systems in cities, Energy Secur Energy Sav. (6) (2019) 13−19.

[13] E.B. Solntsev, A.M. Mamonov, A.N. Fitasov, S.A. Petritsky, A.A. Sevostyanov, Analysis of power quality (voltage fluctuations) in distributed generation, Energy Secur Energy Sav. (3) (2019) 37−40.

[14] V.A. Barinov, A.S. Manevich, A.S. Murachev, Application in power systems of a new class of distributed network controlled devices, Electron. Electr. Eng. Power Eng. Electr. Ind. (3) (2017) 2−7.

[15] N.I. Cherkasova, Schematic diagram of the rural power supply system, differentiated by voltage levels, Electron. Electr. Eng. Power Eng. Electr. Ind. (1) (2017) 8−12.

[16] W. Zheng, Y. Zhang, J. Xia, Cleaner heating in Northern China: potentials and regional balances, Resour. Conserv. Recycl. (2020). T. 160. Article number: 104897.

[17] F. Huang, Z. Wang, J. Liu, Exploring rural energy choice from the perspective of multi-dimensional capabilities: evidence from photovoltaic anti-poverty areas in rural China, J. Clean. Prod. (2021). T.283. Article number: 124586.

[18] D. Satola, A.B. Kristiansen, A. Houlihan-Wiberg, Comparative life cycle assessment of various energy efficiency designs of a container-based housing unit in China: a case study, Build. Environ. (2020). T.186. Article number: 107358.

[19] S.C. Chua, T.H. Oh, Review on Malaysia's national energy developments: key policies, agencies, programs and international involvements, Renew. Sustain. Energy Rev. 14 (9) (2020) 2916−2925.

[20] D.K. Twerefou, J.O. Abeney, Efficiency of household electricity consumption in Ghana, Energy Policy (2020). T.144. Article number: 111661.

[21] L.G. Fernandes, A.A. Badin, D.F. Cortez, Transformerless UPS system based on the half-bridge hybrid switched-capacitor operating as AC-DC and DC-DC converter, IEEE Trans. Ind. Electron. (2021) 2173−2183. T. 68. N. 3.

[22] C. Zheng, T. Dragicevic, Z. Zhang, Model predictive control of LC-filtered voltage source inverters with optimal switching sequence, IEEE Trans. Power Electron. (2021) 3422−3436. T. 36. N. 3.

[23] J. Beiron, R.M. Montanes, F. Normann, Combined heat and power operational modes for increased product flexibility in a waste incineration plant, Energy (2020). T. 202. Article number: 117696.

[24] K. Koch, P. Hoefner, M. Gaderer, Techno-economic system comparison of a wood gas and a natural gas CHP plant in flexible district heating with a dynamic simulation model, Energy (2020). T. 202. Article number: 117710.

[25] A. Bartolini, F. Carducci, C.B. Munoz, Energy storage and multi energy systems in local energy communities with high renewable energy penetration, Renew. Energy (159) (2020) 595−609.

[26] Y. Fang, S. Zhao, Risk-constrained optimal scheduling with combining heat and power for concentrating solar power plants, Sol. Energy 208 (2020) 937−948.

[27] Z. Zhang, C. Dou, D. Yue, High-economic PV power compensation algorithm to mitigate voltage rise with minimal curtailment, Int. J. Electr. Power Energy Syst. (2021). T. 125. Article number: 106401.

[28] C. Alfaro, M. Castilla, A. Camacho, A distributed control for accurate active-power sharing in islanded microgrids subject to clock drifts, IET Power Electron. (2021) 518−530. T. 14. N. 3.

Chapter 4

Optimization of the anaerobic conversion of green biomass into volatile fatty acids for further production of high-calorie liquid fuel

Marina A. Gladchenko and Sergey N. Gaydamaka
Department of Chemical Enzymology, Chemistry Department, Lomonosov Moscow State University, Moscow, Russia

4.1 Introduction

Safe and economically viable biological processing of agro-industrial waste is one of the most serious problems in the world today [1]. Untreated solid wastes such as straw, sawdust, cellulose, lignine, pharmaceutical waste, and liquid food waste and associated industrial wastewater globally pollute the environment. Recycling carbonaceous waste into energy sources such as biofuels based on volatile fatty acids (VFAs) (ethanoic, propanoic, and butanoic) [2,3] appears to be both environmentally and economically attractive.

At the same time, lignocellulose remains a significant part of agro-industrial waste. This material is an object for the biological production of fuel. Bioprocesses with the participation of specialized microbial associations occupy one of the main places in the processing of these wastes. The most studied and energetically beneficial is the process of anaerobic conversion of biomass into biogas [4−6], where the main component is methane (methanogenesis):

$$(CH_2O)_2 \rightarrow CH_4 + CO_2 \tag{4.1}$$

In this case, the difference in chemical nature components of the waste (proteins, carbohydrates, lipids, etc.) is transformed into the combustible gas. The biogas yields can be achieved of up to 90% depending on the process conditions and the threated materials [7].

Advances of Artificial Intelligence in a Green Energy Environment
https://doi.org/10.1016/B978-0-323-89785-3.00007-4
Copyright © 2022 Elsevier Inc. All rights reserved.

68 Advances of Artificial Intelligence in a Green Energy Environment

However, methanogenesis has a number of disadvantages: it is a relatively low activity of the natural methane-generating consortium of microorganisms; the complexity of the biological destruction of lignocellulose polymeric complexes; inhibition of the process of methanogenesis, due to the transition of the system to a state of hyperproduction of organic acids; and also a sharp decrease in the pH of the medium.

As mentioned earlier, lignocellulosic substrates are difficult to decompose and, therefore, this materials are small extent subject to significant transformation to biogas. But the initial stage of the methanogenesis, namely, obtaining VFAs from these substrates, proceeds quite efficiently. Therefore, the hyperproduction of VFAs was of interest by blocking further stages of the methanogenesis process for the further production of liquid biofuel (in the form of ethyl esters of fatty acids), which can be used alone or as an additive, for example, to diesel fuel with little or no engine modification [8]. Esters of VFAs (EVFAs), which have found a noticeable octane-increasing effect in petrol, can be considered as an analog of traditional biological diesel produced by alcoholysis of triglycerides [8].

At low pH values of the medium, the conversion of biomass can proceed with the obtaining of butanoic acid by the stoichiometric equation:

$$2(CH_2O)_6 \rightarrow C_3H_7COOH + 2C_2H_5OH + 4CO_2 + 2H_2 \qquad (4.2)$$

In practice, small amounts of interfering ethanoic and propanoic acids are formed with butanoic acid; nevertheless, under the necessary conditions, this process becomes permanent and rich [9,10].

The transformation of VFAs into the form of ethers can be carried out in different procedures. Thus, VFAs and ethyl alcohol, jointly formed in the biological catalytic action, are straight substrate for the formation of ethyl carboxylates. The processes of esterification of biogenic acids are very efficient at high speed in the presence of supercritical alcohols, for example, ethanol [11,12].

As you know, the process of biotransformation of lignocellulose is restricted by the difficulty of depolymerization of its main substances, such as wood pulp, hemicellulose, lignine, and proteins. One of the modern approaches to intensifying the procedure is premised on the use of preliminary complex organic biomass depolymerization to obtain dissoluble compounds and their following conversion into biofuel [13,14]. There are different pretreatment procedures for solid lignine materials to improve biodegradation processes [13−16].

Earlier, to generate liquid biological fuel, a technology was developed for the oxidative depolymerization of technogenic carbon-containing wastes and a biological consortium was selected for the efficacious transformation of different types of organic mass into VFAs after degradation of biopolymers with subsequent esterification of fatty acids in the supercritical conditions [8,15−17].

Depolymerization of green mass is carried out with alkaline substances with using copper-containing catalyst in the temperature range from 20 to 90 °C [15,16]. This method makes it possible to significantly increase the concentration of oligomers and the amount of carbohydrates in a solution of pretreated biological mass. On the other hand, in order to increase the degree of conversion of the pretreated biomass into alcohols and fatty acids, in addition to mandatory achieving optimal values of the methane-generating consortium, it is important to study the possibility of modification, enrichment with other types of anaerobic microorganisms of the biocatalyst.

Wherein the addition of a consortium inductor such as glycerin intensifies the anaerobic degradation of difficult-to-decompose compounds. In this regard, it was of interest to evaluate the effect of the addition of glycerin, which is also an industrial waste, on the degradation process of the hard-to-decompose lignocellulose substrates under study.

It is known that bacteria of the genus *Clostridium acetobutylicum* have the potential for processing lignine; therefore, it was of interest to study the possibility of enhancing the anaerobic acidogenic consortium with this strain for transforming the processed green biomass. Also, in order to reduce the contact time of the biocatalyst with the pretreated biomass, the subthermophilic (45°C) mode of the process of obtaining VFAs and ethanol was studied.

The presented work shows the possibility of consistent application of the method of oxidative depolymerization of green biomass (straw, sawdust, and lignine) and biological acid hydrolysis by an optimal and enhanced acidogenic consortium with the addition of an inducer, glycerin, to the medium. The data obtained are important for evaluating the efficiency of the technology for the production of biofuels from hardly decomposable lignine-containing ones due to their transformation into VFAs and ethanol (Fig. 4.1).

4.2 Experimental part and methods

4.2.1 Biocatalyst

A previously obtained acidogenic biocatalyst [17] was used; it was isolated via selection from the anaerobic microorganisms of an acting methanogenic bioreactor (the treatment plant of a factory the production of "FritoLay" chips (Kashira town, Russia)). After selection, the biocatalyst mainly contained hydrolytic, enzymatic, and acetogenic microorganisms of the original methanogenic consortium, which prevailed during growth under acidogenic conditions. Methanogens at pH values from 5.4 to 5.6 did not show active growth. The characteristics of the biocatalyst before and after selection were determined according to the methods described in Refs. [18,19]. After selection, acidogenic activity increased from 500 to 627 COD/(g AFB day), ash-free biomass of the substances and dry matter also increased from 36 to and from 58 to 63 g/L, respectively. The ash content remained practically unchanged and remained at the values of 35%−37%.

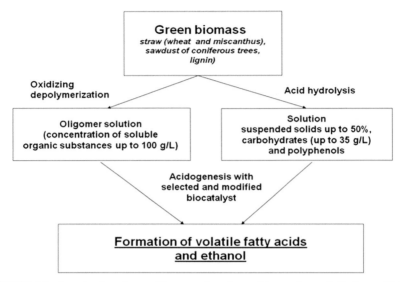

FIGURE 4.1 Investigation of anaerobic conversion of green biomass into volatile fatty acids for further production of high-calorie liquid fuel.

After storage (4 °C, 6 months), the biocatalyst was reactivated in an anaerobic UASB reactor 0.5 m high and 1.25 L in mesophilic (35 °C) and submesophilic (22−28°C) modes. The bioreactor was inoculated with the selected biocatalyst (750 mL, which is 60% of the total reactor volume); model drains of two different compositions (see below) prepared in a mineral medium were used as substrates [19]. A phosphate buffer (Sigma), pH 5.4−5.6, 10 mL of "A" solution, and 1.4 mL of "B" solution were used to prepare 1 L of the mineral medium. "A" solution had the following composition, g/L: NH_4Cl, 100; KH_2PO_4, 37; $CaCl_2 \cdot 2H_2O$, 8; $MgSO_4 \cdot 4H_2O$, 9; "B" solution contained the following components, mg/L: $FeCl_3 \cdot 4H_2O$, 2000; $CoCl_2 \cdot 6H_2O$, 2000; $MnCl_2 \cdot 4H_2O$, 500; $CuCl_2 \cdot 2H_2O$, 30; $ZnCl_2$, 50; H_3BO_3, 50; $(NH_4)_6Mo_7O_2 \cdot 4H_2O$, 90; $Na_2SeO_3 \cdot 5H_2O$, 100; $NiCl_2 \cdot 6H_2O$, 50 (Sigma).

Model wastewater no. 1 and no. 2 comprised glucose (1−4 g/L) or serum (3−6 g/L), respectively, as a source of organic matter. The pH value was measured with a Mettler Toledo potentiometer (Switzerland). The formulas described earlier [17,19] were used to calculate the residence time of the wastewater in the reactor (day), the load on organic matter per liter of bioreactor per day (gCOD/(L * day), and the efficiency of transformation of organic compounds to ethanol (g/L) and various VFAs (g/L).

Measurement of the acid-producing activity of biocatalysts (initial, selective, and reactivation), the calculation of its specific value, and the study of the efficiency and rate of formation of VFAs on various substrates were performed

Optimization of the anaerobic conversion of green biomass Chapter | 4 71

in accordance with the previously described procedures at two temperature conditions (35 and 20°C) [17–19].

We studied the effect of additional introduction of microorganisms *C. acetobutylicum* strain B-1787 (VKPM, Moscow) at a concentration of 9.25 g/L into a reactivated acidogenic biocatalyst at the output of fatty acids and ethanol during the conversion of depolymerized and then hydrolyzed straw, sawdust, and lignine at two temperature modes. Nontarget metabolic products of this strain are butyl alcohol and dimethyl ketone. However, the addition of this microbial strain has proven to be suitable for an acidic catalyst. The rest of the consortium consumed butyl alcohol/dimethyl ketone and even better accumulated the target products of the process—ethanoic and butanoic acids. The biomass of *C. acetobutylicum* was grown under anaerobic conditions at 37°C. The nutrient medium for growing this strain was of the following composition, g/L: peptone (tryptone) 10; yeast extract, 5; and glucose, 25 (Sigma).

4.2.2 Substrate

Milled sawdust and lignine were obtained at the Arkhangelsk woodworking plant (Russia); chopped wheat straw was provided at the Timiryazev Agricultural Academy (Moscow, Russia). Before conversion to VFA and ethyl alcohol, green biomass was pretreated by the previously described methods of catalytic oxidation with copper salts in an alkaline medium as a catalyst in the temperature range 20–90°C [15,16] and acid hydrolysis [17].

Glycerol, analytical grade (Chimmed, Russia) was also used as an inducer substrate. All studies on the bioconversion of the pretreated biomass into VFAs and ethyl alcohol were performed under batch conditions (periodic cultivation in flasks under stationary conditions) at two temperature modes (35 and 20°C). For this purpose, the 50 mL sample under study diluted with mineral medium up to a concentration of the organic substance of 2.5–15.0 g COD/L was introduced to the flasks (120 mL) and 5 mL of an acid-producing biocatalyst was added from the working reactor (9% out of the total volume of the liquid phase); after this, the gas space of the reactor was filled with argon (anaerobic conditions). The efficiency of the conversion of organic substances into ethyl alcohol (EtOH, g/L) and VFAs was calculated according to the equations given in Refs. [17,19].

4.2.3 Methods

Analysis of the products in the gas phase (hydrogen (H_2), methane (CH_4), and carbon dioxide (CO_2)) was analyzed by gas chromatography using an LKhM 8 MD model 3 chromatograph equipped with katharometer (the carrier gas was argon, the flow rate of the carrier gas was 20 cm^3/min, the column was 2 m in length and was filled with porapak Q, and the temperature of the thermostat of the columns was 50°C). Hydrogen comes after 26 s, methane 42 s, and carbon dioxide 65 s; observed chromatographic peaks are quite narrow, so the amount of gas was calculated from the peak height for each assay sample and was taken as 0.2 cm^3 gas [20].

Analysis of the volatile products in liquid phase ethanol, acetic, propionic, and butyric acids determined chromatograph "GC-15A" (Shimadzu, Japan) with a flame ionization detector on a 1 m long packed with Chromosorb 101. Carrier gas—argon, carrier gas velocity—30 cm^3/min, and the temperatures of the thermostat of the columns, detector, and evaporator were 190°, 210°, and 220°C, respectively. The exit time (minimum) described above products: ethyl alcohol 1.43, ethanoic acid 3.45, propanoic acid 6.55, and butanoic acid 12.35. Observed chromatographic peaks were not always sufficiently narrow, so the amount of volatiles is calculated by an integrator. For each test, sample was taken 1 mkL liquid phase; chromatograph calibration was performed as follows. The flasks were prepared solutions of ethyl alcohol, ethanoic acid, propanoic acid, and butanoic acid in a concentration range from 0.01% to 0.1% by volume. From each bottle, 1 mcL and fixed volatile component amount of analyte on chromatograph indications integrator was sampled three times. The values obtained were averaged and a calibration graph of the readings from the integrator component concentration.

The concentration of glucose in the model stream no. 1 was measured using a standard set of reagents (corporation "Impakt," Russia).

The concentrations of organic substances in model stream no. 2 and subsequent samples of pretreated biomass (COD, g/L) were determined according to the method [20].

The content of the sum of reducing sugars was determined according to the procedure described [19].

4.3 Results and discussion

4.3.1 Specific acid-producing activity of the biocatalyst

Previously, we studied two separate methods of pretreatment of lignine-containing biomass. The method of oxidative depolymerization shows the yield of oligomers and carbohydrates up to 10% and 1%, respectively. The method of acid hydrolysis increases the carbohydrate content to 35% and the concentration of the suspended compound to 50% [17]. In order to increase the availability of the substrate, this work also studied the possibility of sequential application of oxidative depolymerization and acid hydrolysis methods for hardly degradable green biomass materials. This procedure, carried out on sawdust of coniferous trees, lignine, and wheat straw, made water-soluble organic matter in concentrations of 44, 94, and 48 gCOD/L, respectively. Also, such pretreatment led to an increase in the concentration of reducing sugars up to 22.0%−36.0% in the medium.

To prepare for the next stage of biological acidogenic hydrolysis, the consortium was selected by incubation for 2 months (1 month on medium

Optimization of the anaerobic conversion of green biomass **Chapter | 4** **73**

TABLE 4.1 Specific acid-producing activity (A, g COD/gAFB/day) of the biocatalyst prior to and after selection and after reactivation.

Temperature (°C)	Prior to selection	After selection	After reactivation
35	0.50±0.01	0.63±0.01	0.70+±0.02
30	0.42±0.01	0.58±0.01	0.66±0.01
20	0.29±0.01	0.39±0.01	0.42±0.01
15	0.20±0.01	0.25±0.01	0.32±0.02

containing glucose and then 1 month on medium containing milk whey). The reactivation of the biocatalyst after storage was carried out in a flow-through bioreactor and included a session use of glucose as a substrate (2 weeks) and then milk whey (3 weeks) under optimal conditions for the production of VFAs. As a result of such cultivation, the activity of the consortium in the production of acids increased by 1.28 times in comparison with the methanogenic consortium (Table 4.1).

4.3.2 Biomass pretreatment

As a result of the research, several methods of biomass pretreatment were tested in order to obtain the maximum amount of organic substances available for the biocatalyst and to increase the yield of fatty acids and ethanol. We obtained substrates with different concentrations of organic substances and the content of RS (Table 4.2) after pretreatment of biomass (wheat straw and miscanthus straw, lignine, and sawdust of coniferous trees) by means of oxidative depolymerization and/or hydrolysis. The technological scheme, including oxidative depolymerization followed by acid hydrolysis of sawdust of coniferous trees, lignine, and wheat straw, made it possible to obtain not only a high concentration of organic substances (44, 94, and 48 gCOD/L, respectively) but also led to an increase in the amount of RS from 1.0% to 22.0%−36.0% (Table 4.2).

4.3.3 Obtaining VFAs and ethyl alcohol during the conversion of pretreated green biomass with a biocatalyst in submesophilic and mesophilic modes

The effect of joint preliminary physicochemical treatment (depolymerization > acid hydrolysis) of solid agro-industrial complex wastes (wheat straw, coniferous sawdust, and lignine) on the yield of VFAs during fermentation with a biological catalyst was studied (Fig. 4.2).

The use of a new scheme for pretreatment of straw made it possible to increase the content of organic compounds in the solution to 5.61 gCOD/L.

74 Advances of Artificial Intelligence in a Green Energy Environment

TABLE 4.2 Characteristics of pretreated biomass.

Green biomass	Pretreatment method	Concentration of soluble organic substances (gCOD/L)	Concentration of reducing sugars (%)
Straw (wheat)	Hydrolysis	6.9	35.0
	Oxidative depolymerization with hydrolysis	4.0	36.0
	Oxidative depolymerization	48.0	1.4
Sawdust	Oxidative depolymerization	44.0	1.0
	Oxidative depolymerization with hydrolysis	44.0	31.0
Lignin	Oxidative depolymerization	94.0	1.0
	Oxidative depolymerization with hydrolysis	94.0	22.0

This, in turn, led to an improvement in the qualitative composition of straw biological acidogenesis products (Fig. 4.2A). Biological acidogenesis in this case lasted 6 days at a temperature of 35 °C and 10 days at a temperature of 20 °C. The yield of by-products decreases: ethanoic acid by 1.82%−5.32%, butanoic acid by 8.84%−16.83%, and ethyl alcohol by 2.21%. At the same time, there is an increase in the production of propanoic acid (at least 10 times at an ambient temperature of 35°C and at least 5 times at an ambient temperature of 20°C).

Acidogenic fermentation of pretreated sawdust (biological transformation 12−16 days) and lignine (biological transformation 14−18 days) allowed in this case to achieve an increase in the yield of all products regardless of the selected temperature (Fig. 4.2B and C). In this case, the increase in the yield of propanoic acid during biological transformation of sawdust was two times and in the case of lignine three times. It is noted that using such a procedure, it is possible to achieve an increase in the yield of VFAs by 18.2% and an increase in the yield of ethyl alcohol by an average of 2.1%.

4.3.4 Increase in the yields of volatile fatty acids and ethanol alcohol during the transformation of pretreated green mass with a biological catalyst modified by bacteria *Clostridium acetobutylicum* in submesophilic and mesophilic modes

The selected biological acidic catalyst was modified by the introduction of bacteria *C. acetobutylicum* at 1/4 of the consortium volume. The experiments

Optimization of the anaerobic conversion of green biomass Chapter | 4 **75**

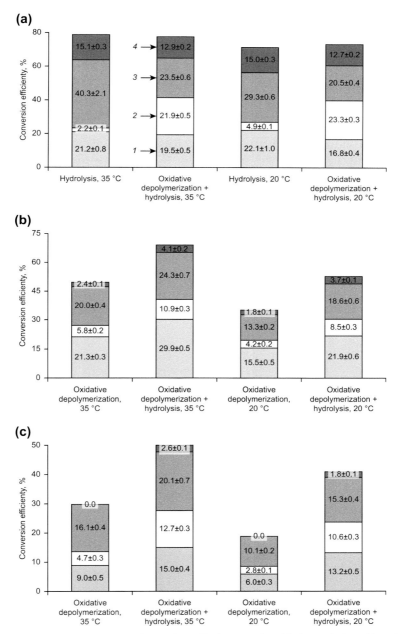

FIGURE 4.2 Yields of volatile fatty acids (VFAs) (1, acetic acid; 2, propionic acid; and 3, butyric acid) and (4) ethanol from pretreated straw (a), sawdust (b), and lignin (c) during their conversion at two temperature regimes by reactivated biocatalyst depending on the pretreatment method.

76 Advances of Artificial Intelligence in a Green Energy Environment

were carried out on samples of pretreated green biomass (straw, sawdust, and lignine, 6−10, 12−16, and 14−18 days of fermentation, respectively) in submesophilic (20°C) and mesophilic (35°C) modes. The concentration of soluble organic matter was in this case varied from 5.34 to 5.82 gCOD/L (Fig. 4.3).

The addition of the specific culture into the biocatalyst significantly increases the yield of ethanoic acid, butanoic acid, and ethyl alcohol from all studied substrates in both temperature modes. The content of butanoic acid and ethyl alcohol during straw processing was higher than in other hydrolysates (27.01% (0.82 g/L) versus 17.41% (0.44 g/L), respectively) (Fig. 4.3).

4.3.5 Increase in the yields of volatile fatty acids and ethyl alcohol with adding a cosubstrate of glycerin to pretreated green biomass during fermentation with the selected biocatalyst in submesophilic and mesophilic modes

It was studied the effect of partial replacement of a complex substrate (depolymerization products of green biomass samples) with glycerol at a concentration of 1/4 and 1/2 of the total volume of soluble organic substances (about 5.63 gCOD/L). The introduction of the cosubstrate should stimulate the metabolism of the consortium and affect the yield of the target substance. Fig. 4.4 shows the results obtained at two different temperature regimes (35 and 20°C) depending on the time of fermentation of each sample of green biomass. It was found that a twofold dilution of a solution of pretreated straw with glycerin at a temperature of 35°C contributed to an increase in the yield of ethanoic and butanoic acids from 19.3% to 25.1% and from 23.1% to 32.2%, respectively. At the same time, there is a significant decrease in the yields of ethyl alcohol and propanoic acid from 13.1% to 9.0% and from 22.3% to 14.1%, respectively. The same dependence was observed at a temperature mode of 20°C. Dilution of the solution of pretreated sawdust with glycerin by 2 times made it possible to increase the yield of ethanoic acid by 38.1% at a temperature mode of 35 °C and by 29.4% at a temperature mode of 20°C. Nevertheless, the yield of butanoic acid did not increase when the sawdust biomass was diluted with glycerol in all studied ratios (yield 30.2% at 35°C and yield 22.4% at 20°C). The use of glycerin in the case of pretreated biomass of lignine made it possible to achieve an increase in the yield of ethanoic and butanoic acids up to 24.2% and 23.3%, respectively. At the same time, an increase in the yield of propanoic acid up to 14.0% was found. The use of the submesophilic temperature regime of acidogenesis (20°C) made it possible to increase the propanoic acid content to 16.2%. It is important to note that the content of ethanol by-product in the fermentation procedures of pretreated green biomass using glycerin dropped sharply at any temperature regime (Fig. 4.4).

Optimization of the anaerobic conversion of green biomass **Chapter | 4** **77**

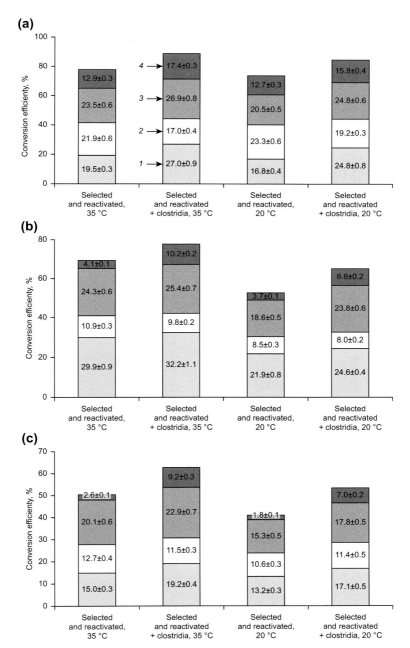

FIGURE 4.3 Effect of the addition of *Clostridium acetobutylicum* bacteria to reactivated selected biocatalyst on the yields of volatile fatty acids (VFAs) (1, acetic acid; 2, propionic acid; and 3, butyric acid) and (4) ethanol during conversion of depolymerized straw (a), sawdust (b), and lignin (c) at two temperature regimes.

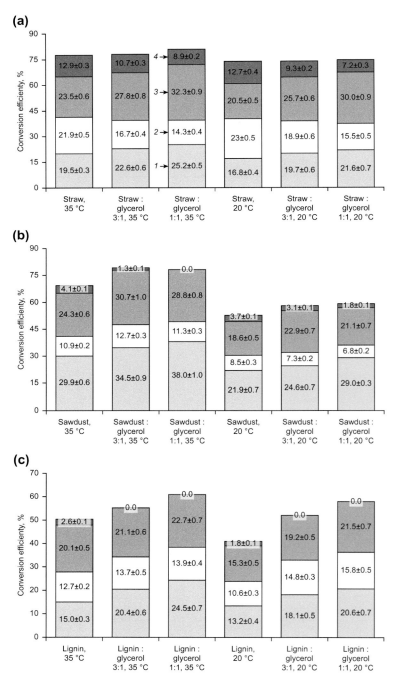

FIGURE 4.4 Effect of partial substitution of glycerol for pretreated straw (a), sawdust (b), and lignin (c) on yields of volatile fatty acids (VFAs) (1, acetic acid; 2, propionic acid; and 3, butyric acid) and (4) ethanol during their conversion by reactivated selected biocatalyst at two temperature regimes.

4.3.6 The efficiency of conversion of substrates into volatile fatty acids and ethanol, depending on the temperature and time of their contact with the biocatalyst

Earlier, it was found that the contact time for each of the substrates with the biocatalyst ranged from 6 to 14 days at a temperature of the acidogenesis process of 35°C. In order to optimize the contact time and reduce it, the subthermophilic (45°C) mode of the process of obtaining VFAs and ethanol on two varieties of depolymerized straw (wheat straw and miscanthus straw) was studied. Table 4.3 shows the results for two substrates obtained at two temperature conditions.

The maximum efficiency of OSs conversion to butyric acid (33.7% and 22.5%) on both substrates (wheat straw and miscanthus straw) was observed at initial concentration—5 gCOD/L on 6 days at mesophilic (35 °C) and on 4 days at subthermophilic (45°C) modes. However, the temperature rises to 45°C, and a more intense formation of acetic acid was observed. The formation of propionic acid changed insignificantly, while the efficiency of formation of butyric acid decreased on average by 23% and 9% in wheat straw and in miscanthus straw, respectively. An increase in the concentration of initial substances from 5 to 15 gCOD/L on both substrates led to a decrease in the efficiency of the formation of all three VFAs. In this case, the formation of ethanol changed insignificantly. A decrease in the efficiency of the conversion of green biomass into VFA at high substrate concentrations is probably associated with both substrate inhibition and inhibition of the process by acidogenesis products. However, despite the decrease in the efficiency of the conversion of green biomass into VFA, the yield of acidogenesis products increases with an increase in the initial concentration of straw (wheat straw and miscanthus straw). The maximum yield of butyric acid of 3.9 gCOD/L was observed at the conversion of OSs in wheat straw at a concentration of 15 gCOD/L at 35 °C (Table 4.3).

The obtained results show that the process of conversion of the organic matter of the investigated straw can be carried out with good efficiency both at mesophilic (35°C) and at subthermophilic (45°C) modes; however, the duration of the conversion process and the high yield of unnecessary by-product (acetic acid) should be taken into account.

4.4 Conclusions

Ethyl alcohol together with VFA—ethanoic acid, propanoic acid, butanoic acid—produce as a result of biocatalytic processing plant of polysaccharide raw material, preferably waste agriculture and wood industry (straw, sawdust, etc.).

Oxidative depolymerization followed by acid hydrolysis of sawdust of coniferous trees, lignine, and wheat straw under optimal conditions made it possible to obtain organic compounds at a concentration of 44, 94, and 48 gCOD/L, respectively, which in turn led to an increase in the amount of reducing sugars from 1.0% to 22.0%—36.0%.

TABLE 4.3 The efficiency of conversion of substrates into volatile fatty acid and ethanol, depending on the temperature and time of their contact with the microbial association after selection (biocatalyst).

Temperature (°C)	Substrate concentration (gCOD/L)	Substrate contact time, day	The efficiency of conversion of the substrate into (%)			
			Ethyl alcohol	Ethanoic acid	Propanoic acid	Butanoic acid
			Substrate—wheat straw			
35	5	6	7.0	22.1	11.6	33.7
	10	6	8.7	21.4	9.7	29.8
	15	6	7.7	17.7	8.2	25.9
45	5	4	11.0	30.7	15.1	25.1
	10	4	9.8	27.4	12.9	23.1
	15	4	9.7	24.0	9.5	20.7
			Substrate—miscanthus straw			
35	5	6	5.1	19.3	14.2	22.5
	10	6	5.8	19.7	12.3	18.6
	15	6	5.6	17.7	9.5	16.5
45	5	4	7.0	27.3	16.4	20.6
	10	4	7.8	25.7	14.9	16.5
	15	4	7.5	22.9	11.6	15.3

The process is implemented using specialized microbial associations (biocatalyst). A new biocatalyst is created as a catalytic system consisting of a consortium of free and immobilized cells acidogenic microorganisms that can withstand considerable loads on the organic matter in the flow reactor. The addition of *C. acetobutylicum* microorganisms to the biocatalyst increases the yield of VFA and ethyl alcohol, and the indicators for butanoic acid and ethyl alcohol on biomass from straw are higher (27.01% and 17.41%, respectively) than on other substrates in a similar process.

The use of glycerol as a substrate inductor makes it possible to increase the yield of butyric and acetic acids during the conversion of pretreated biomass, while straw diluted 2 times with glycerin showed the maximum yields of butanoic acid and VFA, which were 32.2% and 72.2%, respectively.

A temperature of 35°C with a duration of contact of the substrate with the biocatalyst of 6 days seems to be the most acceptable for obtaining the maximum amount of butanoic acid and ethyl alcohol. However, it should be noted that the ethyl alcohol and VFA formed in the process are in the state of low-concentration aqueous solutions, which necessitates the development of methods for their effective extraction.

Abbreviations

AFB ash-free biomass of the substances
COD chemical oxygen demanded—rate of content of organic substances
OSs organic substances
RS reducing sugars
VFAs volatile fatty acids

References

[1] R. Du, C. Li, P. Pan, C.S.K. Lin, J. Yan, Characterization and evaluation of a natural derived bacterial consortium for efficient lignocellulosic biomass valorization, Bioresour. Technol. 329 (2021), https://doi.org/10.1016/j.biortech.2021.124909.

[2] C. Moretti, A. López-Contreras, T. de Vrije, et al., From agricultural (by-)products to jet fuels: carbon footprint andeconomic performance, Sci. Total Environ. 775 (2021), https://doi.org/10.1016/j.scitotenv.2021.145848.

[3] D. Kumar, B. Singh, J. Korstad, Utilization of lignocellulosic biomass by oleaginous yeast and bacteria for production of biodiesel and renewable diesel, Renew. Sustain. Energy Rev. 73 (2017) 654−671, https://doi.org/10.1016/j.rser.2017.01.022.

[4] A.N. Nozhevnikova, A.Y. Kallistova, Y.V. Litti, M.V. Kevbrina, in: A.N. Nozhevnikova (Ed.), Biotekhnologiya i mikrobiologiya anaerobnoi pererabotki organicheskikh kommunal'nykh otkhodov (Biotechnology and Microbiology of Anaerobic Processing of Organic Municipal Waste), Universitetskaya Kniga, Moscow, 2016 (in Russ).

[5] M.A. Gladchenko, D.A. Kovalev, A.A. Kovalev, et al., Methane production by anaerobic digestion of organic waste from vegetable processing facilities, Appl. Biochem. Microbiol. 53 (2) (2017) 242−249, https://doi.org/10.7868/S055510991702009X.

[6] O.V. Sen'ko, M.A. Gladchenko, I.V. Lyagin, et al., Transformation of the biomass of phototrophic microorganisms into methane, Al'ternat. Energ. Ekol. 107 (3) (2012) 89−94 (in Russ).

[7] A.E. Maragkaki, M. Fountoulakis, A. Gypakis, et al., Pilot scale anaerobic co digestion of sewage sludge with agro-industrial by products for increased biogas production of existing digesters at wastewater treatment plants, Waste Manag. 59 (2017) 362−370.

[8] S.V. Mazanov, A.R. Gabitova, R.A. Usmanov, R.R. Gabitov, Experimental study of the process of obtaining biodiesel in the presence of the heterogeneous catalyst Al_2O_3, Vestn. Tekhnol. Univ. 18 (7) (2015) 159−161 (in Russ).

[9] S.V. Kalyuzhnyi, D.A. Danilovich, A.N. Nozhevnikova, in: S.D. Varfolomeev (Ed.), Anaerobnaya biologicheskaya ochistka stochnykh vod (Anaerobic Biological Wastewater Treatment), Itogi Nauki Tekhn., Ser. Biotekhnol., vol. 29, VINITI, Moscow, 1991 (in Russ).

[10] S.D. Varfolomeev, S.V. Kalyuzhnyi, D.Y. Medman, Chemical basis of biotechnology for producing fuels, Usp. Khim. 57 (1988) 1201−1230 (in Russ).

[11] E. Santacesaria, F. Trulli, G.F. Brussani, Oxidized glucosidic oligomers − a new class of sequestering agents preparation and properties, Carbohydr. Polym. 23 (1994) 35−46.

[12] S.A. Biktashev, R.A. Usmanov, R.R. Gabitov, et al., Transesterification of rapeseed and palm oils in supercritical methanol and ethanol, Biomass Bioenergy 35 (2011) 2999−3011.

[13] P. Kumar, D.M. Barrett, M.J. Delwiche, P. Stroeve, Methods for pretreatment of lignocellulosic biomass for efficient hydrolysis and biofuel production, Ind. Eng. Chem. Res. 48 (8) (2009) 3713−3729.

[14] V.B. Agbor, N. Cicek, R. Sparling, A. Berlin, D.B. Levin, Biomass pretreatment: fundamentals toward application, Biotechnol. Adv. 29 (6) (2011) 675−685.

[15] I.P. Skibida, R.M. Aseeva, P.A. Sakharov, A.M. Sakharov, Intumescent Coke-Forming Antiperen, Method for its Production, Method for Fire Protecting Processing of Combustible Substrate, and the Method for Extinguishing the Hearth of Combustion, RF Patent no. 2204547, 2003 (in Russ).

[16] S.D. Varfolomeev, S.M. Lomakin, V.N. Gorshenev, et al., Antiperen, Method for its Production, Method for Fire-Protecting Processing of Materials, and Method for Extinguishing the Hearth of Combustion, RF Patent no. 2425069, 2011 (in Russ).

[17] M.A. Gladchenko, S.N. Gaidamaka, V.P. Murygina, S.D. Varfolomeev, Optimization of the conversion of agro-industrial complex waste into volatile fatty acids under anaerobic conditions, Vestn. Mosk. Univ. Seriya 2 Khimiya 55 (4) (2014) 241−248, https://doi.org/10.3103/S0027131414040026.

[18] C.M. Hooijmans, S. Veenstra, H.J. Lubberding, Laboratory course process parameters and microbiology, in: G. Lettinga (Ed.), International Course in Anaerobic Waste Water Treatment, Agricultural University, Delft, Wageningen (Holland), 1990.

[19] M.A. Gladchenko, Development of Biotechnological Methods of Recycling Waste Wine (Cand. Sci. (Eng.) Dissertation), Ross. Khim. Techn. Univ. im. D.I. Mendeleeva, Moscow, 2001 (in Russ).

[20] S.V. Kalyuzhnyi, M.A. Gladchenko, E.A. Starostina, et al., Integrated biological (anaerobic-aerobic) and physico-chemical treatment of baker's yeast wastewater, Water Sci. Technol. 52 (10−11) (2005) 273−280.

Chapter 5

Life cycle cost and life cycle assessment: an approximation to understand the real impacts of the Electricity Supply Industry

Joaquina Niembro-García[1], Patricia Alfaro-Martínez[2] and Jose Antonio Marmolejo-Saucedo[1]

[1]*Universidad Panamericana, Facultad de Ingeniería, Mexico City, Mexico;* [2]*Universidad Panamericana, Facultad de Gobierno y Economía, Ciudad de México, México*

5.1 Importance of Electricity Supply Industry

Advances in the quality of life of societies are linked to technological development. One of the first technological advances was the conservation of fire and since then the use of energy has been one of the engines of social development. It would be impossible to imagine our current daily life without electric power; we have created so many needs around electricity that running out of it is unthinkable. Since its discovery, first to enlighten us, then to help us: in industry, communications, transport and households, in our daily lives, and finally, to improve our quality of life and our knowledge. Energy is constantly cited in speeches of politicians and decision-makers from local level to the highest level. As an example, the former Secretary General of the United Nations [1], said that in the United Nations, "Energy is the golden thread that connects economic growth, increased social equity and a healthy environment. Sustainable development is not possible without sustainable energy."

While electricity provision is commonly regarded as a basic utility that is noticeable in the most developed economies only when it fails, in developing countries electricity provision remains a core aspiration and development indicator ([2]; p.10). The energy is the fundamental ingredient of modern society because the economic growth and energy consumption go in parallel. Electric energy has allowed the development of a large number of appliances and

84 Advances of Artificial Intelligence in a Green Energy Environment

devices that make our lives easier, improve our communications, and increase our productivity. If there is any technology that has influenced a change in humanity and in the world, it is electricity. Electricity has changed our world; mainly we use it to improve the quality of our life. We may make our life more comfortable, continue to have activities at night, facilitate our daily activities, communicate, and entertain ourselves more and more immediately.

Electricity is a service that society may not part with. It is part of "common" life and is a standard in modern society. The operation of our homes is dependent on it (cooling, heating, cooking, cleaning, recreation, etc.), and so do our medical and diagnostic services, our spaces of entertainment and culture, our means of transport, industry, financial services, etc. The list is endless and as big as the catastrophe of losing service. An example is the economic loss of 60 million euros generated by the blackout in Barcelona in 2007. A cosmopolitan city in the north of Spain, with 323,000 homes and businesses affected, was without electricity for up to 4 days. The financial penalty imposed on the companies for not having provided the service was 20 million euros to be paid to the government; however, the compensation to homes and businesses was 150 million euros [3,4]. Today, we do not imagine an activity that does not require electricity, even doing yoga, today it is done accompanied by a device that dictates exercise instructions (app) that is charged with electricity. On the other hand, since the availability of energy is synonymous with development, the difference is between electrical/energy consumption and availability of energy between developed and developing world. Low-income countries have serious deficiencies to satisfy their primary needs, one of them being electricity. Therefrom that working on Sustainable Development Goal 7: "Affordable and Clean Energy" is relevant. In this sense, electricity is "one of the energy forms where renewable sources can more economically and efficiently substitute traditional greenhouse emitting energy sources" ([5]. p.3). That is, an energy clean source may satisfy the same electricity demand as a traditional fossil source without the environmental negative impacts. Nevertheless, it will depend on government politics and regulations and renewable sources availability.

The electricity is the most sophisticated energy form. Everyone uses electricity: industry, service providers, and households. It has taken a relevant role in our daily life. Nowadays, according to World Bank [6], we consume three times more energy per inhabitant compared to the consumption of 1970 and also the population growth was 2% each year, in 1970 population was 3.7 inhabitants billion, and in 2020 reached 7.8 inhabitants billion. "More people and more development suggest even higher future needs for energy" ([7], p. 4). Electricity has become an integral part of commercial exchange; in parallel its use has expanded rapidly at the residential level, as both the population and per capita consumption have grown.

Primary energy is the one we find in nature, before being converted or transformed. This energy is transformed to take advantage of it, and to improve the society quality of life, 1 kWh of final energy from conventional electricity is equivalent to 2461 kWh of primary energy [8] as shown in Fig. 5.1. The world electricity consumption in 2014 was 3,132,788 kWh per capita [6]. It is divided into two groups: renewable (solar, hydro, wind, etc.) or nonrenewable (coal, gas, oil, uranium). According to data from the International Energy Agency (IEA) in 2018, 24,738.9 TWh of electrical energy was consumed in the world (2018), see Fig. 5.2. In addition, in that same year, the intensity of energy carbon in the industry was 51.8 gCO2/MJ; this has decreased by 2% each year since 2010, as shown in the table below [10]. The World Bank [6] in the last 4 years contributed to improve access to electricity for more than 45 million people, between 2014 and 2018, and provided more than 11.5 billion dollars in financing for renewable energy and energy efficiency initiatives. Although one might think that energy is universal, this is not the case; one billion people do not have access to electricity, which is 13% of the world's population. Three billion, 40% of the planet's inhabitants, continue to cook with polluting fuels (coal or wood). Only 17.5% of all the energy consumed in the world is renewable. Hence the importance of being able to work from many fronts to improve the situation. Making the shift to cleaner energy sources generates greater demand for materials required to manufacture solar, wind, and storage technologies.

Electric power is a commodity, that is, a good that has value and cannot be differentiated, which is why its profit margin is small. This commodity has special characteristics [11]:

1. It has great variations throughout the day and throughout the year. In other words, the generation and transmission capacity necessary to guarantee

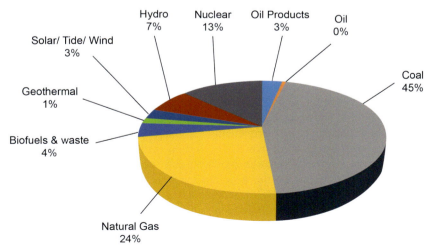

FIGURE 5.1 2018 world power generation. *Data from IEA, World Balance, 2021. https://www.iea.org/sankey/#?c=World&s=Balance.*

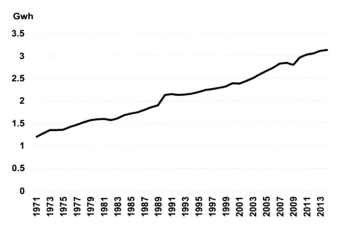

FIGURE 5.2 World electricity consumption between 1975 and 2018. *Create with information from World Bank, Energy in 2018, 2021. World bank sitio web: https://datos.bancomundial.org/indicator/EG.USE.ELEC.KH.PC/.*

peak demand is not used at other times. This implies that its price should vary subject to daily, weekly, and seasonal fluctuations in which it is consumed.

2. It must be satisfied almost instantly, because for technical and economic reasons, it may only be stored in limited quantities. Storage is a central point in the supply of electricity and makes the difference with respect to other commodities, where the variations in production and demand may be covered with the stock [2].
3. In order to be satisfied, it is necessary to use a diversified portfolio of generation technologies so that they have the lowest possible cost.

These characteristics of electric power make it a unique good, a basic product for which storing is almost unviable. Throughout each day, week, and season, we as consumers use different amounts of electricity and then balance between supply and demand must be guaranteed at all times. "Therefore, electricity generation and electricity consumption must be synchronized continuously. In order to secure the supply of electrical energy for all costumers, it is necessary to permanently maintain an amount of capacity in power generation which exceeds the maximum load" ([12]; p.269).

Until now electricity must be produced and consumed simultaneously in equal quantities. However, the next challenge is to be able to accumulate energy efficiently. Storage is a fundamental pillar of the energy transition because, in periods of high production and low demand, it would help accumulate excess energy produced. This would reduce the use of less efficient generating plants that would otherwise only operate during peak hours. Therefore, storage would help integrate more renewable energy into Electricity

Supply Industry (ESI), which would delay or prevent the construction of new power plants, achieving a cost reduction and an increase in the efficiency of the plants.

5.2 The economics of Electricity Supply Industry

Electricity markets are a consequence of the liberalization, privatization, deregulation, and restructuring of the electricity sector. These markets are made up of a large number of participants who must operate in a competitive environment, with what would be expected, an increase in the efficiency of the system and a reduction in prices, as a consequence of the end of the monopoly.

For many years, the economics classified the electricity system as a natural monopoly, that is, a market situation, where it is preferable that the good be produced by a single company. Considering that it requires technology or infrastructure whose investment is so high, it would be difficult for more than one competitor to offer it at the same price and would have capacity restrictions in the short term.

Vertical integration seeks to have the following effects:

1. Greater production efficiency. Reduction in production costs due to economy of scale; greater ability to negotiate long-term contracts to ensure the continuity and quality of raw materials that allow a competitive advantage; and decrease in transaction costs to guarantee the supply of the required resources in the production chain.
2. Greater allocative efficiency. Decrease in the price charged to the final consumer due to the price elasticity of demand, the ability to set prices and existing regulation. For that reason, in order to avoid market distortions via prices, in several countries some government entity took on the role of producer.
3. Improvement in the quality of the service. Decrease in the interruption of supply, and improvement information, attention, and relationship with the end user.

In light of the foregoing, this allowed competition to begin to make sense for electricity generation. Starting in the 1980s, some countries began to experiment with long-term contracts to purchase electricity from independent generating companies.

In the 1990s, the process of deregulation of generation activities and price liberalization began in the world. Thus, the new sectoral model considered that power generation and supply and marketing activities can be developed under a competition regime; whereas the activities related to the transmission and distribution network would be regulated in order to guarantee access to the network, because they have characteristics as a natural monopoly. "The objectives of these reforms differed slightly from one country to another, but one of the most important objectives was improving the efficiency and productivity of these industries. There was a common concern that government-owned vertically integrated electricity supply companies had become inefficient, over-manned, unproductive and unresponsive to customer desires" ([13], p. 78).

In accordance with Kirschen [14], competition in ESI generally means competition in the power generation and supply and marketing:

1. **Single buyer:** A single company is responsible for buying electricity wholesale, it is the intermediary. In most cases, this entity is also the system operator and the electricity retailer. To introduce a more competitive industry structure, other companies were allowed to participate as an independent power producer (IPP). Single buyer is forced by law to buy whole the power produced by IPPs. See Fig. 5.3 when the vertical integration is showed.
2. **Wholesale competition:** Generator companies produce electricity, which is freely traded; the Transmission and Distribution networks are in the hands of a central operator, and there are large consumers (companies) and some companies called traders (DisCo) that have local monopolies to supply electricity to small consumers. In other words, consumers cannot choose their supplier; retailers and large consumers can enter into long-term contracts with generators companies. Wholesale market is operated by a centralized way; there is an independent system operator (ISO). See Fig. 5.4.
3. **Retail competition:** There is competition between generators companies, which compete to establish supply contracts with retailers and large consumers. However, in this market, traders are not local monopolies, but rather compete with each other to sell energy to end consumers. In this market, end consumers can choose their supplier considering its price and quality. See Fig. 5.5.

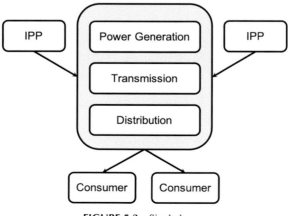

FIGURE 5.3 Single buyer.

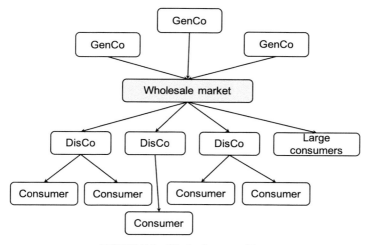

FIGURE 5.4 Wholesale competition.

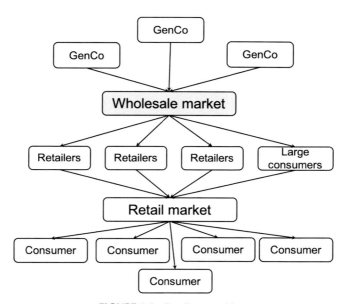

FIGURE 5.5 Retail competition.

In this point, it is important to explain interplay demand and supply of electricity. It is the main model of price determination used in economic theory. In basic economic analysis, all factors except the price of the commodity, in this case electricity, are held constant; the analysis then involves examining the relationship between various price levels and the maximum quantity that would potentially be purchased by consumers and be offered by producers at each of those prices.

Typical main model demand-supply is shown in Fig. 5.6; regarding the figure it is important to consider the following:

- The demand curve for a good electricity is almost always downward-sloping, reflecting the willingness of consumers to purchase more electricity at lower price level (light gray).
- The supply curve for electricity is usually upward-sloping, reflecting the willingness of producers to sell more of electricity they produce in a market with higher prices (dark gray).
- Then the price of electricity is determined by the interaction of supply and demand in a market. The resulting price is referred to as the "equilibrium price" or the "market clearing price" and represents an agreement between producers and consumers of electricity. Therefore, in equilibrium the quantity of electricity supplied by producers equals the quantity demanded by consumers.

Furthermore, since the demand for electricity is cyclical, there is not a single demand curve. There are numerous demand curves related to time. A daily load profile is a graph of the variation in the electrical load vs. time during the day. In other words, the total demand is for electricity during the day.

Typical daily load profile is shown in Fig. 5.7A. Winter demand is usually higher than summer demand, also daily demand fluctuates depend on hour. The comparison between loads is shown in Fig. 5.7B. Base load is the minimum level of demand on an electrical supply system over 24 h. Base load

FIGURE 5.6 Demand-supply model.

Life cycle cost and life cycle assessment **Chapter | 5** **91**

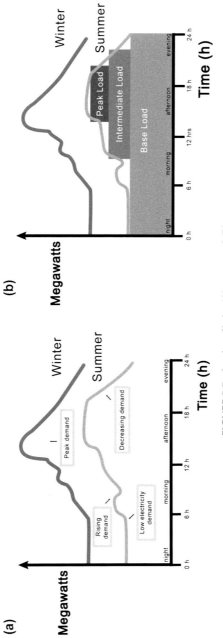

FIGURE 5.7 Load profile by (A) season and (B) type.

power sources are those plants that can generate dependable power to consistently meet demand. But, at some time, it will not suffice to provide all the electricity, we will need additional generators, which is called middle load. In addition, during the hours with the highest demand, we have the peak load. As a result, we also need to have generators providing electricity during these hours. Despite technological progress, we still do not have technology that allows us to store electricity efficiently and at an affordable cost. Consequently, the electricity demand must be satisfied in real time.

To achieve this, Bhattacharyya [15] indicates that three types of plants are required. The first is the one that "running around the year—all the time they are available—to meet the base demand; these plants normally do not have the capability to vary the supply depending on the demand." The second type is "having the capability to follow the demand and vary output frequently during their operation." The third type of a generation plant is "running only during the peak period" ([15]; p. 230). Hence, the load curve to satisfy demand is a function of a plant operation, including the energy source, the cost of service, and the system efficiency.

Therefore, the demand curve shifts from left to right depending on the time of day and day of the week and the ability to meet demands based on the technology available for it (Fig. 5.8).

Because electricity must be generated and transmitted as it is consumed, it is necessary to have generation capacity installed to satisfy peak load demand and to have too some generation capacity reserve in case of an unexpected event. Therefore, each electric grid system must have plants for every type of load: base, intermediate, and peak, for the goal of supply to the grid in a dependable and efficient way.

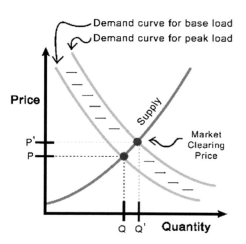

FIGURE 5.8 Load curve family.

The quality of life that developed nations enjoy today is due in large part to the availability of abundant and affordable fossil fuels over the past centuries. As we have pointed out, more technologies have emerged from electricity that could not survive without the first one; therefore, the current challenge for electricity is to promote cleaner generation in order to allow sustainable use over time through sustainable energy technologies, and this is sustainable energy.

Gro Harlem Brundtland, Prime Minister of Norway, in 1987 defined sustainable development as, "The humanity's ability to make development sustainable to ensure that it meets the needs of the present without compromising the ability of future generations to meet their own needs."

As we have pointed out, more technologies have emerged from electricity that could not survive without the first one; therefore, the current challenge for electricity is to promote cleaner generation in order to allow sustainable use over time through sustainable energy technologies, and this is sustainable energy. To Tester, mentioned to achieve a more sustainable energy future; we need a set of factors: (1) increased efficiency, (2) reduced consumption, and (3) used new generation of renewable energy technologies [7].

There are several types of renewable energy technologies:

- Solar—obtained from solar energy
- Wind—obtained from the wind.
- Biomass—obtained from organic matter that is converted into energy when applying chemical processes.
- Hydraulic—obtained from the use of kinetic energy.
- Geothermal—obtained from the heat inside the Earth.
- Tidal—obtained from the mechanical energy generated by the movement of the waves.

Bhattacharyya [15] points out five advantages of using renewable energy: (1) reduction of CO_2 emissions and mitigation of climate change; (2) security of energy supply; (3) improving energy access; (4) employment opportunities; and (5) decreasing energy dependence.

Each renewable energy source is performing differently; one could be the best option for one location/purpose/season and could not perform with that efficiency at another location/purpose/season ([16], p. 2). However, it should also be considered that the most common forms of renewable energy (wind, solar, geothermal, and tidal) are intermittent in their natural state, and since electricity may not be stored in large quantities, this type of energy may only be used during the time in which they are available. They are also local, so they depend on endemic factors. Most of these energies are available abundantly and the humankind has been using them for various purposes from time immemorial. The direct cost to the consumer remains low in their traditional form of use (such as drying). However, modern ways of using these energies

94 Advances of Artificial Intelligence in a Green Energy Environment

require sophisticated conversion processes, which in turn increase the cost of supply ([15]; p.249).

The competition created from the introduction of large amounts of renewable energy sources brought lower prices, by putting downward pressure on the profit margins of generators and suppliers. Due to the costs of renewable energy, generation units are very low; they are at the beginning of the merit order; hence, they will always run then if they meet electricity demand, the most expensive plants would not be needed.

It would be very efficient that the market would provide incentives for electricity generators to build new plants based on renewable sources. However, there is a problem with this kind of product, electricity demand is not flexible enough, and this is known in economics as inelasticity. The electricity price does not have enough impact to change the consumption preferences of end users; therefore, the quantity demanded does not change as much as the price. As a result, they do not reduce the need for dispatchable generation capacity in certain time by much. In addition, this efficiency does not pass on directly to end consumers.

5.3 The life cycle of Electricity Supply Industry

According to ISO 14001:2015, life cycle is a "consecutive and interlinked stages of a product (or service) system, from raw material acquisition or generation from natural resources to final disposal." The main objective is to prevent the transfer of environmental impacts caused by third parties along the supply chain. This approach helps us recognize that all our decisions and actions are one part of a whole system. Life cycle perspective brings powerful insights for decision-makers, because it starts premise that the purchase price reflects only a part of cost production, the externalities (costs or benefits) are not reflected, and therefore the real cost is superior. A life cycle perspective supports the proposition that energy provides a useful service and therefore usually does so with costs (direct or indirect), which may or may not be known immediately.

The life cycle of the ESI includes each and every one of the steps from the production and extraction of raw materials, processing, transportation, storage, distribution, and use. Each of these may have an impact on different dimensions: environmental, economic, and social. To transform energy into electricity from any renewable or nonrenewable source, certain stages may be identified. Table 5.1 shows a correlation between the ESI stages and economic characteristics.

Electricity rates or prices depend on several factors: amount of demand, technology used in its generation, the capacity of the transmission and distribution lines, the location, and the fuels. Reforms to the electricity industry typically have focused on the use of competition to achieve efficient use of, and investment in, generation resources. However, different reforms differ in

TABLE 5.1 Life cycle of the Electricity Supply Industry.

Stages	Some details	Economic characteristics	Implications
Energy sourcing (primary source)	There are seven approaches of transforming energy into electrical energy. The mix of energy sources is chosen by each country based on its endowment, certain political/social/cultural aspects, its climate, among others.	Property rights	Potentially competitive between different sources
Power generation	Generating plants are classified by merit order. 1. Base load plants are those able to run continuously and consistently, without being shut down. These have the lowest marginal cost of production. 2. Peak load plants are those able to run at certain times of day or seasons for the purpose of turned on and off suddenly to cover demand spikes. These have the highest price. 3. Mid-merit plants are those able to cover demand between base load demand and peak load demand.	Limited economies of scale at the plant level Coordination economies at the system level Complementarity with transmission	Potentially competitive
Transmission network	Transmission usually refers to transport large volumes of electric power with high voltages over relatively long distances. It is convenient to specify that although transmission has economies of scale of great magnitude, that is, it is economically more efficient to have a single line operating than two or more with less capacity, this is at the level of the transmission line, not necessarily in the entire network. Although this activity is not a natural monopoly, the scope for competition is limited.	Network of externalities In general, it is not a natural monopoly Big sunk costs	Investment incentives require special attention One network but possibly multiple owners

Continued

96 Advances of Artificial Intelligence in a Green Energy Environment

TABLE 5.1 Life cycle of the Electricity Supply Industry.—cont'd

Stages	Some details	Economic characteristics	Implications
Distribution network	Distribution is transportation at lower voltage levels. It is considered a natural monopoly due to large fixed cost of the investment and because it would be inefficient to duplicate the physical network.	It is often a natural monopoly Big sunk costs	Without competition
System operation	It ensures the reliable delivery of electricity to users. It is responsible to guarantee the functioning of system: 1. Optimization, balancing, and dispatch 2. Applying technical access criteria to networks 3. Ancillary services (activities beyond generation and transmission that are required to maintain grid stability and security)	Monopoly (due to technical restrictions)	Without competition
Supply and marketing	It refers all services around end user. It includes the commercial operation, procurement of energy, metering, billing of consumption, and customer relationship.	Limited economies of scale	Potentially competitive
Metering	It has the most relevant role for increasing the issues of sustainability and the environment. With the implementation of more and more accurate meters, that users know how much they must pay for an amount consumed would seem a simple process; however, in practice, it is quite complicated.	Big sunk costs	Without competition

Generated from the information of Organisation for Economic Co-operation and Development, International Energy Agency, Competition in Electricity Markets. Competition in Electricity Markets, 2001 (p. 18) and C. Harris, Electricity Markets. Pricing, Structures and Economics, John Wiley & Son Ltd., 2006 (p. 12–14).

TABLE 5.2 Traditional model of electricity supply.

Technical	Power generation	Transmission network	Distribution network	Supply and marketing
Regulation	Functioning market	Regulation needed		Functioning market
Economics	Market price	Tariffs	Tariffs	Consumer price

Source: How to play with market players by TU Delft (Rudi Hakvoort with https://www.edx.org/es/bio/rudi-hakvoort and Laurens de Vries https://www.edx.org/es/bio/laurens-de-vries-2) is licensed under CC-BY-NC-SA 4.0. This material was created by or adapted from material posted on Energy Markets of Today — https://courses.edx.org/courses/course-v1:DelftX+EnerTran1x+3T2020/course/.

the degree of competition introduced on the "buyer" side of the market ([13] p. 79). Given its economic and regulatory characteristics, the ESI has adopted a different market design at each stage. By economic characteristics, we refer to the way of establishing costs and charges and by regulatory characteristics to focus on the supervision and protection of the final consumer. Table 5.2 shows the traditional model of the ESI. There the sale regulation attended the technical and the economic variables.

The power generation and supply and marketing stages have been left under a liberalization model, that is, they are activities where the market allows free competition. Regarding the transmission and distribution stages, they continue to be a centralized system since fixed costs are very large, due to the construction of the transmission network. That is, the construction of a parallel network owned by a new competitor would lead to a doubling of costs, which would only increase prices. Therefore, in general, it is not economically viable to introduce competition in the transmission and distribution networks.

However, in some cases, two transmission lines running more or less in parallel may increase dependability and ensure security of supply. For the specific case of distribution, this is classified as a natural monopoly, where it would be better if only was one competitor, due to the large fixed costs of the investment. It is more efficient in these stages to have a network operator that guarantees the operation of the same, with what connection and investment are made efficiently. This operator's main function is providing stability to the network. It controls and coordinates the injection of electricity to the grid in order to avoid overloads, at the same time; this operator must avoid shortages

98 Advances of Artificial Intelligence in a Green Energy Environment

to end consumers due to insufficient injections. The grid operator decides which and how long the generating plants must generate electricity in order to achieve a balance between supply and demand.

Hence, in order to ensure access to the network, in some countries, a regulatory system has been established, which has the mission of ensuring that any member of the market has access to the network and achieving a fair rate by not allowing that costs are multiplied or the network is monopolized.

5.4 Importance of life cycle assessment of Electricity Supply Industry

The world is not yet heading toward a sustainable energy future. The IEA in 2006 remarks about the record highs reached by oil prices raise concerns about the balance of the 20% increase in CO_2 emissions during 1995–2005 during the last decade and anticipate ground in the oil demand over the next 25 years even taking into account improvements in energy efficiency and probable technological progress [17]. Therefore, it has happened, the demand is increasing as previously presented in this chapter. Unfortunately, our development concept includes an environmental load, and in this sense, the energy systems are not the exception. The necessary exploitation of the primary energy pollutes the world and the evil relationship between energy, greenhouse gases (GHG), and anthropogenic climate change is already a fact. Weisser [18] refers to the importance of a deep knowledge of life cycle GHG as an important indicator for mitigation strategies to the energy sector. The research work of Weisser [18] revolves around the life cycle of GHG emission of the electricity generation chains and the type of generation technology and its thermal efficiency including its operation mode, determining the GHG emissions, has been found. Parallel that and for a better approach to studying the problem, GHG emissions can be broken down by the economic activities associated to their production, as shown in Fig. 5.9, with information of IPCC [19]. There you can see that electricity and heat production systems represent the 25% of the GHG emissions at 2010. In the same way, Hondo [20] recommends an appropriate understanding of the power generation GHG emission characteristics of the power generation systems and in this sense the life cycle assessment (LCA) studies have a roll to include the environmental perspective to attend the concerns over anthropogenic climate change. Moreover, like Hondo [20] comments, the technologies are associated at the society and depend crucially on "the characteristics of the society wherein they exist" and "therefore, it is essential to clearly define the scope with regard to time and space for any technology assessment" ([20] page 2). In other words, it is a key issue to include in the integral GHG study the technological component, as the power generation systems.

Sustainability has passed the environmental dimension and has permeated the discourse of business and social decision-makers. Constructive measures

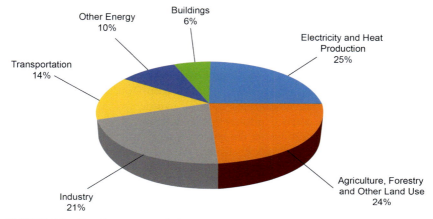

FIGURE 5.9 Greenhouse gas emissions by economic sector. *Adapted from IPCC, in: O. Edenhofer, R. Pichs-Madruga, Y. Sokona, E. Farahani, S. Kadner, K. Seyboth, A. Adler, I. Baum, S. Brunner, P. Eickemeier, B. Kriemann, J. Savolainen, S. Schlömer, C. von Stechow, T. Zwickel, J.C. Minx (Eds.), Climate Change 2014: Mitigation of Climate Change. Contribution of Working Group III to the Fifth Assessment Report of the Intergovernmental Panel on Climate Change, Cambridge University Press, Cambridge, United Kingdom and New York, NY, USA, 2014.*

such as "end of pipe" are obsolete; evolution leads a proactive measure like a cleaner production (including distribution). In the industry challenge, it is increasingly common to provide a solution in terms of sustainability measures and best if this measure is inherent to the system design. As discussed by Niembro [21], a methodology specifically developed to visualizing a product with a system perspective is the LCA. The methodology allows showing and understanding the impacts associated with products or services. The LCA is an excellent tool for both prognoses and diagnosis of system performance. Its application as an aid in the design and manufacturing is a smart move that promotes the eco-efficiency. According to the ISO 14040 standard [22, 23], the LCA treats the potential environmental aspects and impacts throughout the life cycle of a product from the acquisition of the raw material, through production, use, final treatment, recycling, until its final disposition, and this may include the entire supply chain, that is, from the cradle to the grave. In general, the economic and social impacts are beyond the scope of the LCA. Although, they can be combined with other analysis techniques. In this chapter, we explore the possibility of a parallel work to the best understanding of the ESI using LCA and LCC.

LCA is different from other environmental assessment tools such as environmental performance assessment, environmental impact assessment, and risk assessment. The LCA approach is based on a functional unit, not so the other techniques, although they can use information collected by them. The context and environment under study affected the results. The LCA is an

100 Advances of Artificial Intelligence in a Green Energy Environment

excellent tool to make a comparative assertion in situ. In different systems, isolated from one another, with different conditions, the LCA of each system offers an approximate idea of each situation, but not be compared; the comparison requires the analysis of the entire system. In order to apply a multi-objective optimization of a process, one of the main purposes of using LCA is obtaining input data. Jacquemin et al. [24] discuss the confluence between LCA and "process systems engineering" tools in order to improve a production analysis with more detailed information concerning to the influence of the process operating conditions on the environment.

LCA is an objective tool for evaluating the environmental loads associated with a product, process, or activity. It is carried out by

- identification and quantification of energy, materials used, and waste streams of all kinds;
- qualification of the impacts of the use of energy, materials and emissions, waste, and discharges to the environment;
- evaluation of actual or potential impacts and implementation of appropriate environmental improvements.

In addition, as Finnveden et al. [25] comment in a mandatory reference article, the LCA is useful to avoiding problem-shifting from one life cycle stage to another stage of the process, or of another system stage, from one region to another or from one environmental problem to another different problem.

Fig. 5.9 shows the framework of an LCA according to the ISO 14040 and ISO 14044 standards. As can be seen in the figure, the four phases of LCAs are

- Phase 1: Definition of the objectives that is the agreed goal and scope, including functional unit and references flows.
- Phase 2: Inventory analysis, when the recollected date is processed to generate useful information for the study.
- Phase 3: Environmental impact assessment, when the distribution of environmental burdens is made based on scientific models already validated to obtain a list of different potential environmental impacts and their magnitudes.
- Phase 4: Interpretation, to determine the importance of the results, their real magnitude beyond the amount, give a context of the result and be able to make decisions about the hot points of the analyzed process or system.

Usually the LCA studies study and use the so-called electric mix, which is the aggregated information of the different sources of technologies that generate electricity, annual average of a certain geography of a specific past year in the life cycle of the product. However, considering that, "electricity demand fluctuates over the years, over the season, over the week, and over the day [...] can also highly influence the consumption pattern" ([26]; p. 2). That is why the approach we propose in our work is relevant.

The negative bias is that energy (ESI) infers in the calculation of the environmental impacts of any system recognized by the practitioners of the LCA methodology. However, the methodology continues to be simplified for the sake of practicality and quick information, although robustness is sacrificed and potential for richness in the interpretation of results is lost. It is known that there is a much more attractive indicator in terms of its simplicity of calculation and therefore a faster application is the carbon footprint. However, it is recognized as a partial indicator, extremely simplified, and lacking a framework that allows a comprehensive information analysis of the system under study. It could be said that the carbon footprint is only one impact category of the many others that are studied in the LCA. Several investigations have focused on studying the "decarbon" scenarios after the implementation of certain measures that seek these effects; these are works that have allowed to calculate the future results of the decarbonization of electrical networks [27–30] and even LCA studies [26,31,32]. Although they are very valuable works, they do not include a comprehensive assessment of the actions or measures. That is when political and governmental decisions on electricity and GHG are made and implemented, in the end, their environmental results are evaluated, but the economic and financial implications are not considered, not the social part. In our research, outlined in this chapter, a methodological path is offered to comprehensively evaluate the economic and environmental aspects of the ESI and thereby offer information that allows us to have real data on the current situation. We are not yet considering social impacts. However, the approach already is an advance in the sustainable track. Our research will allow evaluating potential measurements of "decarbon" less idealized and closer to the real system.

A 167 LCA case studies review of electricity is developed by Ref. [33]. The main objective was to provide an overview of emissions from electricity generation technologies based on published LCA studies until that date. In the Turconi review, the results of the LCA studies were compared according to the GHG. In addition, they "evaluated and categorized according to contributions from the following three life cycle phases: (1) fuel provision (from the extraction of fuel to the gate of the plant), (2) plant operation (operation and maintenance, including residue disposal), and (3) infrastructure (commissioning and decommissioning)" [33]. For this chapter, that is so relevant, because apply this "three life cycle phases" as the stages to include in own properly proposal LCA methodology and we include the "supply and marketing." The cycle mentioned earlier can be seen in Table 5.2. On the other hand, Turconi recommends paying special attention to the formulation of objectives and scopes, as well as the Functional Unit proposal to avoid incomparability setbacks of results between similar studies, which we attend to [26]. develop a method for linking an hourly economic model with LCA, it is an excellent approach to the problem. We are working in another way in the same route, taking into account life cycle costing in the equation.

5.5 Life cycle cost of Electricity Supply Industry

Life cycle costing is the economic analysis process that assesses the total cost of acquisition, ownership, and disposal of a product over a period of time. It includes a comparison between alternatives projects (including their externalities). Life cycle cost (LCC) and life cycle costing are not the same. Life cycle costing is a methodology to develop a systematic economic evaluation of LCCs in a determinate time. The LCC is the cost of an asset, and assets system, or a part of them; in other words, it is the cost of some resource, resource system, or a partial system, with intrinsic economic value. According to ISO:15686-5, "The purpose of life-cycle costing should be to quantify the life-cycle cost (LLC) for input into a decision-making or evaluation process, and should usually also include inputs from other evaluations." Life cycle costing includes all costs of the product or service throughout its life cycle. This technique considers both direct costs (planning, design, construction, operation, and disposal) and the indirect costs related to investment, use, maintenance, or end of life. The classical graphic representation of the life cycle costing methodology is shown in Fig. 5.10.

The relevance of the LCC has even led the petroleum, petrochemical, and natural gas industry to develop its own standard, the ISO 15663:2021 petroleum, petrochemical, and natural gas industries—life cycle costing. It is relevant to mention that the standard does not include its own methodology for

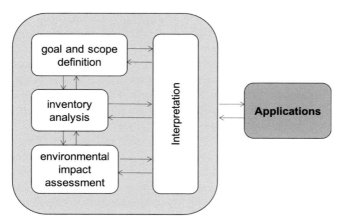

FIGURE 5.10 Framework of life cycle assessment according to the ISO 14040 standard. *Source: AENOR, UNE EN ISO 14040 gestión ambiental. Análisis de ciclo de vida. Principios y marco de referencia, Asociación Española de Normalización y Certificación, Madrid, 2006.*

"electricity." The ISO 15663:2021 is applicable when making decisions between competing options that are differentiated by cost and/or economic value. The methodology presented in ISO:15686-5 and ISO 15663:2021 is similar and is based on the same principles (Fig. 5.11).

In an ESI study of life cycle costing, the objective is the obtaining of the net present value of all financial flows. It will incorporate all costs and inversions produced, from the "cradle to the grave" of all supply chain of energy. It is an ambitious study with a large amount of information whose modeling has a certain degree of complexity. On the other hand, the income generated during the operation of the ESI is accounted for as the current of opposite flows. The life cycle costing is a methodology that is widely accepted in certain sectors of construction in the world. So much so that there is the ISO standard: 15686-5: 2017 Buildings and constructed assets—Service life planning—Part 5: Life-cycle costing. On it, the applicable methodology for the construction industry is described. Additionally, European directives define the road to implement the methodology, such as Directive 2010/31/EU that to define the scope of the costs and benefits of a life cycle costing study ([34], page 5), includes both the perspective of financial character (immediate costs and benefits of the investment decision) and macroeconomics (indirect

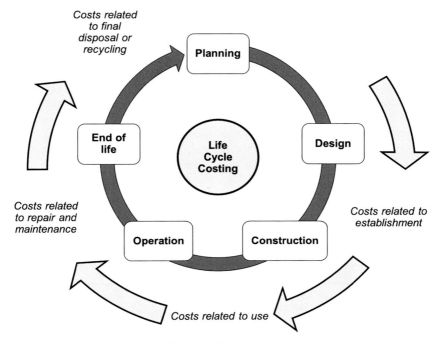

FIGURE 5.11 Stages of life cycle.

economic concepts that affect market players other than the investor). The ISO: 15686-5: 2017 standard can be used as a reference for the life cycle costing concept.

The costs of a product throughout its life cycle can be easily visible, such as direct costs of production, or they can have less visibility, such as indirect costs for the manufacturer or costs for society. That is, the long-term perspective helps us to show some costs that would otherwise be hidden despite having an impact on the organization.

Performing an LCC is important for ESI because one main objective is the level of efficiency energy or energy consumption having the minimum LCC for end users.

There are four phases to work out an LLC (Fig. 5.12):

In accordance with García-Erviti there are four main elements into the LCC Analysis: the costs or monetary flows; the period of time over which these costs are incurred (project lifetime considering each project element life); the discount rate (the interest rate that would make a company or investor indifferent to receive a payment now or a greater payment in the future) applied to future costs to equate them with present day costs (this present value allows the addition of costs along the time); and finally results analysis method [34]. This author explains that the discount rate is the risk-free return plus a premium that compensates the risk of each project.

A whole life cost typical is shown in Fig. 5.13 where in x-axis is time and y-axis is cost expressed in percentage. Usually, we could distinguish two kinds of cost, committed and incurred. A committed cost is a cost that will be incurred in the future but is not possible to change it. Costs are incurred only when a resource is done. This means that both of them are realized at different times. Murman points out that in accordance with many programs, the chance to influence the LCCs is greatest in its earliest stages and quickly decreases when the planning stage is completed. This showed the importance to add all life cycle attributes into early stages of planning [35].

García-Erviti mentioned that the relevant issue for determining the whole life of project is getting a comparison between alternative projects that offer different technological solutions [34].

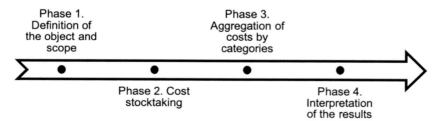

FIGURE 5.12 Phases to work out a life cycle cost.

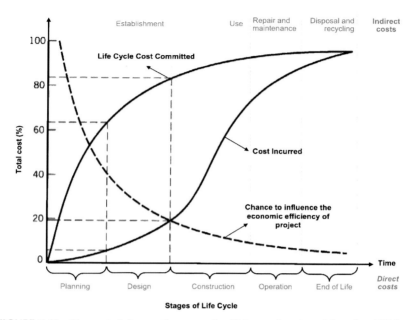

FIGURE 5.13 Chance to influence the economic efficiency of project. *Adapted to ESI from: Roger France, Jean-Francois. Life-cycle costs in construction. (2003) 96 pages http://hdl.handle.net/2078.1/171424.*

Besides all costs should be carefully classified and grouped by category. Each cost should be assigned to three categories: first to stage of life cycle of ESI, then to stage of life cycle, and finally to cost category. So, we get a graph like shown in Fig. 5.9, where x-axis has stage of life cycle, y-axis has stage of life cycle of ESI, and z-axis has cost category.

Fig. 5.13 shows that each cost should be associated to one cost category, for example, labor, energy, equipment, and transport.

The interpretation of the results will also be carried out taking into consideration the three dimensions. García-Erviti points out that this analysis also requires long-term perspective, an uncertainty and risk estimation, and finally a sensitivity analysis that allows to evaluate benefits and impacts in variables as a result as premises adopted [34].

5.6 Mobilizing industry for a clean and circular economy

Usually, the production system of industries works by designing products to be used, consumed, and discarded after a single life cycle. While in the circular economy, a system for the use of resources where the reduction, reuse, and recycling of elements is privileged, where waste becomes materials, seeking the use of all resources and evaluating the environmental impact in all phases of the life cycle.

Industry consumes almost the same energy as transport, making it one of the largest energy users. Industry uses large amounts of electricity, natural gas, and fossil fuels; for example, it consumes about 40% of worldwide electricity. Industry may increase its energy efficiency in several ways because manufacturing processes offer endless possibilities to save energy.

Modernizing industrial facilities may lead to a significant reduction in emissions by introducing new technologies. Increase the implementation of energy innovation and technology to achieve an energy transition committed to production systems that respect the environment, with active prosumers.

With the implementation of renewable technologies for electricity generation (technologies: hydraulic, photovoltaic, solar, wind, biomass, and geothermal), the negative impact that the consumption of electrical energy has caused on the planet has been reduced. However, these forms of renewables are intermittent in their natural state, and since electricity may not be stored in large quantities, this type of energy may currently only use for as long as it is available. In addition, most of them are local, so it depends on endemic factors.

These energies are available in abundance and humanity has been using them for a long time. So, the direct cost to the consumer remains low in its traditional form of use. However, modern ways of using these energies require sophisticated conversion, which in turn bought the cost of supply. Despite this, it would be expected that beyond the immediate future, when renewable energies will be preponderant, the cost of these energies with respect to fossil fuel energy would be lower, given that their marginal cost is zero.

This leads us to reflect, among other things, if the rate or price we would pay corresponds to all materials used and energy consumed during all stages of the life cycle of the ESI.

A possible solution would be to adopt the LCC to calculate the price of electricity in order to help raise awareness that our consumption decision is not isolated, it has consequences on the environment and the economy; so, it is important to consider that production systems do not exceed the capacity of the ecosystem, in addition to the environmental impact of manufacturing, use, and disposal, thus promoting a more sustainable production rate and consumption.

Due to the growing use of energy consumption that is demanded worldwide, they must consider all the impacts that its generation will produce. The sustainability of energy must be understood with the purpose of benefiting human beings and the ecosystem.

If consumers of electrical energy would participate in electricity markets through price, it may benefit not only consumers also would help to market operate more efficiently. Energy efficiency is minimizing the amount of energy needed to meet demand without affecting its quality.

In summary, giving the consumer a more active role within decision-making and focusing the vision on sustainability concepts provide a framework for approaching the assessment of energy technology and policy options and their trade-offs.

Finally, the importance of the environmental component in the systems is no longer an issue to be considered, but rather a required issue. If, in addition, the system is an energy supply system, the importance of the issue is irrefutable. Combining environmental and economic components to analyze the performance of the ESI is of vital importance now when production systems are not at odds with efficiency and the preservation of nature.

Our research is not finished; we are validating the models; and the results that we only sketch (see Figs. 5.14 and 5.15) still need more work to show that our methodology is quick and effective to apply.

Throughout this chapter, we discuss the generalities of the ESI and the importance of LCA studies and the life cycle costing methodology. All of the above, as an approach to solving the problem of the characterization of an ESI from a joint perspective, allows obtaining more and better information on what exists in the field of sustainable ESI.

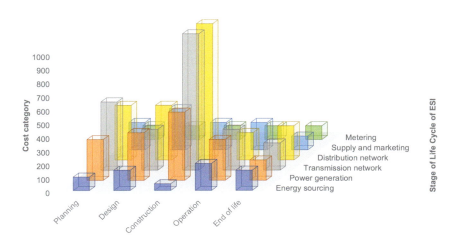

FIGURE 5.14 Stages of the life cycle costing for the ESI.

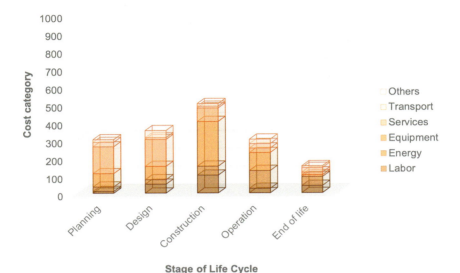

FIGURE 5.15 One cost category in life cycle costing.

References

[1] K.-m Ban, Energy is the Golden Thread, United Nations, 2012.
[2] C. Harris, Electricity Markets. Pricing, Structures and Economics, John Wiley & Son Ltd., 2006.
[3] D. Cordero, Red Eléctrica y Endesa pagarán 20 millones por el apagón de 2007, 2015. Febrero 26, 2021, de Periódico El País Cataluña Sitio web: https://elpais.com/ccaa/2015/08/03/catalunya/1438606204_930577.html.
[4] A. Monge, Aniversario De Una Crisis Energética 10 años del gran apagón de Barcelona, 2017. Febrero 26, 2021, de El Periódico, Barcelona Sitio web: https://www.elperiodico.com/es/barcelona/20170723/10-anos-del-gran-apagon-de-barcelona-6185930.
[5] A. Gómez-Expósito, A. Conejo, C. Cañizares, Electric Energy Systems, CRC Press, 2018.
[6] World Bank, Energy in 2018, World Bank Sitio web:, 2021 https://datos.bancomundial.org/indicator/EG.USE.ELEC.KH.PC.
[7] W. Tester, M. Drake, J. Driscoll, A. Peters. Sustainable Energý, MIT Press Books, 2012.
[8] Twenergy, Qué es la energía primaria?, 2019. Febrero 26, 2021, de Twenergy Sitio web: https://twenergy.com/energia/que-es-la-energia-primaria-1792/.
[9] IEA, World Balance, 2021. https://www.iea.org/sankey/#?c=World&s=Balance.
[10] International Energy Agency, 2021. International Energy Agency. 2018 Energy information de IEA Sitio web: https://www.iea.org/data-and-statistics/data-tables.
[11] Organisation for Economic Co-operation and Development, International Energy Agency, Competition in Electricity Markets. Competition in Electricity Markets, 2001.
[12] P. Zweifel, A. Praktiknjo, G. Erdmann, Energy Economics. Theory and Applications, Springer International Publishing AG, 2017.

[13] R. Biggar, H. Reza. The Economic of Electricity Markets, IEEE Press and Jonh Willey & Son Ltd., 2014.

[14] D. Kirschen, Fundamentals of Power System Economics, John Wiley & Son Ltd, 2019.

[15] C. Bhattacharyya, Energy Economics. Concepts, Issues, Markets and Governance, Springer-Verlag London Limited, London, 2011.

[16] A. Singh, D. Pant, S. Olsen. Life Cycle Assessment of Renewable Energy Sources, Springer, 2013.

[17] IEA, Perspectivas Sobre tecnología energética, 2006 de International Energy Agency Sitio web: http://aceer.uprm.edu/pdfs/etp_spanish.pdf.

[18] D. Weisser, A guide to life-cycle greenhouse gas (GHG) emissions from electric supply technologies, Energy 32 (9) (2007) 1543−1559.

[19] IPCC, in: O. Edenhofer, R. Pichs-Madruga, Y. Sokona, E. Farahani, S. Kadner, K. Seyboth, A. Adler, I. Baum, S. Brunner, P. Eickemeier, B. Kriemann, J. Savolainen, S. Schlömer, C. von Stechow, T. Zwickel, J.C. Minx (Eds.), Climate Change 2014: Mitigation of Climate Change. Contribution of Working Group III to the Fifth Assessment Report of the Inter-governmental Panel on Climate Change, Cambridge University Press, Cambridge, United Kingdom and New York, NY, USA, 2014.

[20] H. Hondo, Life cycle GHG emission analysis of power generation systems: Japanese case, Energy 30 (11−12) (2005) 2042−2056.

[21] I. Niembro, S. García, C. Sierra, M. González, Actualización de la herramienta informática CICLOPE para Análisis de Ciclo de Vida, in: XIII Congreso Internacional De Ingeniería De Proyectos, 2009, pp. 1069−1078.

[22] International Organization for Standardization, Environmental Management-Life Cycle Assessment-Principles and Framework (ISO 14040), 1997.

[23] AENOR, UNE EN ISO 14040 gestión ambiental. Análisis de ciclo de vida. Principios y marco de referencia, Asociación Española de Normalización y Certificación, Madrid, 2006.

[24] L. Jacquemin, P.Y. Pontalier, C. Sablayrolles, Life cycle assessment (LCA) applied to the process industry: a review, Int. J. Life Cycle Assess. 17 (8) (2012) 1028−1041.

[25] G. Finnveden, M. Hauschild, T. Ekvall, J. Guinée, R. Heijungs, S. Hellweg, S. Suh, Recent developments in life cycle assessment, J. Environ. Manag. 91 (1) (2009) 1−21.

[26] B. Kiss, E. Kácsor, Z. Szalay, Environmental assessment of future electricity mix−Linking an hourly economic model with LCA, J. Clean. Prod. 264 (2020) 121536.

[27] M. Arbabzadeh, R. Sioshansi, J. Johnson, G. Keoleian, The role of energy storage in deep decarbonization of electricity production, Nat. Commun. 10 (1) (2019) 1−11.

[28] A. Mileva, J. Johnston, J. Nelson, D. Kammen, Power system balancing for deep decar-bonization of the electricity sector, Appl. Energy 162 (2016) 1001−1009.

[29] M. Pehl, A. Arvesen, F. Humpenöder, A. Popp, E.G. Hertwich, G. Luderer, Understanding future emissions from low-carbon power systems by integration of life-cycle assessment and integrated energy modelling, Nat. Energy 2 (12) (2017) 939−945.

[30] C. Woo, A. Shiu, Y. Liu, X. Luo, J. Zarnikau, Consumption effects of an electricity decarbonization policy: Hong Kong, Energy 144 (2018) 887−902.

[31] D. Burchart-Korol, P. Pustejovska, A. Blaut, S. Jursova, J. Korol, Comparative life cycle assessment of current and future electricity generation systems in the Czech Republic and Poland, Int. J. Life Cycle Assess. 23 (11) (2018) 2165−2177.

[32] A. Gargiulo, M. Carvalho, P. Girardi, Life cycle assessment of Italian electricity scenarios to 2030, Energies 13 (15) (2020) 3852.

[33] R. Turconi, A. Boldrin, T. Astrup, Life cycle assessment (LCA) of electricity generation technologies: overview, comparability and limitations, Renew. Sustain. Energy Rev. 28 (2013) 555–565.

[34] F. García-Erviti, J. Armengot-Paradinas, G. Ramírez-Pacheco, Life cycle cost analysis as an economic evaluation tool for sustainable building. State of the art, Informes de La Construcción 67 (537) (2015) 1–8, https://doi.org/10.3989/ic.12.119.

[35] E.M. Murman, M. Walton, E. Rebentish. Challenges in the Better, Faster, Cheaper Era of Aeronautical Design, Engineering and Manufacturing, MIT, MA, USA, 2002.

Chapter 6

Comparison of open access multi-objective optimization software tools for standalone hybrid renewable energy systems

Mahesh Wagh[1], Purshottam Acharya[1] and Vivek Kulkarni[2]
[1]*Department of Technology, Shivaji University, Kolhapur, Maharashtra, India;* [2]*Sanjay Ghodawat Group of Institutions, Atigre, Kolhapur, Maharashtra, India*

6.1 Introduction

With the rapid growth of the world's economy, energy requirements have increased remarkably especially in developing countries. So to fulfill these requirements, we need energy which should be more sustainable and economically cheaper, especially when we are going to supply the electricity for remote areas. To connect the remote areas directly with the grid is a very costlier option. So to solve this problem, a standalone hybrid renewable energy system gives a solution, which includes PV, wind, diesel generator, and sometimes a hydrogen fuel cell. In the case of remote areas, the battery is a must part for the standalone hybrid renewable energy system, for example, in the case of remote bases of defensive forces which are situated at mountain regions or sometimes on an island. The design and optimization of hybrid renewable energy systems plays a crucial role in the installation of the system. If not properly designed, the system will result in a financial, technological, and finally a time loss. The proper utilization of the system components according to their necessity and a proper energy mix for minimization of the total project cost and in case of multiobjective optimization, decrease in CO_2 emissions, and increase in HDI (Human Development Index) should be fulfilled. So due to this situation, it becomes a complex problem of operations research, so the conventional methods with smaller iterations will not give the proper results. This problem requires sophisticated algorithms, which can be

---

Advances of Artificial Intelligence in a Green Energy Environment
https://doi.org/10.1016/B978-0-323-89785-3.00010-4
Copyright © 2022 Elsevier Inc. All rights reserved.

111

112 Advances of Artificial Intelligence in a Green Energy Environment

developed by using the computer languages, such as C, C $^{++}$, Windows Fortran, Visual Basic, and Pascal. These algorithms have also different types which are discussed below in detail. Here the term multiobjective software tool comes which can directly intake the energy-related data and perform various operations according to the algorithm defined and give the result as an output in form of text, bar chart, histogram, pie chart, schematic diagram, and energy flow diagrams [8]. Though there are multiple tools available for the optimization of the hybrid renewable energy systems, some open access or freely available software tools are there, which can be freely downloaded and are capable to optimize the multiobjective renewable energy application problem. This software includes mainly HYBRID-2, iHOGA, and HOMER. In this chapter, a case study has been conducted for the comparison of these software tools and to find out the best software tool among the three.

6.2 Literature review

In 1997, G. C. Selling-Hochmuth developed a method for the general optimization and sizing of the hybrid PV system using genetic optimization technique; this strategy contains mainly two algorithms: one is main and another one is the subalgorithm; the main algorithm deals with the sizing of the hybrid system and subalgorithm deals with the optimization of the system. The final result is a combined output of both operations [1]. Turcotte D. et al. in 2001 in their paper categorized available software tools according to their capability as prefeasibility, sizing, simulation, and open architecture tools, respectively, suggested the improvements needed in software tools, and categorized HOMER as a sizing tool and HYBRID-2 as a simulation tool [2]. Dufo-LÓpez R. et al. in 2005 in "Design and control strategies of PV-Diesel systems using genetic algorithms" used genetic algorithms for the optimization of the hybrid renewable energy system and they developed a tool named HOGA which is becoming a basic software tool for the base of iHOGA in the future. They compared results obtained by using the HOGA with those which are obtained by the HOMER. They observed the characteristic of SOC (state of charge) point for the batteries, i.e., they compared Ah of batteries and found out that HOMER lacks in finding the optimal set point for SOC, but HOGA optimizes SOC point [3]. Jose′ L. Bernal-Agustín and Rodolfo Dufo-LÓpez both in 2009 compared multiobjective optimization software tools used for simulation of hybrid renewable energy systems, but there is no such information or case study conducted. In this chapter, the comparison was done which states that control strategies are available for only three software, namely HYBRID-2, HOMER, and HOGA [4]. Conolly D. et al. in their paper "A review of computer tools for analyzing the integration of renewable energy into various energy systems" studied 37 software tools and identified the required area of application [5]. Luis Arribas et al. also categorized software tools for hybrid PV systems as a dimensioning tool, simulation tool, research cannot use the fuel cell and electrolyzer. It is the software used for the optimization of hybrid renewable systems utilizing genetic algorithms; it optimizes the system by using the genetic algorithms to reduce the data time step for calculations when a user requires quicker simulation of his/her system than if he enters the time required for simulation which is less than the standard calculation time required for the simulation of all estimated cases based on input given by the user, then it uses a genetic algorithm. This could give the feasible solution for the energy mix problem of hybrid renewable; this solution is probably 98%−97% matching with the exact one, but this value can be varied with the increasing number of constraints and components included in the project which affect the complexity of the problem. The iHOGA uses mainly two algorithms.

6.3.2.1 Primary algorithm

The primary algorithm gets optimal energy mix of PV panels, batteries, diesel generator sets, and other components; it minimizes the total net present value of the system, which includes capital, replacement, operation and management, fuel, salvage, and total costs.

6.3.2.2 Secondary algorithm

The secondary algorithm defines load dispatch strategy, by reducing the total operation and maintenance cost.

6.3.3 Load dispatch strategies

iHOGA has multiple options for defining load dispatch strategy, which includes the following.

6.3.3.1 Demand following

If batteries are unable to generate electricity to meet the demand, the diesel generator operates at a value that generates the energy which is only sufficient to fulfill the net demand. Batteries will get charged when renewable energy overcomes the necessary demand, but will not be energized by a diesel generator.

6.3.3.2 Cycle charging

Cycle charging includes running DG set at a rate such that it will charge the batteries with the excess power if batteries are unable to reach the demand. If the SOC point is set already, then the generator will charge the batteries until the SOC point is not reached. Another point which is a complementary point of SOC is depth of discharge (DOD) point, which indicates the SOC in Ah units, e.g., if $0 =$ full, then $300 =$ empty, whereas SOC is given in percentage (%), if $0\% =$ empty, then $100\% =$ full.

6.3.3.3 Mixed strategy

This strategy combines both the options mentioned earlier; if the demand is lower than the charging load, then the cycle charging strategy is applied; if net demand is higher than the charging load, then the demand following strategy is applied.

The genetic algorithm technique evaluates N number of the separate solutions, and each solution has a unique number given, and each solution is shown by a vector, and components of that vector represent the components and costs related to each solution. The iHOGA has an option of importing the database from the NASA RET screen, we can input the same data which appear on the screen after entering the latitude and longitude of the location, it includes daily solar radiation in kWh/day, for PV panel, speed of wind in m/s for wind turbine, and finally the air temperature for the calculation of battery performance. The schematic diagram which is generated by iHOGA is very simple, robust, and easy to understand, as compared to the line energy diagrams generated by HOMER and HYBRID-2.

iHOGA can be operated on both modes, either it can be mono-objective, which means it can only be used to minimize the net present cost along with the CO_2 emissions, or it can be operated on the multiobjective model, which means it can be used to estimate the minimum net present value, CO_2 emissions, along with the minimization of the total unmet demand. Sometimes it becomes necessary to find out the percentage of the total unmet load; such unmet load is not desired, when it is necessary to match the demand always, for example, a colony situated on a remote island, or a military airbase situated on a plateau, or a remote SAM (surface to air missile) site. Fig. 6.2 explains the algorithm flowchart of iHOGA software.

6.3.4 HYBRID-2

The HYBRID-2 software was developed by the National Renewable Energy Laboratory (NREL) and the University of Massachusetts at Amherst with funding from the US Department of Energy. The latest version of this software is version 1.3f. HYBRID-2 was constructed using MS Visual Basic language. It can be used for the smaller load simulation; it has the advantage that we can input the probabilistic/time series data from 10 min to 60 min range. There are separate options available for economic simulation and performance simulation. Three basic inputs are required for the resources: wind speed, insolation, and ambient temperature. Power generators are limited to wind turbines, PV arrays, and diesel generators. The energy storage options are available as the pseudo grid, heat storage, and battery. The inputs required are mainly categorized as load, resource, power system, and economics. Loads are categorized mainly as AC and DC; further category includes the primary load, deferrable load, heating load, and optional load. In each input, the time series

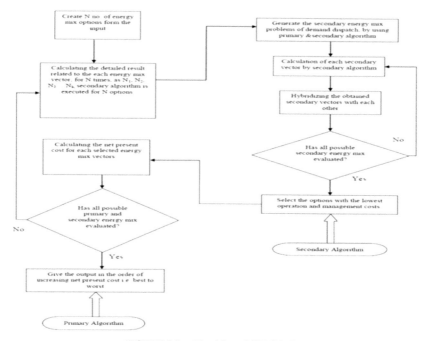

FIGURE 6.2 Algorithm of iHOGA [2].

data must be given for the optimization of the problem. The inputs required by this software are lesser as compared to the HOMER and iHOGA software. This software is also free but requires a password for the installation. The load dumping is also possible in HYBRID-2 when there is an excess load. In the energy diagram, the AC and DC buses are separated; in between the two buses, there are power converters. The standard diagram is already drawn by default by clicking on the inputs or by adding the data we are just connecting or interlinking the two components of the whole power system, and when all the data are added, the line diagram is generated which shows the interconnection between the two entities. However, only one of the multiobjective optimizations tools, the HYBRID-2, can be used for the optimization of the smaller grids because it lacks in generating a detailed report about the optimization result. HYBRID, which is the base case of HYBRID-2, is one of the oldest multiobjective hybrid renewable energy software. Fig. 6.3 reveals the algorithm flowchart of HYBRID-2 software.

HYBRID-2 has different options available for the load dispatch strategy which are peak shaving, load following cycle charging, renewable battery only system, and renewable genset only system. We have studied earlier about the load following and cycle charging; along with these two, there are also other options which are necessary for the study.

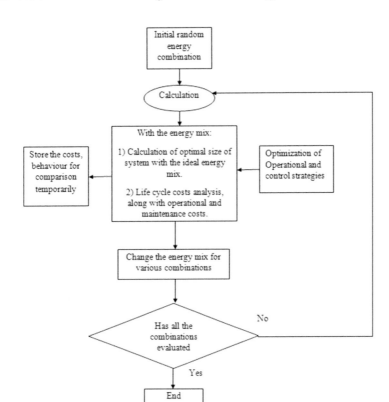

FIGURE 6.3 Algorithm flowchart of HYBRID-2.

6.3.4.1 Peak shaving

Batteries are charged regularly for every 2 weeks only by using the genset, and the peak demand only is supplied by the batteries.

6.3.4.2 Renewable battery only

In this system, the demand is fed by only renewable energy and battery systems; the generator is absent in this system.

6.3.4.3 Renewable genset only

In this system the batteries are absent and the whole demand is fed by using the renewable energy options and gensets only.

6.3.4.4 Customized load dispatch strategy

This is the unique feature of HYBRID-2, which gives it the benefit over other software tools because the operator on his/her own can enter the new strategy of load dispatch according to load requirement and trend of demand. In this strategy, we can set the optimal SOC point (boost level) for the battery and the time interval between the successive boosts.

6.3.5 Comparison of multiobjective optimization algorithm

Table 6.1 represents the comparison between the various multiobjective algorithms, which are generally utilized to run the considered software.

6.4 Case study

The case study presented here was performed to compare the iHOGA, HOMER, and HYBRID-2. This case study includes the simulation of a residential energy modeling problem for a house located in Maharashtra, at Warananagar District: Kolhapur latitude $= 16.876°$N and longitude $= 74.190°$E. The height of the place is about 458 m above sea level. This system typically includes a load of 4.4 kWh/day; the house is considered to be completely dependent on the hybrid; renewable energy only means it is considered as separated from the grid. The system includes the power generators typically as a DG set, PV modules, and wind turbine. Energy storage is fulfilled by using batteries. Fig. 6.4 represents the iHOGA software-generated hybrid standalone renewable energy system.

(1) When a new project has been started in all the three software the HOMER gives, the best method to the input project location, HOMER allows user by just clicking on the world map to go to the project site or by a simple entry of the name of that particular place where the project is going to be installed, and here it becomes best among the others by automatically relocating the server to gather the particular resource data related to the wind, solar irradiance, temperature, etc., of that particular chosen place.
(2) HYBRID-2 when started asks for the data time step required rather than the location and meteorological information at first; after that, the load should be entered.
(3) iHOGA EDU opens a standard case which has a consumption of 3.63 kWh/day AC load with a system of PV-diesel-battery; the user has to alter the values for the base case to give his/her input.

6.4.1 Input of load

(1) In HOMER, there are only two peak months considered for the peak load, but in our case, the load is maximum in April when the summer is on its high, and our load is 4.4 kWh/day.
(2) In iHOGA, a curve can be seen instantly after entering all the necessary data, which can be customized, by varying number of days; it gives a

TABLE 6.1 Comparison of multiobjective algorithms.

Algorithms	Genetic algorithm	Memetic algorithm	Particle swarm optimization algorithm	Ant colony optimization algorithm	Simulated annealing	Differential evolution
Optimization method	Metaheuristic population-based	Metaheuristic population-based	Metaheuristic population-based	Metaheuristic population-based	Metaheuristic trajectory-based	Metaheuristic population-based
Influence of population size on solution processing time	Exponential	Exponential	Linear	Linear	Linear	Linear
Requires ranking of solution	Yes	Yes	No	No	No	No
The tendency of premature convergence	Medium	Medium	High	Low	Low	Low

Open access multi-objective optimization software tools **Chapter | 6** **121**

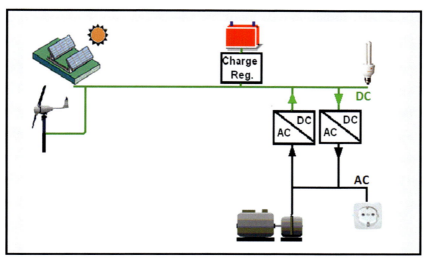

FIGURE 6.4 iHOGA software-generated hybrid standalone renewable energy system starting a project.

load profile of a whole year or only 1 day. Fig. 6.5 represents the daily load profile in iHOGA software.
(3) In HYBRID-2, there is also a provision of load profile generation but it depends upon the sensitivity of load data.

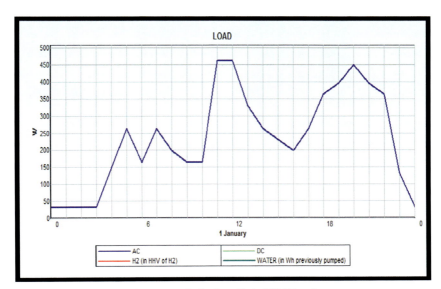

FIGURE 6.5 Daily load profile in iHOGA software.

6.4.2 Solar radiation data input

(1) In HOMER, there is no need for this step because at first, you are entering the data indirectly by giving the location coordinates.
(2) In iHOGA, solar energy—related data can be imported from the NASA RET screen website. Fig. 6.6 represents the solar radiation data at Warananagar site generated in iHOGA.
(3) In HYBRID-2, the load profile also can be generated the same as iHOGA, but cannot be compared with the profile generated by the surface whose azimuth angle is zero degrees.

6.4.3 Wind speed data input

(1) In HOMER, there is no need for this step because at first, you are entering the data indirectly by giving the location coordinates. Fig. 6.7 reveals the Weibull wind speed distribution at the site location.
(2) Fig. 6.8 represents the wind speed distribution in HYBRID-2.

Total annual irradiation comes out to be 1886.55 kWh/m^2, and the average wind speed is estimated as 3 m/s about 10 m from the ground, so this location is not suitable for large wind power generation, but except iHOGA, no software considered this; in the first case itself, HOGA calculated the system output and optimized it without considering much wind power generator, though it includes the wind turbines that have the rated speed nearly about 3.5 m/s.

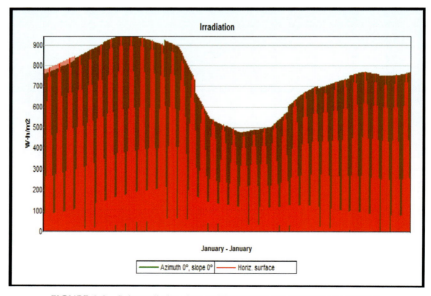

FIGURE 6.6 Solar radiation data at Warnanagar site generated in iHOGA.

Open access multi-objective optimization software tools Chapter | 6   123

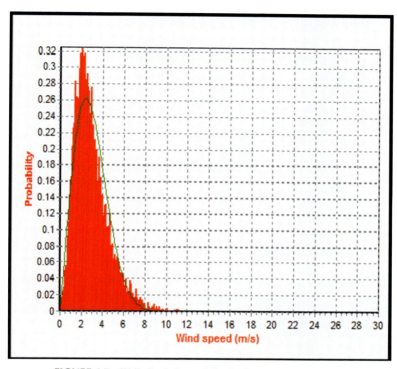

FIGURE 6.7 Weibull wind speed distribution at the site location.

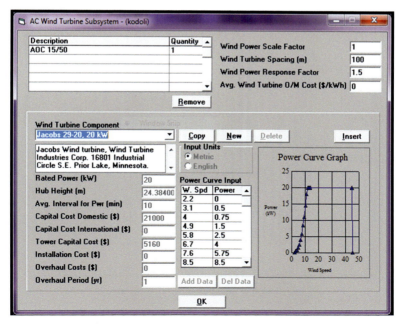

FIGURE 6.8 Wind speed distribution in HYBRID-2.

6.4.4 Selecting other components data

(1) In iHOGA, there is a component zero in the standard component database which is included in the datasheet of each of the components and is used to show the probability of that component, not being added in the system; so whenever there is no need in the particular mix selected by the system, every component each time neglected gets added for various possible combinations.
(2) In HOMER, there is a lack of setting for giving account for shadows while specifying the solar data.
(3) HYBRID-2 will give the output as per the time step series data entered. HYBRID-2 software has multiple load dispatch strategies as compared to the remaining two. Fig. 6.9 reveals the graph of total power supplied and time of the day.

6.4.5 Selection of battery and inverter

(1) In iHOGA, the battery life cycle calculation is done through the main flow (cycle counting) method. However, we can use the Schiffer model because in summer, the ambient temperature is greater than the manufacturer-specified temperature.

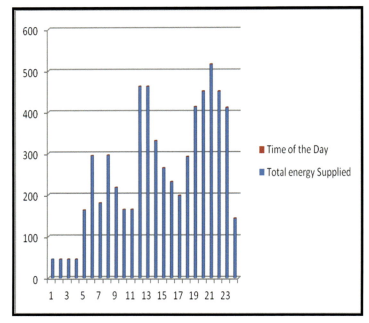

FIGURE 6.9 Graph of total power supplied and time of the day.

(2) HOMER lacks in setting the battery at optimum set point, i.e., SOC point, or it should at least indicate the battery DOD. The maximum power demanded was 577.5 VA, so the average power of the inverter selected to 1000 VA comes out to be 23% of total capacity, so the efficiency comes out to be 91%.

6.4.6 Calculation time

Among the three software, the HOMER requires much time for the simulation as it has no probabilistic approach added. iHOGA, when dealing with the simulation time given by the user greater than the time required for simulation of all cases acts like HOMER, and gives result. But, when such condition is not satisfied, it uses the genetic algorithm. Specific output can be selected for the renewable fraction in HOMER only while such a case is not possible in HOGA. Both HOMER and iHOGA calculate for minimum net present value. Both HOMER and iHOGA require an internet connection for the operation. Figs. 6.10 and 6.11 reveal the simulation results from HOMER and iHOGA software, respectively. Fig. 6.12 represents the power generation share of wind and solar PV versus total demand in iHOGA. Similarly, Fig. 6.13 reveals the time series analysis results in HYBRID-2.

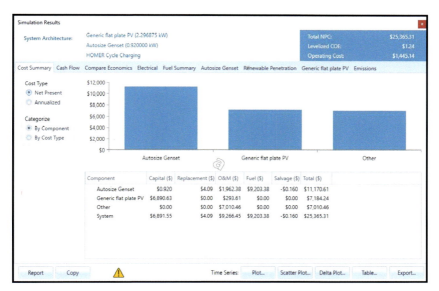

FIGURE 6.10 Simulation result of HOMER.

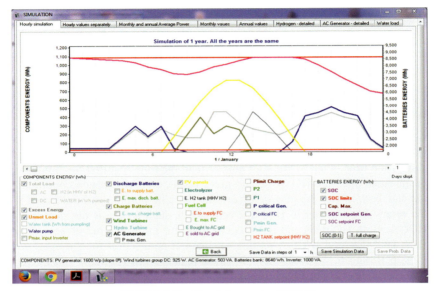

FIGURE 6.11 Simulation result of iHOGA.

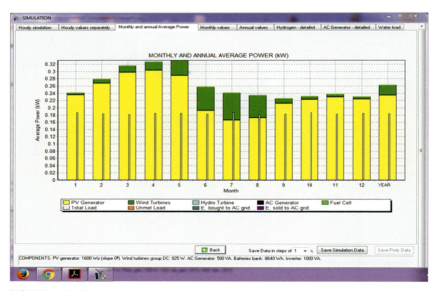

FIGURE 6.12 The power generation share of wind and solar PV versus total demand in iHOGA.

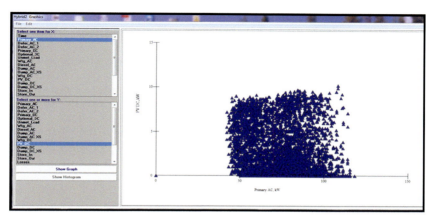

FIGURE 6.13 Time series analysis results in HYBRID-2.

6.5 Conclusion

From the comparison carried out for three software, it is found that the iHOGA, though lagging in some features like direct importing the resource data and ready to match demand curves, proved to be more beneficial for the evaluation of large projects related to the society and having larger capital costs. It gives the results through various dimensions which cannot be estimated perfectly in the case of HYBRID-2 and HOMER. It is required that iHOGA should give a simple and robust report like HOMER, in tabular forms and bar charts rather than giving it in numeric values and pie charts, which are difficult to understand. iHOGA should also apply the various load fulfilling strategies, such as HYBRID-2, the renewable battery only system, and renewable genset only system. The specific output selection, for the minimum renewable energy fraction, should be considered for the improvement in iHOGA, as it is present in HOMER.

References

[1] G.C. Seeling-Hochmuth, A combined optimisation concept for the design and operation strategy of hybrid-PV energy systems, Sol. Energy 61 (2) (1997) 77−87, https://doi.org/10.1016/S0038-092X(97)00028-5.

[2] D. Turcotte, M. Rossb, F. Sheriffa, PV Horiz. Work. Photovolt. Hybrid Syst., Photovoltaic Hybrid System Sizing and Simulation Tools: Status and Needs, vol. 13, September 10, 2001, pp. 1−10.

[3] R. Dufo-López, J.L. Bernal-Agustín, Design and control strategies of PV-diesel systems using genetic algorithms, Sol. Energy 79 (1) (2005) 33−46, https://doi.org/10.1016/j.solener.2004.10.004.

[4] J.L. Bernal-Agustín, R. Dufo-López, Simulation and optimization of stand-alone hybrid renewable energy systems, Renew. Sustain. Energy Rev. 13 (8) (2009) 2111−2118, https://doi.org/10.1016/j.rser.2009.01.010.

128 Advances of Artificial Intelligence in a Green Energy Environment

[5] D. Connolly, H. Lund, B.V. Mathiesen, M. Leahy, A review of computer tools for analysing the integration of renewable energy into various energy systems, Appl. Energy 87 (4) (2010) 1059—1082, https://doi.org/10.1016/j.apenergy.2009.09.026.

[6] L. Arribas, G. Bopp, A. Lippkau, World-wide overview about design and simulation tools for hybrid PV systems, in: IEA PVPS Task 11, 2008, p. 28. Rep. IEA-PVPS T11-012011, no. 1.

[7] I. Tegani, et al., Optimal sizing study of hybrid wind/PV/diesel power generation unit using genetic algorithm, Int. Conf. Power Eng. Energy Electr. Drives (May 2013) 134—140, https://doi.org/10.1109/PowerEng.2013.6635594.

[8] S. Sinha, S.S. Chandel, Review of software tools for hybrid renewable energy systems, Renew. Sustain. Energy Rev. 32 (2014) 192—205, https://doi.org/10.1016/j.rser.2014.01.035.

Chapter 7

Optimization of the organic waste anaerobic digestion in biogas plants through the use of a vortex layer apparatus

Andrey A. Kovalev[1], Dmitriy A. Kovalev[1], Yuriy V. Litti[2], Inna V. Katraeva[3] and Victor S. Grigoriev[1]

[1]*Federal Scientific Agroengineering Center VIM, Moscow, Russia;* [2]*Federal Research Centre "Fundamentals of Biotechnology" of the Russian Academy of Sciences, Moscow, Russia;* [3]*Nizhny Novgorod State University of Architecture and Civil Engineering, Nizhny Novgorod, Russia*

7.1 Introduction

The negative impact of human activity on the environment is associated not only with the increasing consumption of natural resources but also, to a greater extent, with the formation of liquid and solid waste from agricultural and processing industries [1,2]. In recent years, society's attention has been increasingly drawn to solving two inextricably linked problems—preventing the depletion of natural resources and protecting the environment from anthropogenic pollution. The rapid consumption of reserves of fossil fuels and the limitation of the construction of hydro and nuclear power plants have aroused interest in the use of renewable energy sources, including the huge masses of organic waste generated in agriculture, industry, and urban communal services. In this regard, the use of methods of biological conversion of organic waste with the production of biogas and high-quality organic fertilizers while solving a number of issues of environmental protection from pollution is very promising [3,4].

130 Advances of Artificial Intelligence in a Green Energy Environment

7.2 Anaerobic processing of organic waste: general characteristics of fermentation

7.2.1 General characteristics of the methanogenesis

Anaerobic bioconversion is a complex multistage decomposition of various volatile solids at conditions without oxygen access under the influence of anaerobic microorganisms, the end result of which is the formation of methane and carbon dioxide.

Lacking access to either oxygen or other energy-preferable electron acceptors (nitrate, sulfate, sulfur, etc.), microorganisms are forced to use the carbon of organic substances for this purpose, which ultimately leads to the most strongly reduced of the existing in the nature of carbon compounds—methane. At the same time, microorganisms in this process also use, as a rule, carbon of organic substances as an electron donor, oxidizing it to carbon dioxide (the second most important component of biogas).

According to modern concepts, the anaerobic bioconversion of volatile solids into biogas goes through four stages, proceeding sequentially, but simultaneously:

- hydrolysis;
- fermentation;
- acetogenesis;
- methanogenesis.

Hydrolysis is splitting of complex biopolymer molecules (proteins, lipids, polysaccharides, etc.) into simpler oligo- and monomers: amino acids, carbohydrates, fatty acids, etc.

Hydrolysis is carried out by various hydrolytic microorganisms by secretion of exogenous enzymes into the intercellular environment. The action of these enzymes leads to the production of relatively simple products that are effectively used by the hydrolytics themselves and by other microorganisms in the subsequent stages of bioconversion. Anaerobic bioconversion hydrolysis is closely related to acidogenesis, and hydrolytic microorganisms perform both stages (hydrolysis and acidogenesis).

The stage of fermentation (acidogenesis) of the formed monomers to even simpler substances—lower acids and alcohols, while carbon dioxide and hydrogen are also formed. The group of fermentative bacteria (fermentative bacteria) is a complex mixture of the many types of bacteria, most of which are strict anaerobes and function best in the pH range 4.0—6.5. The dominant organisms can be anaerobic mesophilic bacteria such as *Bacteroides*, *Clostridium*, *Butyrivibrio*, *Eubacterium*, *Bifidobacterium*, *Lactobacillus*, and many others. The main feature of fermentors is the ability to use the same substrates as hydrolytics, that is, the products of hydrolysis of polymeric compounds and monomers supplied with the effluent, but in lower concentrations. Enzymatic bacteria, along with proteins, carbohydrates, and fats, metabolize phenolic nitrogen and sulfur-containing compounds.

At the acetogenic stage, the direct precursors of methane are formed: acetate, hydrogen, and carbon dioxide. The decomposition of the products of the acidogenic stage is carried out by obligate proton-reducing or obligate-syntrophic bacteria, as a rule, requiring hydrogen-using partners, for which they are often called syntrophic bacteria. The bacterium *Syntrophomonas wolfei* anaerobically oxidizes butyrate in co-culture with *Methanospirillum hungatei*. The bacterium *Syntrophomonas buswelii*, which decomposes benzoate in a binary culture with methanogen to methane, acetate, and carbon dioxide, is also described. In the methane biocenosis, there is a close symbiosis between acetogenic bacteria and hydrogen-using methanogens; interspecies hydrogen transfer plays a decisive role in it. The nature of the synergism described above has not yet been fully clarified, since the group of acetogenic bacteria has not yet been sufficiently biochemically and physiologically characterized.

The methanogenic stage leads to the end product of the decomposition of complex organic substances—methane. Methane-forming bacteria are obligate anaerobes that are sensitive to oxygen. Representatives of only two genera *Methanosarcina* and *Methanosaeta* are capable of decomposing acetate with the formation of methane and carbon dioxide. Methane suction is capable of using acetate at very low concentrations and providing deep wastewater treatment. Acetate-using methanogens have morphological features that allow them to form shaped structures with other bacteria or mixed colonies. These properties are extremely important for the formation of fouling and pellets in reactors. In the absence of limiting at the stage of hydrolysis, the second limiting factor of anaerobic conversion is the methanogenic stage. This is due to the low growth rates of acetate-using methanogens (bacteria of the genus *Methanosarcina* double in 20−30 h, of the genus *Methanosaeta* in 200−300).

In a methanogenic community, there are close and complex relationships between groups of microorganisms. In view of the substrate specificity, the development of methanogens is impossible without their interaction with bacteria of the previous stages.

The low increase in biomass during anaerobic waste treatment is explained by the low energy yield of reactions (89% goes into methane) carried out by the methane biocenosis. The low energy yield in the process of methane fermentation, in particular, explains an important property of microorganisms of the methane biocenosis: the ability to do without a substrate for a long time (months) with a rapid recovery of activity after refeeding [3,5,6,8].

7.2.2 General characteristics of the hydrolysis process

An important role in the methanogenic community is played by microorganisms that hydrolyze polysaccharides present in the effluents of various branches of the food and pulp and paper industries. Isolated from anaerobic reactors fermenting effluents with suspended solids containing cellulose,

132 Advances of Artificial Intelligence in a Green Energy Environment

strains of cellulolytic bacteria, as a rule, are strict anaerobes and their concentration with a significant content of cellulose components in the effluent is usually 10^5-10^6 cells/mL of sludge, reaching 10^{11} cells/mL for livestock drains. Currently, more than 20 species of anaerobic cellulolytic bacteria belonging to the genera *Clostridium, Bacteroides, Butyrivibrio, Ruminococcus, Eubacterium, Acetivibrio*, and *Micromonospora* have been described. In anaerobic reactors, cellulose-forming bacteria are represented mainly by *Clostridial* forms. In particular, *Clostridium thermocellum* is the leading form under thermophilic conditions. Anaerobic hydrolytic bacteria decomposing hemicellulose, pentosans, polysaccharides of bacterial cell walls, starch, and pectin have been isolated and described. They belong to different genera: *Clostridium, Bacteroides, Lactobacillus, Bacillus*, etc. The number of bacteria-hydrolyzing polysaccharides in anaerobic reactors depends on the composition and concentration of the fermented effluent and the design of the reactor, so the number of amylolytics is especially high on wastes from starch production. The biochemistry of polysaccharide fermentation has been studied quite well in pure cultures. The main products are various fatty and carboxylic acids, alcohols, hydrogen, and carbon dioxide. It was shown that in the methanogenic community, the metabolism of hydrolytics, in particular, cellulose-destroying bacteria, is subject to regulation by methane and homoacetate bacteria that consume hydrogen.

Most of the proteolytic bacteria isolated from anaerobic reactors are also *Clostridia*, capable of growing by fermenting peptides and amino acids formed during the hydrolysis of protein compounds. This is their difference from the proteolytic microflora of the rumen, a characteristic feature of which is the need for carbohydrates and the inability to use amino acids. Bacteria of the genera *Peptococcus, Bifidobacterium, Staphylococcus*, and *Eubacterium* found in digesters also have proteolytic activity. The nonspore thermophilic proteolytic bacterium *Thermobacteroides proteolyticus* was isolated from the reactor fermenting the waste streams of the tannery. Thermophilic proteolytic bacteria, developing at temperatures up to $80-85°C$, are found in thermophilic fermented manure runoff. The presence of psychrophilic proteolytics developing on protein substrates at temperatures up to $0°C$ was found in the sediment of the manure storage and the silt of a hypereutrophied reservoir. In some proteolytic *Clostridia*, high proteolytic activity is associated with virulence. Most proteolytic bacteria are also capable of fermenting carbohydrates, which confirms the high plasticity of microorganisms included in the methane biocenosis.

The least studied is the process of anaerobic lipid hydrolysis. It is carried out by special enzymes—lipases. Most lipases of microbiological origin attack lipids, which are glyceride esters of long-chain fatty acids (FALCs) at positions 1 and 3 of glycerol. And only some lipases have the ability to carry out complete lipid hydrolysis with the formation of glycerol and FALC. It has been shown that in anaerobic reactors, lipolytic microflora is represented by

Organic waste anaerobic digestion in biogas plants **Chapter | 7 133**

clostridia and micrococci. Individual strains of bacteria of the genera *Ruminococcus*, *Eubacterium*, and *Butyrivibrio* are capable of hydrogenating insoluble fatty acids.

Anaerobic hydrolysis of lignin practically does not occur; however, it has been shown that its fragments with a molecular weight of up to 330 decompose in enrichment cultures with the formation of CO_2, CH_4, and acetate. Monomeric aromatic lignin derivatives are unstable under anaerobic conditions and decompose to form fatty acids. There are data on the biodegradation of one of the components of lignin—coniferyl alcohol under anaerobic conditions along one of the side metabolic pathways [6—8]

7.2.3 The impact of external factors on the intensity of the gas formation process

The main factors affecting the performance of anaerobic reactors are

- phase and chemical composition, as well as the particle size of the loaded substrate,
- the hydraulic retention time (HRT) in the reactor,
- the concentration of microorganisms inside the reactor,
- the efficiency of mass transfer of the reaction medium,
- reactor loading speed,
- temperature regime.

Temperature is one of the most important parameters that determine the speed of the process and the productivity of industrial plants. Temperature modes of operation of anaerobic reactors are subdivided into psychrophilic, mesophilic, and thermophilic, the temperature values at which are 20°C, 20—45°C, and 50—65°C, respectively. Naturally, the higher the temperature, the higher the rate of biochemical processes; therefore, as a rule, the thermophilic mode of fermentation is the most productive. But despite the higher fermentation rates in thermophilic conditions, the effect obtained is not large enough to compensate for the heat costs required to maintain it. In addition, under these conditions, the species composition of the microflora is rather poor, which explains the lower stability of the thermophilic regime compared to others. In this regard, most anaerobic reactors currently operate in a mesophilic mode, as a rule, 30—40°C, which provides an acceptable rate of methanogenesis, the relative energy advantage of maintaining the temperature regime, and the stability of the process due to the existence of a sufficient number of microorganism species. Currently, there are two fundamentally different methods of maintaining the temperature: with preheating of the original manure and without preheating. Heating devices of various types are used to heat the liquid manure supplied to the digester and compensate for heat losses into the environment.

When fermenting waste in a batch mode, two phases of the process are clearly distinguished—acidic or hydrogen and methanogenic itself. These phases can be clearly traced at the start of the anaerobic reactor. First, simple compounds decompose with the formation of organic acids and hydrogen, and then hydrogen begins to be consumed with the formation of methane, at about the same time the hydrolysis of polymers increases. A decrease in hydrogen concentration allows syntrophic bacteria to develop, and an increase in the number of acetate-using methane bacteria leads to a decrease in acid concentration. The system gradually becomes balanced. It sometimes takes tens of days to reach steady-state concentrations of various groups of microorganisms. This time can be shortened by using sludge from a well-functioning digester, selected associations of microorganisms or specially grown methane, and syntrophic bacteria as starter cultures [6–8].

In practice, methane fermentation of cattle or pig manure often develops spontaneously in places where manure is accumulated or stored, for example, in manure storage facilities. The launch of digesters for processing manure, especially cattle, can also be carried out without the use of any seed. However, the period when the unit reaches the operating mode can be quite long; it can take up to several months, often during this period "acidification" is observed—the accumulation of propionic and other volatile fatty acids, which leads to an irreversible stop of the process. All this is associated with the formation of a methanogenic microbial community, which differs from the microflora of the gastrointestinal tract and is adapted to development on a given raw material and in a given mode [6].

It is believed that cattle manure contains a sufficient amount of the complete set of microorganisms necessary for its fermentation, since it is known that methanogenesis occurs in the rumen and intestines of ruminants. However, there the anaerobic decomposition of volatile solids ends at the stage of formation of fatty acids, which are absorbed in the gastrointestinal tract, and the formation of methane serves as a protective reaction for binding hydrogen and carbon dioxide. It is carried out by hydrogen-using methane bacteria. The process of digestion of food in the gastrointestinal tract is at $39°C$. In order for the fermentation of manure to proceed at a high rate, it is necessary to accumulate the appropriate microflora, primarily acetate-using methane and syntrophic bacteria using volatile fatty acids. Fermentation of manure requires a microbial system that is significantly different from the cicatricial one, especially for the thermophilic process. Such associations can be obtained by the autoselection method both in laboratory and in large installations, which, as a rule, requires a longer period.

In practice, two temperature regimes of anaerobic conversion are usually used: mesophilic and thermophilic, at $35–40°C$ and $50–55°C$, respectively. In these temperature regimes, there are optima for the development of the main groups of mesophilic and thermophilic anaerobic microorganisms involved in the decomposition of volatile solids of organic waste [6].

7.3 Methods of preprocessing

The limiting stage of the anaerobic bioconversion of organic waste is the hydrolysis of the solid phase, in particular activated sludge, which consists mainly of cells of microorganisms and a small amount of dissolved volatile solid. The sediment of the primary settling tanks contains more dissolved volatile solids and fewer active microorganisms than activated sludge. The efficiency of biogas production during the conversion of organic waste is directly dependent on their biodegradability and, accordingly, the rate of hydrolysis. One of the technological methods for increasing the rate of hydrolysis is their preliminary treatment before anaerobic conversion in biogas plants. Pretreatment of the sludge allows to

- destroy microbial cells;
- partially dissolve the solid particles of the sediment;
- partially decompose organic polymers into simpler organic compounds [8].

A number of methods for pretreatment of precipitates are known, among which there are acid, alkaline, and thermal alkaline hydrolysis, mechanical and ultrasonic pretreatment, and thermohydrolysis, or combinations thereof. In general, methods of pretreatment of sediments are divided into

- thermal, in which the sediment (excess sludge) is exposed to high temperatures ($100-180°C$);
- chemical, in which the destruction of cells of microorganisms occurs under the influence of acids or alkalis, as well as oxidants, such as ozone or peroxides;
- thermochemical, combining two destructive factors: exposure to high temperatures and alkalis;
- mechanical, of which the most widespread is the treatment of sludge with ultrasound, as well as processing in ball mills or high-pressure pumps, lysis centrifuges;
- biological, in which for the lysis of the cell walls of the sludge (including the cell walls of pathogenic microorganisms present in the sludge), preparations of enzymes or hydrolytic microorganisms are used [9−12].

Many of the listed methods are still at the testing stage, but the most promising ones have already been commercialized. Thermohydrolysis of sludge is one of the most common methods that ensure the destruction of stable OM of sludge before it is processed in digesters. At treatment plants in Europe and America, thermohydrolysis has been implemented in the Cambi and BIOTHELYS (Veolia Water Solution and Technologies) processes since the mid-1990s. The Cambi Process (Oslo, Norway), based on the thermohydrolysis of activated sludge at a temperature of $130-180°C$ and a pressure of over 6 bar [13], is in great demand. Application of this method allows to increase the decomposition depth of BW and biogas yield by 30%. However, to use this method, it is necessary to build an expensive capital thermohydrolysis unit, consisting of three reactors, a heat exchanger, and a steam supply system, with the help of which the activated sludge is heated, cooled, and recirculated. In recent years, Krüger (Veolia) has introduced a new modification of the thermohydrolysis technology—EXELYS at a full-scale wastewater treatment plant in Hillerod (Denmark). This technology, in contrast to Cambi and BIOTHELYS, provides a continuous thermal process. Two modifications of the process are possible: "fermentation−hydrolysis−fermentation" and "hydrolysis−fermentation." The question of the optimal temperature regime, as well as the expediency of treating the entire sediment or only activated sludge, remains open [14]. In fact, the thermohydrolysis method is a simplified version of the outdated method of thermal conditioning of the sludge, one of the drawbacks of which was the formation of a large amount of biodegradable colored compounds (refractory compounds)—the products of interaction between proteins and carbohydrates released during thermohydrolysis [15].

Despite the high degree of destruction of activated sludge cells and high solubilizing ability, thermochemical treatment is inferior in prevalence to thermohydrolysis due to high costs of reagents. Mechanical pretreatment focuses on the size reduction of solids and can be implemented in ball mills or high-pressure homogenizers. The use of this technology entails significant energy costs. World practice shows that ultrasonic pretreatment of WWS is one of the new promising methods for increasing the biodegradability of sediments and the yield of biogas in the process of anaerobic digestion [16]. According to the literature, the use of this pretreatment method for the fermentation of activated sludge in laboratory and pilot reactors makes it possible to increase the decomposition depth of VS by 10%−56%, depending on the processing conditions [10−12]. The absence of the need to use reagents and the possibility of simple integration of ultrasonic generators into existing technological schemes make this method attractive for future use. However, this is an energy-intensive technology, and the feasibility of its application requires a careful calculation of the economic efficiency for each specific treatment plant. At JSC "Mosvodokanal," a comparative analysis of methods for pretreatment of sludge (activated sludge, primary sludge, and their mixture) of WWTP was carried out. There was a comparison of the effect of acidic, alkaline, and thermo-alkaline hydrolysis, thermohydrolysis, mechanical grinding with a homogenizer and ultrasonic treatment on the increase in the concentration of soluble COD in the sediment after treatment, which was a criterion for the efficiency of destruction of microbial cells and transformation of VS into a soluble state [9]. The transformation of VS from a bound to a soluble state increases the bioavailability of the sediment for methanogenesis, which is accompanied by an increase in the biogas yield. In the course of the conducted studies, it was shown that activated sludge is exposed to the greatest effect, which is due to the presence of bound VS in its composition. To check the effect of treatment on the depth of sediment fermentation, laboratory anaerobic reactors operating in a thermophilic (53°C) mode were used. As a criterion for evaluating the efficiency, the ratio of the received energy (in terms of biogas yield) to the energy expended with various methods of preprocessing (costs for the actual processing, heating of digesters) was chosen [9]. The most effective were the methods of thermohydrolysis and ultrasonic treatment [16].

Preliminary aerobic treatment: The functioning of these systems is based on fundamentally different biotechnological processes. The main advantage of anaerobic systems is the production of a high-calorie gaseous energy carrier, which allows not only to minimize operating costs but also in some cases to obtain commercial energy. The main advantage of aerobic-type systems is the relatively high intensity of microbiological processes, which makes it possible to reduce the volume of structures by a factor of 2—3 in comparison with the anaerobic process. The decision to use a type of recycling system is made on a mutually exclusive basis using a technical and economic analysis of options and taking into account the individual characteristics of waste source objects.

It is known that with the degree of decomposition of VS of the initial substrate by a consortium of aerobic thermophilic microorganisms not more than 10%—15% within 0.5—1.5 days, the following are achieved:

- preliminary heating of the substrate to thermophilic temperatures, which makes it possible to avoid the most energy-intensive and material-intensive operation of the anaerobic stage—heating the initial substrate to operating temperatures, and thereby increase the yield of commercial biogas;
- intensification of the coupled anaerobic conversion due to dissolution (hydrolysis) of volatile solids of the initial substrate;
- increasing the reliability of the anaerobic conversion process by increasing the pH of the treated substrate.

Aerobic thermophilic treatment of organic polysubstrates before biological gasification in digesters was developed to a certain extent abroad (USA, Germany, Switzerland, and South Africa) in relation to sludge and sludge of domestic waste water. It is also known that research and development of the corresponding direction in relation to manure were carried out in Germany.

At present, interest in obtaining energy from renewable energy sources, such as organic waste of a wide range of industries and consumption (agricultural waste from plant growing and animal husbandry, waste from the food and alcohol industry, organic fraction of MSW), otherwise called biomass, is still great.

Biomass is composed of water and nonvolatile and volatile solids. In this case, biomass can be converted into energy carriers of various phase states (solid [granules, briquettes], liquid [bioethanol], or gaseous [hydrogen, methane]) [17—20].

By anaerobic bioconversion of volatile solids contained in biomass, biogas can be obtained in the form of hydrogen and/or methane. Calculations show that the degree of bioconversion of volatile solids into hydrogen does not

138 Advances of Artificial Intelligence in a Green Energy Environment

exceed 20%−30%, while about 80% of volatile solids are converted into methane [21,22]. Almost any organic substrate, with the exception of waxes and untreated lignin, can undergo anaerobic bioconversion to hydrogen and/or methane [23−25].

One of the ways to enhance the efficiency of the anaerobic conversion of volatile solid of organic waste to biogas is to increase the solid retention time (SRT) in the reactor [26].

In industrial tests, it was found that the thickened digestate recirculation allows to achieve the decomposition degree of volatile solids on average 56.6%, while the value in the control variant was 42.2%. The increase in the degree of decomposition led to a significant improvement in the properties of the digestate, affecting the efficiency of separation into fractions. At the same time, the biogas yield increased by 3.0%. During the experiment, no negative impact of the thickened digestate recirculation on the technological operation of the digester was recorded [27].

In the technology of anaerobic conversion of organic waste, an important stage is the preparation of waste for bioconversion. The main tasks of preparing waste for bioconversion are

1. separation of large nonvolatile solid inclusions;
2. homogenization of the initial substrate;
3. heating the initial substrate to the fermentation temperature.

For this, various equipment is used, which has significant energy consumption; therefore, the development of new technical solutions for the preliminary processing of organic waste before anaerobic bioconversion in biogas plants is an urgent scientific and technical task [9,28].

One of the most promising methods for the pretreatment of organic waste before anaerobic bioconversion is its processing in an electromagnetic mill (vortex layer apparatus [VLA]), in which the waste is subjected to a number of actions (rotating electromagnetic field [EMF], cavitation, impact action of working organs, etc.) [28].

7.4 Vortex layer apparatus

The main direction for the development and implementation of environmental protection measures is to improve the state of water resources and waste disposal.

The creation of effective technologies for wastewater treatment in order to reduce the negative impact on water bodies is an urgent task.

The problem of intensifying chemical transformations in various media is extremely relevant for many technologies. One of the methods for intensifying technological processes is the processing of materials in a VLA, in which the main effect is created by a pulsating magnetic field [29].

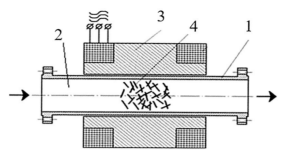

FIGURE 7.1 Schematic diagram of the vortex layer apparatus (VLA). (1) Control unit, (2) cooling system, (3) housing, and (4) insert pipe.

The main elements of the VLA are a rotating magnetic field inductor, a working chamber, and ferromagnetic particles.

Fig. 7.1 shows a schematic diagram of the VLA.

Case (1), which is a hollow cylinder made of nonmagnetic material, is placed inside the inductor (3), which creates a rotating magnetic field. In this case, the inductor must be located in a metal jacket that protects against the effects of a magnetic field, and inside which a cooling liquid circulates (for example, dry transformer oil). Ferromagnetic particles (4) are located inside the working chamber (2) [29].

The principle of operation of the VLA is quite simple. The set of working bodies rotating and colliding in a magnetic field forms a so-called vortex layer, in which it is possible to carry out the following processes with a sufficiently high degree of efficiency:

(1) mixing of liquids and gases;
(2) mixing of solid bulk materials;
(3) dry grinding of solids;
(4) grinding of solids in liquid dispersion media;
(5) activation of the surface of solid particles;
(6) implementation of chemical reactions;
(7) change in the physical and chemical properties of substances.

It should be noted that in most cases, many of the listed processes are carried out simultaneously.

Fig. 7.2 shows a laboratory version of the VLA, and Fig. 7.3 shows ferromagnetic particles used in the processing of various media and materials.

In recent years, VLA has been widely used in various industries for the intensification of many physical and chemical processes, including wastewater treatment.

Main ideas of talented scientist D.D. Logvinenko were presented by him in co-authorship with O.P. Shelyakov in the book [29] published in 1976 by

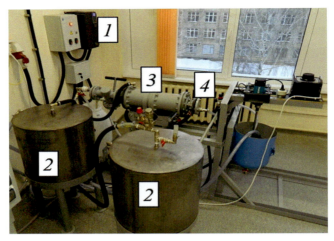

FIGURE 7.2 External view of the vortex layer apparatus. 1) control panel; 2) coolant tanks; 3) coil of the vortex layer apparatus; 4) the outer part of the working chamber of the vortex layer apparatus.

FIGURE 7.3 Ferromagnetic particles.

academician N.P. Vershinin. As a result of many years of scientific research, Academician Vershinin managed not only to confirm the conclusions made in the 70s of the last century but also to develop new, more productive types of devices, significantly expanding the areas of their application. He deeply analyzed the processes occurring in the apparatus of the vortex layer to clarify the nature of the effect of EMFs and their derivatives on matter. This allowed him to make a bold assumption that the enormous productivity and the observed anomalies in the course of processes, in all likelihood, are a consequence of the release of the internal energy of matter, and the devices themselves are its generators. With this, he explains the wide possibilities of installations for activating processes in all areas of technology, agriculture, everyday life, and ecology. This very important fact, perhaps, will serve as a starting point for a new leap in the development of the theory and practice of using this energy. After all, it is believed that an increase in the productivity of technological lines or apparatuses by 10%−15% is a very good result and by 2−3 times is a qualitative breakthrough in the development of production technology. However, the devices created by Logvinenko are capable of increasing productivity tens and thousands of times in comparison with technologies of the same purpose. It is not surprising that the scientific world accepted such information with distrust. Therefore, the creator of highly efficient installations for activating processes, D.D. Logvinenko, had to make a lot of efforts to popularize his ideas [29].

At the present stage of development of the manufacturing industry, the interest of engineers in this area is fueled by the fact that when organizing modern production, one of the mandatory requirements is its environmental safety. At the same time, the share of energy spent on neutralizing the waste of an enterprise using traditional technologies can be commensurate with the energy spent on manufacturing products. Considerable prospects for using devices with a rotating EMF are opening up in the field of preparing raw materials with minimal energy and other costs. Thus, the task of developing highly efficient methods of material processing is extremely urgent. Taking into account the more complicated ecological and energy situation in the world every year, it is worth paying attention to highly efficient methods of material processing using the processes taking place in the VLA. However, despite the fact that devices with rotating EMF have been used in industry for several decades, the explanation of the course of processes and energy sources capable of providing an unusually high performance of devices with negligible external energy consumption and very small material consumption and dimensions is still presented in the form of scientific hypotheses, the conclusions of which often go beyond the framework of classical physics. This state of affairs, obviously, indicates the presence of a wide scope for scientific research and a significant potential for the further development of this technology for engineers in the field of its improvement.

A typical installation for activating processes to create a vortex layer uses an electric power supply with a three-phase current of an industrial network, which makes it possible to easily generate a rotating EMF, with an industrial frequency, as is done in electrical machines. However, static current converters of high power that have already become quite available allow to significantly expand the range of used field parameters, for fine-tuning the operating mode of devices for specific technological requirements. Structurally, the device is similar to an asynchronous electric motor with a removed rotor, in the place of which the working area is located. The main unit that creates the rotating EMF of the process is an inductor, which includes an inductor core and a three-phase winding. The rotating magnetic field of the inductor is closed in the area of the working area of the apparatus, limited by the nonmagnetic body. The processed material is pumped into the working area of the apparatus, together with it needles made of ferromagnet are placed there, which interact with the rotating magnetic field of the inductor.

Devices with rotating EMF actually exist (only D.D.Logvinenko produced about 2000 of them); the technological lines, the basis of which they are, are operating successfully, all forecasts of the technical feasibility of their use have been confirmed. It would seem that in connection with the predicted high economic effect of using installations for activating processes in various sectors of the national economy, it is necessary to actively develop these technologies. But so far, unfortunately, this is not the case [30].

All processes of the devices under consideration take place in the same working space, in the same mode. Substances in any state are suitable for processing in installations for activating processes, as long as their dimensions (for solid phases) correspond to the dimensions of the working space and needles, the liquids are sufficiently mobile, and the powders are freely poured.

Until recently, the productivity of one apparatus was comparatively low, and in order to ensure, for example, the neutralization of effluents of a large enterprise or city, a large number of units of installations were required. However, at present, as shown earlier, apparatuses and designs of apparatuses for $100-1000 \text{ m}^3/\text{h}$ have already been created. Therefore, all performance problems are removed.

A wide area of application of process activation units is the treatment of aqueous solutions. Consider the issue of water purification. It was shown above that the devices are capable of carrying out the kinetic regime in the working zones when implementing various technological methods. At the same time, high travel speeds of a wide variety of technological processes are found with significantly lower capital and energy costs. Despite this, chemical processes remain almost unchanged, and at the same time, some physical phenomena undergo very noticeable changes [30].

Processes for obtaining powders in vortex bed apparatus can also be effectively implemented. The idea of obtaining iron and other metals by direct reduction of preliminarily prepared ores has always been tempting, and in a number of cases it received an industrial embodiment back in the 20th century. However, the technological difficulties, the complexity of the equipment, and the low reduction rate did not make it possible to obtain iron powders on a large scale. However, the direct reduction of niobium, tantalum, tungsten, and some other metals from their compounds is still the main method for obtaining metal powders.

The peculiarities of the effect of magnetic fields on the substance in the working zone of the process activation units give grounds to assume that the reduction of iron ore in this case will be technically expedient [30,31].

One of the most topical directions of using installations for activating processes can be the processing of waste from human activities— neutralization and utilization of industrial, household, and agricultural effluents, effluents, and wastes. The existing traditional technologies for the neutralization and utilization of industrial and domestic wastewater are technically imperfect, consume a lot of energy and materials, occupy vast areas, and are environmentally hazardous, because the neutralization products are always buried wherever they have to. It is easy to calculate that if you completely neutralize the effluents of any enterprise using modern methods, you will have to spend not much less energy than the production process itself, and the reliability in the completeness of neutralization is low [29,30].

To date, a gigantic arsenal of all kinds of technologies and equipment supporting them has been created in the world. Of course, they, albeit not fully enough, fulfill their functions. The challenge is to improve their efficiency without significant capital expenditures and energy consumption. And now it can be done [30,32].

The process activation units, such as the vortex layer devices, have a very important feature: they can be integrated into existing technological lines without much effort and expense, significantly increasing their productivity and improving product quality. This circumstance opens up wide opportunities for quick and inexpensive reconstruction of already existing treatment facilities of any type. But the main attention in this work is focused on the creation of new highly efficient treatment facilities using the technical features of the process activation unit [33].

Currently, a great variety of types of effluents, discharges, and wastes of different composition and in very large quantities are formed and it is very difficult to understand them and even more so to classify them. Therefore, a kind of concept of approach to solving these problems is proposed. It should take into account the actual state and give some recommendations [33].

There is practically no enterprise that could provide data on the successful purification of industrial waters from oil products. A typical traditional system for the neutralization of oily effluents, which does not provide for their utilization, is difficult and not always within the MAC standards. Such technologies have disadvantages: a multistage system of a technological line, high material and energy costs, large production areas, the consumption of expensive additives, the presence of filter systems, and a relatively low productivity.

The use of installations for activating processes, which make it possible to carry out instead of the diffusion kinetic type of matter transfer, makes it possible to almost completely eliminate the indicated disadvantages [30,33].

With regard to the kinetics of the processes occurring in the working area of devices with rotating EMF, physicochemical prerequisites for stimulating these processes, a number of theoretical assumptions regarding their nature are considered.

As a result of the interaction of the rotating EMF created by the inductor of the installation with the vortex layer of the material being processed and the ferromagnetic needles introduced there, a number of effects arise that, along with the mechanical and thermal effects of needles, directly affect the substance, changing its physicochemical properties. The considered effects arising in the working area of the VLA are distinguished by very high energetics, the source of which is still a mystery. To reveal the prerequisites for their course, it is necessary to consider them one by one [30].

The most important component of the process is the direct impact of the rotating needles on the processed material, as well as the magnetostriction of the needle body. The process of magnetostriction inevitably leads to the appearance of acoustic phenomena. Given the complex nature of electromagnetic interactions, the frequency range of sound waves is quite wide, and as can be concluded from the above data, it can be from tens of hertz to tens of megahertz; however, the potential inability of the system of chaotically emitting needles to provoke any resonance phenomena casts doubt on the significance of the purely acoustic component during mechanical processing of raw materials [30].

In the context of the energetic effect on the material, the most interesting is the acoustic radiation of the ultrasonic frequency range. More often than others, materials in a liquid and mixed phase are subjected to processing in vortex bed installations. The passage of high-intensity, high-frequency acoustic waves through an incompressible liquid medium serves as a source of cavitation. The resulting cavitation bubbles during collapse serve as secondary sources of ultrasound. Studies indicate that the share of acoustic vibrations accounts for no more than 2% of all energy costs; however, cavitation phenomena arising in this process have a huge impact on the course of many physicochemical processes [30,34].

According to Faraday's law, the effect of an alternating magnetic field in the working area of the apparatus, on metal needles, which are conductors, leads to the appearance of induction currents in them. Considering that the liquid component of the processed raw material is most often water with salts dissolved in it, it is difficult to underestimate the role of electrolysis processes in such systems. Indeed, the share of energy that falls on electrochemical processes is significant, and according to some estimates, it is close to 15% of the total costs. Clear signs of the presence of these processes are a decrease in the acidity of the treated solutions and the appearance of hydrogen in a molecular form. In addition, it was found that from hydrochloric acid, sulfuric acid, and nitric acid solutions containing metal ions, metal compounds precipitate in the form of hydroxides, and iron and nickel, which are the material of the needles, dissociate into solution. It is known that the amount of electric current consumed for the process and the amount of the reacted substance must strictly obey Faraday's laws, but some anomalies are observed here. So, for example, 6-valent chromium is reduced to 3-valent chromium without the

Organic waste anaerobic digestion in biogas plants **Chapter | 7 145**

addition of a reducing agent; during the process in the working area of the apparatus, it is possible to even completely drop metals in the form of hydroxides from the electrolytic solution, when the amount of reducing agents is less than 70% of the stoichiometrically required amount. In addition, the spatial configuration of the electrochemical cell remains questionable, since, as is known, in the classical view, electrolysis is carried out in the presence of at least two electrodes of different potential, placed in an electrolyte. In the case of needles in the working area of the process activation unit, it is assumed that the electrolytic cell can be the proximity of two needles with different charges. The second electrode can also be a solid fragment of the material being processed. The electrochemical process is most likely of an impulsive nature, since it is obvious that such a neighborhood is rather short. Thus, in the working area of the VLA, there is a huge number of short-lived electrolyzers; due to their high intensity and despite the short duration of action, they are capable of giving an impulse to the formation of products characteristic of electrolysis, as indicated earlier [30,35].

The physicochemical prerequisites for the intensification and stimulation of the processes occurring in devices with a rotating EMF are multifaceted and complexly interrelated. The energy interaction of the primary rotating magnetic field and the secondary magnetic field induced in the vortex layer with the processed materials has been little studied.

The paradox of the conclusions of experimental studies of the energetics of these processes indicates, first of all, the imperfection of the methods of their implementation and requires careful applied and theoretical scientific study.

7.5 Application of the vortex layer apparatus in biogas plants

One of the topical directions of using the VLA is the processing of household and agricultural waste, effluents, and waste [36,37]. The energy consumption in the VLA, during its operation, increases significantly and is spent on heating the ferromagnetic elements, as well as on the processes that occur in the vortex layer (cavitation, electrolysis) during the processing of components. A significant part (up to 35%) of the power of the vortex layer is spent on heat, in the process of mixing and grinding (up to 48%), up to 15% of the power of the vortex layer is spent on electrolysis, and only 2% of the energy is spent on obtaining a high-frequency magnetic field and acoustic vibrations in the medium. The flow rate of the processed medium affects the mode of operation of the vortex layer during mixing and dispersion of the components of the medium. Moreover, the vortex layer of ferromagnetic elements can exist up to a certain flow rate, and with its further increase it leads to the removal of ferromagnetic elements from the working chamber.

The main units of electromagnetic devices with a vortex layer are an EMF inductor with a cooling system and a working chamber with ferromagnetic elements.

The set of ferromagnetic working bodies, rotating and colliding in a magnetic field, forms a so-called vortex layer, in which it is possible to carry out the following processes with a sufficiently high degree of efficiency:

1. Fine grinding, which allows to improve the availability of nutrients for microorganisms with simultaneous partial heating of the substrate and partial hydrolysis of complex organic compounds.
2. Introduction of ferromagnetic particles of abraded working parts into the substrate, which increases both the volumetric methane production rate and methane yield [38].

The pretreatment in electromagnetic mill results in a significant enhancement in the performance of the subsequent anaerobic bioconversion of the organic wastes. Thus, the biogas yield increased by 25% and methane yield by 80%. The degree of anaerobic decomposition of the volatile solids after treatment in electromagnetic mill increased by an average of 16%. The rate of methane formation during anaerobic fermentation of the substrate after treatment in electromagnetic mill increased by 50%. Thus, pretreatment in electromagnetic mill (VLA) is an effective way to intensify the anaerobic bioconversion of organic waste. Based on the results obtained, it seems reasonable to treat the substrate in VLA for no more than 0.5 min, since with a longer treatment, the characteristics of the anaerobic fermentation process improved slightly [36,40].

1. Treatment of activated sludge from the SBR reactor, which treats the wastewater of a brewery, in VLA for 2 minutes allows the destruction of activated sludge flakes and protozoan cells, which, in the case of subsequent anaerobic treatment, will be a breeding ground for a consortium of anaerobic microorganisms.
2. Treatment of a mixture of activated sludge and raw sludge of treatment facilities in the VLA for 37.5–300 s allows the destruction of activated sludge flakes; however, their disinfection does not occur.
3. Treatment of substrates of anaerobic bioreactors in VLA is not capable of providing disinfection of substrates without their subsequent anaerobic treatment.
4. Pretreatment of liquid organic substrates in the VLA will enhance the effectiveness of anaerobic bioconversion due to the following factors:
   - introduction of ferromagnetic microparticles into the substrate;
   - destruction of protozoan cells and the formation of a nutrient medium from them, which does not require additional retention time in anaerobic bioreactors;
   - destruction of activated sludge flakes without their disinfection, which will lead to improved mass transfer in anaerobic bioreactors [37,39].

The positive effect of pretreatment of various organic wastes in the VLA on the characteristics of methanogenesis, in particular on the kinetics of methanogenesis, the completeness of decomposition of organic matter, the content of methane in biogas, and waste decontamination, is also shown [36−40].

Integration of pretreatment of organic substrates before anaerobic bioconversion into biogas plants makes it possible to increase both the energy efficiency of biogas plants and the degree of decomposition of volatile solid organic waste.

The result of the experiment is in comparison with anaerobic treatment in traditional digesters, the use of pretreatment of organic waste in a VLA before fermentation in anaerobic bioreactors allows increasing the specific yield of biogas by at least 1.18 times with a simultaneous increase in the methane content in biogas and, as a result, its calorific value, which corresponds to an increase in the specific amount of energy of the produced biogas by no less than 1.69 times. At the same time, the specific energy consumption increases by 30% with a decrease in the specific heat consumption by 8%, and the total energy consumption is no more than 60% of the energy of the obtained biogas. Thus, the specific amount of biogas energy obtained in the process of fermentation in anaerobic bioreactors with the use of substrate pretreatment in the VLA completely compensates for the energy consumption for the pretreatment of the substrate in the VLA. At the same time, the specific yield of commercial energy increases by 70%, compared with anaerobic treatment in traditional digesters [38].

7.5.1 Pretreatment of the substrate in the vortex layer apparatus together with biomass recycling

For a more complete and intensive decomposition of dry matter of organic waste, it is proposed to apply the treatment of the return flow in the VLA.

A method has been developed to intensify the anaerobic bioconversion of volatile solids to obtain methane through the combined use of recirculation of solid digestate (to increase SRT and preserve biomass) and pretreatment of the solid digestate together with the raw wastes in VLA.

Fig. 7.4 shows a block diagram of the proposed method to intensify the anaerobic bioconversion of volatile solids.

7.5.1.1 Complex application of microwave and VLA

Also, in the laboratory of bioenergetic and supercritical technologies of the Federal Scientific Agroengineering Center VIM, research was carried out on the process of preheating the substrate of digesters due to complex electrophysical treatment—in a VLA with direct supply of ultrahigh-frequency waves into the working chamber of the apparatus.

The obtained experimental data partially confirm the theoretical prerequisites for using a microwave heater together with VLA: preliminary complex processing of the initial substrate leads to a decrease in operating costs [41].

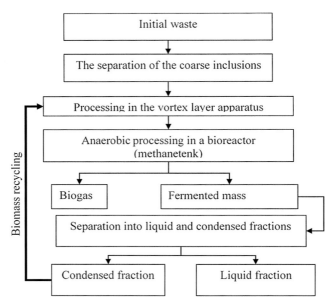

FIGURE 7.4 Block diagram of a method for increasing the efficiency of anaerobic bioconversion of organic waste to obtain a gaseous energy carrier and organic fertilizers based on recycling biomass using a vortex layer.

7.6 Conclusion

Thus, the use of the VLA in the systems of anaerobic processing of organic waste allows

- to reduce the cost of preparing waste for anaerobic digestion;
- to increase the availability of nutrients to anaerobic microorganisms, which increases the processing speed;
- to carry out partial hydrolysis of complex organic substances in waste, which reduces the HRT in anaerobic bioreactors;
- to introduce into the processed waste fine particles of ferromagnet, which are a catalyst for the anaerobic bioconversion;
- to increase the stability of the processes taking place in the anaerobic bioreactor.

Thus, the combination of positive effects of the use of the VLA in the systems of anaerobic processing of organic waste, coupled with low specific energy consumption for own needs and ease of operation, makes it possible to optimize the complex anaerobic bioconversion of volatile solids when processing organic waste in biogas plants.

References

[1] A.Y. Izmaylov, Y.P. Lobachevskiy, A.V. Fedotov, V.S. Grigoryev, Y.S. Tsench, Adsorption-oxidation technology of wastewater recycling in agroindustrial complex enterprises, Vestnik mordovskogo universiteta = Mordovia Univ. Bull. 28 (2) (2018) 207−221, https://doi.org/10.15507/0236-2910.028.201802.207-221.

[2] A.V. Artamonov, A.Y. Izmailov, Y.A. Kozhevni-kov, Y.Y. Kostyakova, Y.P. Lobachevsky, S.V. Pashkin, O.S. Marchenko, Effective puri-fication of concentrated organic wastewater from agro-industrial enterprises, prob-lems and methods of solution, AMA, Agric. Mech. Asia, Afr. Latin Am. 49 (2018) 49−53.

[3] L.I. Gyunter, L.L. Gol'dfarb, Digesters (Metantenki). M.: Stroiizdat, 1991, 128 p. (in Russ.).

[4] A.A. Kovalev, Increasing the Energy Efficiency of Biogas Plants (thesis Cand. tech. sciences. M.), 2014 (in Russ.).

[5] S.V. Kalyuzhny, D.A. Danilovich, A.N. Nozhevnikova, Results of Science and Technology, ser. Biotechnology, M., VINITI, vol. 29, 1991 (in Russ.).

[6] A. Kovalev, Andreevich Technologies and Feasibility Study of Biogas Production in Manure Utilization Systems of Livestock Farms (theses of Doctoral Dis.), 1998 (in Russ.).

[7] D.A. Kovalev, Improving the Technology of Purification of Manure Runoff from Pig Farms (theses of Cand. Dis., Moscow), 2004 (in Russ.).

[8] A.N. Nozhevnikova, A.Y. Kallistova, Y.V. Litti, M.V. Kevbrina, Biotechnology and Microbiology of Anaerobic Processing of Organic Municipal Waste, University Book, 2016, 320 p. (in Russ.).

[9] M.V. Kevbrina, N.G. Gazizova, V.G. Korobtsova, Comparison of different methods of pretreatment of sewage sludge for intensification of the process of methane fermentation, Vodoochistka 1 (2013) 22−28 (in Russ.).

[10] L. Appels, J. Baeyens, J. Degreve, et al., Principles and potential of the anaerobic digestion of waste-activated sludge, Prog. Energy Combust. Sci. 34 (2008) 755−781.

[11] J. Lu, Optimization of Anaerobic Digestion of Sewage Sludge Using Thermophilic Anaerobic Pre-treatment, Technical University of Denmark, Denmark, 2006, 59 p.

[12] D.Q. Zhang, S.K. Tan, R.M. Gersberg, Municipal solid waste management in China: Status, problems and challenges, J. Environ. Manag. 91 (2010) 1623−1633.

[13] P. Walley, Optimizing thermal hydrolysis for reliable high digester solids loading and performance, in: 12th European Biosolids & Organic Resources Conference, November 12−14, 2007 (Manchester, UK).

[14] S.V. Khramenkov, A.N. Pakhomov, S.A. Streltsov, et al., Increasing the efficiency of wastewater sludge treatment using high-temperature hydrolysis before fermentation, Water Supply Sanit. Eng. 10 (2012) 55−60 (in Russ.).

[15] R.T. Haug, D.C. Stuckey, J.M. Gossett, et al., Effect of thermal pretreatment on digestibility and dewaterability of organic sludges, J. Water Pollut. Control Fed. 50 (1) (1978) 73−85.

[16] A. Tiehm, K. Nickel, M. Zellhorn, et al., Ultrasonic waste activated sludge disintegration for improving anaerobic stabilization, Water Res. 35 (2001) 2003−2009.

[17] G. Fuasak, P. Plion, V. Fiche, F. Tabe, Briquette on the Basis of a Compressed Lignocel-lulose Body Impregnated with Liquid Fuel: Pat. 2507241 Rus. Federation 2012108893/04; declared 03/07/2012; publ. 02/20/2014, Bul. No. 5, 2014, p. 11 (in Russ.).

[18] V.V. Starshikh, E.A. Maksimov, Method of Briquetting Waste Products of Animals and Poultry and a Device for its Implementation: US Pat. 2507242 Rus. Federation 2012146319/04; declared 10/30/2012; publ. 02/20/2014 Bul. No. 5, 2014, p. 6 (in Russ.).

[19] P. Sommer, T. Georgieva, B.K. Ahring, Biochem. Soc. Trans. 32 (2) (2004) 283.

[20] O.V. Senko, M.A. Gladchenko, I.V. Lyagin, et al., Altern. Energy Ecol. 3 (107) (2012) 89 (in Russ.).

[21] S.D. Varfolomeev, S.V. Kalyuzhny, D.Y. Medman, Adv. Chem. 57 (1988) 1201 (in Russ.).

150 Advances of Artificial Intelligence in a Green Energy Environment

[22] S.D. Varfolomeev, E.N. Efremenko, L.P. Krylova, Adv. Chem. 79 (2010) 544 (in Russ).

[23] M.G. Khamidov, S.A. Streltsov, D.A. Danilovich, Communal Complex of Russia, vol. 2, 2009, p. 56 (in Russ).

[24] E.A. Tsavkelova, A.I. Netrusov, Appl. Biochem. Microbiol. 48 (5) (2012) 469 (in Russ).

[25] S.V. Khitrin, D.S. Meteleva, O.A. Shmakova, et al., Vseros. Scientific. Conf. Theoretical and Experimental Chemistry through the Eyes of Young People, Irkutsk: ISU Publishing House, 2013, p. 151 (in Russ).

[26] M.V. Kevbrina, Y.A. Nikolaev, A.G. Dorofeev, A.Y. Vanyushina, A.M. Agarev, Highly efficient technology for methane digestion of sewage sludge with biomass recycling, Water Supply Sanit. Equip. 10 (2012) 61 (in Russ).

[27] M.N. Kozlov, M.V. Kevbrina, G.A. Kolbasov, A.M. Agarev, P.S. Shashkina, The main results of industrial tests of the technology of sludge digestion with biomass recycling, in: Materials of the International Conference "Treatment and Disposal of Sewage Sludge in Municipal Services and Industry", May 27, 2015 [Electronic resource]. - M.: CJSC "Company EKWATEK", 2015 (in Russ).

[28] D. Kovalev, A. Kovalev, Y. Litti, A. Nozhevnikova, I. Katraeva, Influence of organic matter load on the bioconversion process of pretreated substrates of anaerobic bioreactors, Ecol. Ind. Russ. 23 (12) (2019) (in Russ).

[29] D. Logvinenko, O. Shelyakov, Intensification of Technological Processes in the Vortex Layer Apparatus, Tekhnika, Kiev, 1976, 113 p. (in Russ).

[30] I.N. Vershinin, N.P. Vershinin, Devices with a rotating electromagnetic field, in: Salsk-Moscow: Advanced Technologies of the XXI Century, 2007, 368 p. (in Russ).

[31] Fundamentals of Metallurgy. Heavy Metals, Metallurgy, Moscow, 1962 (in Russ).

[32] Abstracts of reports. Russia. Moscow, in: World Conference on Climate Change, November 29—December 10, 2003 (in Russ).

[33] N.P. Vershinin, Process Activation Settings. Rostov-on-Don: Innovator, 2004, 96 p. (in Russ).

[34] Ultrasound: Small Encyclopedia, Soviet Encyclopedia, Moscow, 1979, 400 p. (in Russ).

[35] Brief Chemical Encyclopedia. T.5. M., Soviet Encyclopedia, 1967, 939 p. (in Russ).

[36] Y. Litti, D. Kovalev, A. Kovalev, I. Katraeva, J. Russkova, A. Nozhevnikova, Increasing the efficiency of organic waste conversion into biogas by mechanical pretreatment in an electromagnetic mill, J. Phys. Conf. Ser. 1111 (1) (2018) 1—8.

[37] D.A. Kovalev, A.A. Kovalev, I.V. Katraeva, Y.V. Litti, A.N. Nozhevnikova, Effect of disinfection of anaerobic bioreactor substrates in the vortex layer apparatus, Chemsafety 3 (1) (2019) 56—64, https://doi.org/10.25514/CHS.2019.1.15004.

[38] A.A. Kovalev, D.A. Kovalev, V.S. Grigoriev, Energy efficiency of pretreatment of the digester's synthetic substrate in the vortex layer apparatus, Eng. Technol. Syst. 30 (1) (2020) 92—110, https://doi.org/10.15507/2658-4123.030.202001 (in Russ).

[39] V. Litti Yu, D.A. Kovalev, A.A. Kovalev, I.V. Katraeva, E.R. Mikheeva, A.N. Nozhevnikova, Using the vortex layer apparatus to increase the efficiency of methane digestion of sewage sludge, Water Supply Sanit. Tech. 11 (2019) 32—40.

[40] A.N. Nozhevnikova, Y.I. Russkova, Y.V. Litti, S.N. Parshina, E.A. Zhuravleva, A.A. Nikitina, Syntrophy and interspecies electron transfer in methanogenic microbial communities, Microbiology 89 (2020) 129—147.

[41] A.A. Kovalev, D.A. Kovalev, Y.A. Sobchenko, A.G. Makarov, Approbation of the method for preheating the substrate of digesters due to complex electrophysical processing, Electrotechnol. Electr. Equip. Agro-Ind. Comp. 4 (37) (2019) 3—7 (in Russ).

Chapter 8

Search of regularities in data: optimality, validity, and interpretability

O.V. Senko[1], A.V. Kuznetsova[2], I.A. Matveev[1] and I.S. Litvinchev[1]
[1]*Federal Research Center "Computer Science and Control" of Russian Academy of Science, Moscow, Russia;* [2]*Institute of Biochemical Physics, Moscow, Russia*

8.1 Introduction

Main goal of recognition and regression techniques is best prediction of the *outcome Y* from informative features or *explanatory variables*. Many modern machine learning methods reduce full set of informative features to a specially selected subset, while the rest of the set is ignored. Assessment of effects of explanatory variables on outcome also plays a great role. However, accessibility and interpretability of solutions generated by a computer algorithm is poor in many machine learning techniques, such as neural networks, support vector machine, or gradient boosting. One of the ways to overcome this problem is approximation of complicated machine learning algorithm by a set of simple and comprehensible models. This approach can be effective if two conditions are satisfied:

- such simple models describe real-life effects and are not completely or partially false;
- number of models in set is not too large.

In fact, both conditions are closely interrelated and may be satisfied if the effect associated with them is statistically significant. Such significant effects will be further referred to as *empirical regularities*.

In many applications it is interesting to enumerate as much empirical regularities as possible, not only those related to informative features selected by machine learning technique. The traditional way to find regularities in data is using a variety of statistical tests, for example, the standard Student's t-test or the nonparametric Mann–Whitney and Kolmogorov–Smirnov tests to compare distributions in independent groups; Student's and Fisher's tests for assessing the significance of linear correlation, ANOVA method, etc. The

152 Advances of Artificial Intelligence in a Green Energy Environment

search for empirical patterns, along with statistical verification, should also include the search for optimal mathematical functions connecting Y with explanatory variables. The use of linear functions alone for this purpose does not allow revealing valuable nonlinear laws in biology, economics, chemistry, and other fields. Therefore, the search must also be carried out within alternative functional families. For example, one can use families of polynomial, piecewise linear functions or families of nonparametric functions defined by partitions of the feature space or families of combined functions that include both linear parametric and nonparametric elements. Standard statistical tests are usually not suitable for assessing the significance of patterns described by functions that were previously found from the same data set. For example, applying standard techniques (F-test, Chi-square, and others) leads to significantly overestimated differences between groups with values of the studied indicator to the left and right of a boundary if this boundary and statistics evaluating differences between groups were calculated from the same data set. Such overestimation arises from multiple testing effect [1].

One of the ways to find an adequate estimate is randomized splitting of initial data on two subsets. The first one is used for calculating boundaries and the second one is employed for evaluation of statistical validity. But such approach leads to loss of both exactness of approximation and statistical significance of found regularities. Another way to assess statistical significance is using permutation tests [2−4]. Permutation test may be implemented regardless of knowing underlying distributions of test statistics. Permutation test is not based on asymptotic probability estimates. So it preserves accuracy for small data sets. There are many examples of successful use of permutation technique in different types of tasks. Method based on testing significance of corresponding partial correlation coefficients may be mentioned in this regard [5]. A permutation test is used to assess the level of significance associated with the optimal split set of longitudinal data into two mutually exclusive groups [6]. Permutation test was used to make a choice between piecewise regression and a simple linear regression [7]. Also permutation tests are widely used to control multiplicity in various applied tasks including high-dimensional tasks related to DNA microarray experiments. Effective methods calculating nonparametric combination of several dependent permutation tests were discussed in book [2].

An essential role in the search of empirical regularities in data is played by the problem of partially false regularities or the problem of redundancy. A regularity that includes a certain set of elements will be considered partially false if its statistical significance is determined by some improper subset of elements. Moreover, elements that are not included in this subset are actually redundant. An example is a regularity described by a linear regression model, in which several regressors are random variables that have nothing to do with the target variable. One way to avoid redundancy is to use model complexity optimization techniques.

Search of regularities in data Chapter | 8 **153**

Today there are several approaches for complexity optimization that allow to suppress overfitting effect. Akaike information criterion [8], Bayesian information criterion [9], Hannan−Quinn information criterion [10], and Rissanen principle [11] may be mentioned. These techniques often allow to find out complexity level with best generalization ability. But in many application tasks it is important not only to find model of optimal complexity but also to estimate associated statistical significance.

In many applications it is important not only to find the model of optimal complexity but to estimate associated significance as well. Here we present an approach in which the statistical significance of empirical regularities is estimated to overcome the redundancy. The significance of an element is understood as the statistical significance of improving the approximation of data after the element is included in the regularity. A regularity will be referred to as *completely valid* if all its elements are statistically significant. Two ways to evaluate statistical significance of a regularity elements were considered previously. The first one is based on Occam's razor principle that is attributed to William of Occam living in the 14th century. The most popular version of razor is formulated as, "Entities should not be multiplied beyond necessity." The principle was adopted by many scientists and another variant were invented. Principle was stated by Isaac Newton in form, "We are to admit no more causes of natural things than such as are both true and sufficient to explain their appearances." Occam's razor principle is discussed in modern scientific literature on machine learning or knowledge discovery. Usually it is considered that razor is a way to improve forecasting ability. Arguments for and against such razors are represented in detail in Ref. [12]. The form of Occam's razor principle here is close to Newton's one.

8.2 Occam's razor principle for verification of parametric regression models

8.2.1 Choice between complex and simple models

Variant of Occam's razor principle aimed to evaluate statistical significance of regression models was previously considered in Ref. [13]. Suppose that the pattern lies in the existence of a relationship between the target variable Y and the variables $X_1, \ldots, X_n$ described by the function $\mathcal{F}(X_1, \ldots, X_n)$. In other words, the model

$$Y = \mathcal{F}(X_1, \ldots, X_n) + \varepsilon$$

exists, where ε is error term. The goal is to find the possibly best approximation of function $\mathcal{F}(X_1, \ldots, X_n)$ Approximating function $\mathcal{F}(X_1, \ldots, X_r)$ with minimal mean ε^2 at data set is searched inside some family $\widetilde{M}$.

Optimal function $\mathcal{F}(X_1, \ldots, X_r)$ may be searched in simple family $\widetilde{M}_s$ and in more complicated family $\widetilde{M}_c$ by data set $\widetilde{S}_0 = \{(y_1, \mathbf{x}_1), \ldots, (y_m, \mathbf{x}_m)\}$,

154 Advances of Artificial Intelligence in a Green Energy Environment

where y_j is value of target variable Y and $\mathbf{x}_j$ is vector variables $X_1, \ldots, X_n$. Family is considered simple if it consists of models characterized by minimal number of parameters. Family $\widetilde{M}_c$ must include models with some additional parameters. For example, family $\widetilde{M}_s$ may consist of simple linear regressions of target Y on single regressors. Family $\widetilde{M}_c$ may consist of piecewise linear regressions with single break points. The approach is aimed to asses which of two families $\widetilde{M}_s$ or $\widetilde{M}_c$ is more appropriate to calculate approximation $F(X_1, \ldots, X_r)$ of true function $\mathcal{F}(X_1, \ldots, X_n)$ by $\widetilde{S}_0$.

Definition 1: It is considered that some function F exhaustively describes dependence of Y from $X_1, \ldots, X_r$ if residuals $\{r_1 = y_1 - F(\mathbf{x}_1), \ldots, r_m = y_m - F(\mathbf{x}_m)\}$ are realizations of mutually independent identically distributed random values $\xi_1, \ldots, \xi_m$ that are independent on vector descriptions $\mathbf{x}$. It is supposed also that $(\xi_i) = 0$, $i = 1, \ldots, m$.

Supposition 1: More complicated function from $\widetilde{M}_c$ should be used only when there is no function in family $\widetilde{M}_s$ that exhaustively describes data.

Supposition 2: It is possible to reject null hypothesis that function F exhaustively describes dependence on X variables with the help of complicated family $\widetilde{M}_c$. In other words, $Y = F(\mathbf{x}) + \varepsilon$, where ε is noise that does not depend on X variables.

Usually in statistics, validity of choice between two hypotheses is evaluated with the help of P-values. The same way of validity evaluating is used in this chapter. It is considered that complicated family must be used then and only then when any simple model fails to describe exhaustively regularity that exists in data. At that null hypothesis about exhaustive explanation of existing regularity by simple predictive function from $\widetilde{M}_s$ is rejected.

8.2.2 Permutation test technique

Let $\widetilde{f}$ be the set of all possible permutations of integers $\{1, \ldots, m\}$ Let $\widetilde{S}_p(f, F)$ be data set received from initial data $\widetilde{S}_0$ by random permutation of residuals $\{r_1, \ldots, r_m\}$.

$$\widetilde{S}_p(f, F) = \left\{ \left[r_{f(1)} + F(\mathbf{x}_1), \mathbf{x}_1 \right], \ldots, \left[r_{f(m)} + F(\mathbf{x}_m), \mathbf{x}_m \right] \right\} \tag{8.1}$$

Definition 2: Two permutations f' and f'' will be called equivalent if data sets $\widetilde{S}_p(f', F)$ and $\widetilde{S}_p(f'', F)$ are equal.

Let $\widetilde{f}_b = \{f_1^b, \ldots, f_{\mathfrak{N}}^b\}$ be such a set of permutations that

- any two permutation from $\widetilde{f}_b$ are not equivalent,
- any permutation is equivalent to one of the permutations from $\widetilde{f}_b$.

Let us note that due to transitiveness of equivalence any permutation may be equivalent to only one element from $\widetilde{f}_b$. Equivalence class $c(f)$ may be defined for each permutation from $\widetilde{f}_b$ that consists of all permutations that are equivalent to f. Equality

$$\widetilde{f} = \overset{\mathfrak{N}}{\underset{i=1}{\cup}} c\left(f_i^b\right)$$

is true by definition of $\widetilde{f}_b$.

Two statements are true.

Statement 1: In case supposition 2 is true for any $f_j \in \widetilde{f}_b$

$$P\{\widetilde{S}_p\left(f_j, F\right)|\mathbf{x}_1, \ldots, \mathbf{x}_m\} = \prod_{i=1}^{m} P(\xi_i = r_i)$$

Proof: Statement 1 may be easily obtained from independence of residuals r on vectors $\mathbf{x}$ and mutual independence of observations corresponding different objects from $\widetilde{S}_0$. It follows from supposition 2 that probabilities of data sets $\widetilde{S}_p(f_1, F), \ldots, \widetilde{S}_p(f_{\mathfrak{N}}, F)$ are equal to each other. Q.E.D.

Statement 2: All classes $c(f_1), \ldots, c(f_{\mathfrak{N}})$ are of the same size.

Proof: Let $\{\widetilde{r}_1, \ldots, \widetilde{r}_k\}$ be such partition of $\{r_1, \ldots, r_m\}$ that residuals inside each element of partition are equal to each other and residuals from different groups are different. Suppose that $\widetilde{J}_q = \{J_q(1), \ldots, J_q[\mu(q)]\}$ is a set of residual numbers inside group $\widetilde{r}_q$ according to some permutation $f_j \in \widetilde{f}_b$, where $\mu(q)$ is the size of group $\widetilde{r}_q$ and $q = 1, \ldots, k$. It is evident that for any permutation $\widetilde{f}_j'$ that is received from f_j by some permutations of numbers only inside sets $\widetilde{J}_1, \ldots, \widetilde{J}_k$, equality of data sets $\widetilde{S}_p(f_j, F)$ and $\widetilde{S}_p(\widetilde{f}_j', F)$ is preserved. For any permutation $\widetilde{f}_j''$ that is received from f_j by some permutation including exchanges between sets $\widetilde{J}_1, \ldots, \widetilde{J}_k$, data sets $\widetilde{S}_p(f_j, F)$ and $\widetilde{S}_p(\widetilde{f}_j', F)$ are not equal. So class $c(f_j)$ must include all permutations that are received from f_j by some permutations of numbers inside sets $\widetilde{J}_1, \ldots, \widetilde{J}_k$. Class $c(f_j)$ does not include any other permutations. Let us note that the amount of such permutations depends only on sizes of groups $\{\widetilde{r}_1, \ldots, \widetilde{r}_k\}$ and does not depend on specific permutation f_j. So size of class $c(f_j)$ does not depend on f_j. **Q.E.D.**

156 Advances of Artificial Intelligence in a Green Energy Environment

Set $\widetilde{S}_b = \{\widetilde{S}_p(f_1, F), ..., \widetilde{S}_p(f_\mathfrak{N}, F)\}$ includes all possible data sets satisfying conditions.

(a) empirical distribution of residuals r from forecasting function F in $\widetilde{S}$ coincides with empirical distribution of residuals r at initial data set $\widetilde{S}_0$ (condition $C_r[\widetilde{S}_0, F]$);

(b) x-descriptions in $\widetilde{S}$ completely coincide with x-descriptions of $\widetilde{S}_0$ (condition $C_x[\widetilde{S}_0, F]$).

Let $\mathcal{P}$ be some predicate that is defined at set of all possible data sets of size m. Let predicate $\mathcal{P}$ be true at some subset $\widetilde{S}_T(\mathcal{P})$ of set $\widetilde{S}_b$. Probabilities of all data sets from $\widetilde{S}_b$ are equal according to statement 2. So probability $P\{\mathcal{P}(\widetilde{S}) = TRUE | \widetilde{S} \in \widetilde{S}_b\}$ may be evaluated as ratio $\dfrac{|\widetilde{S}_T(\mathcal{P})|}{|\widetilde{S}_b|}$.

Supposition 4: Let $\mathcal{P}_{pv}(\widetilde{S}) = Q_{min}(\widetilde{S}, \widetilde{M}_c) < Q_{min}(\widetilde{S}_0, \widetilde{M}_c)$ It is suggested to use conditional probability

$$P\{\mathcal{P}_{pv}(\widetilde{S}) = TRUE | \widetilde{S} \in \widetilde{S}_b\} = \frac{|\widetilde{S}_T(\mathcal{P}_{pv})|}{|\widetilde{S}_b|}$$

as P-value that evaluates validity of null hypothesis about exhaustiveness.

Statement 3: Equality is true.

$$\frac{|\widetilde{S}_T(\mathcal{P}_{pv})|}{|\widetilde{S}_b|} = \frac{|\{f_j \in \widetilde{f}_g | Q_{min}[\widetilde{S}_p(f, F), \widetilde{M}_c] < |Q_{min}(\widetilde{S}_0, \widetilde{M}_c)\}|}{|\widetilde{f}|}$$

Proof: Evidently

$$\mathcal{P}[\widetilde{S}_p(f, F)] = \mathcal{P}[\widetilde{S}_p(f', F)]$$

if f is equivalent to f'. According to statement 2 all equivalence classes are of the same size. Let n_c be the number of permutations in each equivalence class. Then

$$\frac{|\widetilde{S}_T(\mathcal{P}_{pv})| n_c}{|\widetilde{S}_b| n_c} = \frac{|\{f_j \in \widetilde{f}_g | Q_{min}[\widetilde{S}_p(f, F), \widetilde{M}_c] < |Q_{min}(\widetilde{S}_0, \widetilde{M}_c)\}|}{|\widetilde{f}|}$$

Q.E.D.

Thus ratio

$$\frac{|\{f_j \in \widetilde{f}_g | Q_{min}[\widetilde{S}_p(f,F), \widetilde{M}_c] < |Q_{min}(\widetilde{S}_0, \widetilde{M}_c)\}|}{|\widetilde{f}|} \tag{8.2}$$

theoretically allows to calculate exact P-value testing validity of null hypothesis about exhaustive description of existing regularity by simple regularity from $\widetilde{M}_s$. But practically it is impossible to calculate exact P-values because of huge amount of possible permutation. However, it is easy to estimate (2) using relatively small number of random permutations that are generated by random numbers generator. Let

$$\widetilde{f}_g = \{f_j | j = 1, \dots, N_g\}$$

be the set of permutations calculated by random numbers generator. Then P-value may be estimated as ratio

$$\frac{|\{f_j \in \widetilde{f}_g | Q_{min}[\widetilde{S}_p(f,F), \widetilde{M}_c] < |Q_{min}(\widetilde{S}_0, \widetilde{M}_c)\}|}{N_g} \tag{8.3}$$

8.2.3 Choice of simple model

Technique described in previous subsection may be used only if simple model from $\widetilde{M}_s$ has been previously chosen. Supposition 1 declares that complicated model must not be used when there is simple model that exhaustively describes data. Such model may be searched by evaluating all predicting functions from $\widetilde{M}_s$ with the help of described in previous section PT version. But it is practically impossible to implement such approach because number of models inside $\widetilde{M}_s$ may be too great. In this chapter, only two simple predicting functions from $\widetilde{M}_s$ are evaluated. At first, simple predicting function is studied that is searched with the help of standard least squares (LS) technique. Let try to explain why such technique may be useful. It is naturally to expect that LS regression will be close to a model that exhaustively describes data in case such model really exists in $\widetilde{M}_s$.

Suppose that $F_s(\mathbf{x})$ is some predicting function from $\widetilde{M}_s$ and

$$F_c^0(\mathbf{x}) = arg \ min_{F(\mathbf{x}) \in \widetilde{M}_c} Q[\widetilde{S}_0, F(\mathbf{x})]$$

Discussed approach is based on evaluating upper boundary of $Q_{min}[\widetilde{S}_p(f, F_s), \widetilde{M}_c]$, where $f \in \widetilde{f}$.

Using definition of $\widetilde{S}_p(f, F_s)$ in formula (8.1):

$$Q_{min}[\widetilde{S}_p(f, F_s), \widetilde{M}_c] < Q[\widetilde{S}_p(f, F_c^0), \widetilde{M}_c] = \sum_{j=1}^{m} \left[r_{f(j)} + F_s(\mathbf{x}_j) - F_c^0(\mathbf{x}_j) \right]^2 =$$

$$= \sum_{j=1}^{m} \left[r_{f(j)} + \delta(j) \right]^2 = \sum_{j=1}^{m} r_{f(j)}^2 + 2 \sum_{j=1}^{m} \delta(j) r_{f(j)} + \sum_{j=1}^{m} \delta^2(j),$$

where $\delta(j) = F_s(\mathbf{x}_j) - F_c^0(\mathbf{x}_j)$.

On the other hand,

$$Q_{min}[\widetilde{S}_0, \widetilde{M}_c] = Q[\widetilde{S}_0, F_c^0]$$

$$= \sum_{j=1}^{m} \left[y_j - F_c^0(\mathbf{x}_j) \right]^2$$

$$= \sum_{j=1}^{m} \left[y_j - F_s(\mathbf{x}_j) + F_s(\mathbf{x}_j) - F_c^0(\mathbf{x}_j) \right]^2 =$$

$$= \sum_{j=1}^{m} \left[r_j + \delta(j) \right]^2 = \sum_{j=1}^{m} r_j^2 + 2 \sum_{j=1}^{m} \delta(j) r_j + \sum_{j=1}^{m} \delta^2(j).$$

Taking into account that

$$\sum_{j=1}^{m} r_{f(j)}^2 = \sum_{j=1}^{m} r_j^2$$

we receive

$$Q_{min}[\widetilde{S}_p(f, F_s), \widetilde{M}_c] - Q_{min}[\widetilde{S}_0, \widetilde{M}_c] = 2 \sum_{j=1}^{m} \delta(j) \left[r_{f(j)} - r_j \right]$$

$$\leq 2 \sum_{j=1}^{m} |\delta(j)| |r_{f(j)} - r_j|$$

Thus upper bound for $Q_{min}[\widetilde{S}_p(f, F_s), \widetilde{M}_c]$ tends to $Q_{min}[\widetilde{S}_0, \widetilde{M}_c]$ as $\max_{j=1,\ldots,m}|\delta_j|$ tends to 0. It is more probable that inequality

$$Q_{min}[\widetilde{S}_p(f, F_s), \widetilde{M}_c] < Q_{min}[\widetilde{S}_0, \widetilde{M}_c]$$

is true when $\max_{j=1,\ldots,m}|\delta_j|$ is small. So we may hope that P-value that is calculated by ratios (3) will be greater when $\max_{j=1,\ldots,m}|\delta_j|$ is small. Thus small P-value received when F_c^0 is verified relatively closest simple model is a strong argument for absence of simple model from $\widetilde{M}_s$ that cannot be rejected using complicated model. Existence of such argument corresponds to Supposition 1 correctness.

8.2.4 Using Occam's razor principle to evaluate significance of piecewise linear models

Discussed variant of Occam's razor principle was successfully implemented to verify piecewise linear model describing relationship between the secretion of parathyroid hormone (PTH) and vitamin D level [13]. Vitamin D plays a key role in serum calcium homeostasis. Vitamin D deficiency leads to reduced efficiency of intestinal calcium absorption and the body reacts by increasing secretion PTH. Supposition exists that vitamin D correlates with PTH only when its concentration is less than certain threshold level and there is correlation "loss" when concentration is higher than threshold level. Goal of our research was statistical verification of last supposition and search of optimal model that describes dependence of PTH on vitamin D. It must be noted that discussed supposition corresponds to use of piecewise linear model. The study was aimed to compare performance of piecewise linear model and simple linear model in data set of 139 cases. It is supposed that response variable Y is predicted by variable X with the help of piecewise linear model with 2 segments

$$Y = \beta_0^l + \beta_1^l X + \varepsilon_l, \text{when } X < B;$$

$$Y = \beta_0^r + \beta_1^r X + \varepsilon_r, \text{when } X > B.$$

It is supposed that

$$\beta_0^l + \beta_1^l B = \beta_0^r + \beta_1^r B.$$

Let $\widetilde{M}_{pwl}^B$ be the family of all piecewise-linear predicting functions with 2 segments and fixed B. For each B regression coefficients β_0^l, β_1^l, β_0^r, β_1^r are

160 Advances of Artificial Intelligence in a Green Energy Environment

calculated from data set $\widetilde{S}_0 = \{(y_1, \mathbf{x}_1), \ldots, (y_m, \mathbf{x}_m)\}$ with the help of standard LS technique. It is evident that search of coefficients may be reduced to of quadratic programming task:

$$Q(\widetilde{S}_0, \widetilde{M}_{pwl}^B) = \sum_{x_j < B} \left(y_j - \beta_0^l - \beta_1^l x_j\right)^2 + \sum_{x_j \geq B} \left(y_j - \beta_0^r - \beta_1^r x_j\right)^2 \to min,$$

$$\beta_0^l + \beta_1^l B = \beta_0^r + \beta_1^r B \tag{8.4}$$

Functional $Q(\widetilde{S}_0, \widetilde{M}_{pwl}^B)$ value corresponding to solution of problem (4) will be referred to as $Q_{min}(\widetilde{S}_0, \widetilde{M}_{pwl}^B)$. Let $\widetilde{X}_m = \{x_1, \ldots, x_m\}$ be a set of variable X values in data set $\widetilde{S}_0$ and

$$\widetilde{X}_c = \left\{ x_{j'j''}^c = \frac{x_{j'} + x_{j''}}{2} \mid x_{j'} \in \widetilde{X}_m, x_{j''} \in \widetilde{X}_m, x_{j'} \neq x_{j''} \right\}$$

be a set of boundaries separating neighbor points from $\widetilde{X}_m$. To find LS piecewise linear regression, it is sufficient to calculate for all boundary points from $\widetilde{X}_c$ and to select boundary corresponding to minimal $Q_{min}(\widetilde{S}_0, \widetilde{M}_{pwl}^B)$.

Let x_{vd} be the concentration of vitamin D in serum, y_{ph} be the concentration of PTH, $y_{ph}^l = \log y_{ph}$. Optimal piecewise linear regressions calculating y_{ph} and y_{ph}^l were chosen in $\underset{B \in \widetilde{X}_c}{\cup} \widetilde{M}_{pwl}^B$ with the help of a technique described in previous section. Optimal boundary point was equal to 23.95 ng mL for model predicting y_{ph} from x_{vd} and $B = 24.7$ ng mL for piecewise linear regression predicting y_{ph}^l from x_{vd} (task II). Dependence of $Q_{min}(\widetilde{S}_0, \widetilde{M}_{pwl}^B)$ on B is given at Fig. 8.1. It is seen that point 23.95(ng/mL) corresponds to unique expressed global minimum of $Q_{min}(\widetilde{S}_0, \widetilde{M}_{pwl}^B)$. Graphic of piecewise linear function from model I is represented at Fig. 8.2. It is seen that slope of linear predicting function inside left segment significantly exceeds slope of linear predicting function inside right segment. Correlation coefficient between y_{ph} and x_{vd} in group of patients with $x_{vd} < 23.95$ is equal to -0.2934 (significant at $P < .01$). Correlation coefficient in group of patients with $x_{vd} > 23.95$ is close to zero (equal 0.0351). Such results are in good agreement with supposition that vitD correlates with PTH only when vitD concentration is less than certain threshold level.

Search of regularities in data **Chapter | 8** **161**

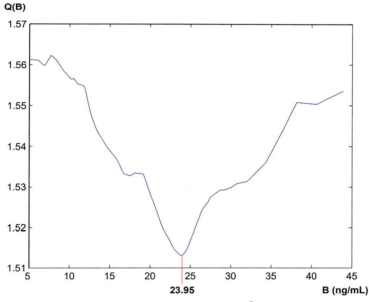

FIGURE 8.1 Dependence $Q_{min}(\widetilde{S}_0, \widetilde{M}_{pwl}^B)$ on B.

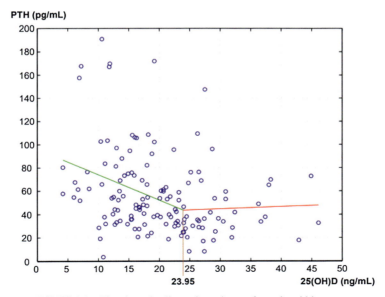

FIGURE 8.2 The piecewise linear dependence of parathyroid hormone.

162 Advances of Artificial Intelligence in a Green Energy Environment

At the first stage null hypothesis about independence of y_{ph} on x_{vd} was tested with the help of previous standard permutation test version. A set of random permutations of integers $\{1, \ldots, m\}$ was formed with the help of random numbers generator. This set $\widetilde{f}_{rng}$ consisted of N_g elements.

Data set $\{\widetilde{S}_p(f_j) \mid f_j \in \widetilde{f}_{rng}\}$ was built from $\widetilde{S}_0$ by random permutation of y_{ph} positions and relatively fixed positions of x_{vd}. Statistical validity of null hypothesis is evaluated with the help of P-value that is equal ratio

$$\frac{|\{f_j \in \widetilde{f}_g \mid Q_{min}[\widetilde{S}_p(f_j), \widetilde{M}_{pwl}] < |Q_{min}(\widetilde{S}_0, \widetilde{M}_{pwl})\}|}{N_g}$$

In other words, P-value is calculated as fraction of random data sets where dependence of y_{ph} on x_{vd} is approximated better than at initial set $\widetilde{S}_0$. Values $Q_{min}(\widetilde{S}_0, \widetilde{M}_{pwl})$ and $Q_{min}[\widetilde{S}_p(f_j), \widetilde{M}_{pwl}]$ are calculated with the help of procedure that is described in Section 8.3.3. Piecewise linear modeling of y_{ph} from x_{vd} allows to reject null hypothesis with P-value equal 0.000041. Piecewise linear modeling of y_{ph}^l from x_{vd} allows to reject null hypothesis with P-value equal 0.000079. At that number of random permutations was equal 10^6. Then piecewise linear regressions were verified with relatively simple regression models. Optimal piecewise linear regression $y_{ph} = F_{pw}^0(x_{vd}) + \varepsilon_{pw}$ was verified by testing null hypothesis about exhaustive description of dependence by simple linear regression $y_{ph} = \alpha_0 + \alpha_1 x_{vd} + \varepsilon_1$. Piecewise linear regression $y_{ph}^l = F_{pw}^l(x_{vd}) + \varepsilon_{pwl}$ was verified by testing null hypothesis about exhaustive description of dependence by simple linear regression $y_{ph}^l = \alpha_0^l + \alpha_1^l x_{vd} + \varepsilon_2$.

Two ways of regression coefficients α_0, α_1, α_0^l, α_1^l calculating were considered:

(1) simple regression coefficients were searched with the help of standard LS procedure,
(2) such simple regression coefficients were chosen that the distance between verified piecewise linear regression and simple regression was minimal.

Let us suppose that x values in $\widetilde{S}_0$ belong to some interval (a_l, a_r). Then the distance between two predicting functions $F_1(x)$ and $F_2(x)$ is calculated by formula.

TABLE 8.1 Results of verification of piecewise linear model relatively of linear models, based on Occam's razor principle.

Target	Type of simple linear model	P-value
y_{ph}	Standard LS	.022
y_{ph}^l	Standard LS	.026
y_{ph}	Most close to $F_{pw}^0(x_{vd})$	.015
y_{ph}^l	Most close to $F_{pw}^l(x_{vd})$	.0218

$$D[F_1(x), F_2(x)] = \int_{a_l}^{a_r} [F_1(x) - F_2(x)]^2 dx$$

It is seen from Table 8.1 that P-values for null hypotheses about exhaustive description of data by simple regressions do not exceed 0.026. This result is a strong argument that simple regressions are not sufficient to explain data and more complicated piecewise linear regression models are really necessary. Thus supposition that vitD correlates with PTH only when vitD concentration is less than certain threshold level is statistically valid.

8.3 Occam's razor principle for verification of regression models based on optimal partitioning

8.3.1 Optimal partitioning

The method of optimal valid partitioning (OVP) [14−16] is aimed to explore dependence of binary variable $Y \in \{a, b\}$ on explanatory variables $X_1, ..., X_n$.

OVP implements best partitioning of one-dimensional intervals of values of independent (explanatory) variables or two-dimensional joint areas of values of pairs of such variables by the training sample having the standard form $\widetilde{S}_0 = \{(y_1, \mathbf{x}_1), ..., (y_m, \mathbf{x}_m)\}$. The boundaries are searched to achieve the best separation of groups with different values of the binary Y. Quality of separation for partition $r = \{q_1, ..., q_k\}$ of the feature space is described by functional

$$Q(r, \widetilde{S}_0) = \frac{1}{v_0(1 - v_0)} \sum_{i=1}^{k} m_i(v_0 - v_i)^2,$$

where v_0 is the share of objects with $y_j = 1$ in the entire training sample $\widetilde{S}_0$, v_i is the share of objects with $y_j = 1$ inside region q_i, m_i is the number of objects

164 Advances of Artificial Intelligence in a Green Energy Environment

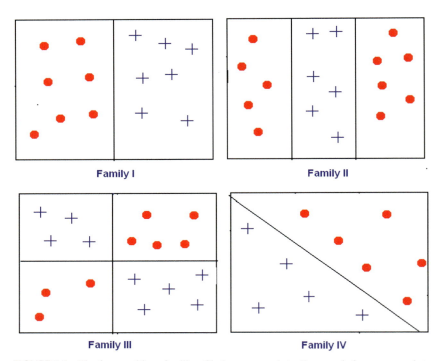

FIGURE 8.3 The four partitions families. Circle corresponds to $Y = a$ and plus corresponds to $Y = b$. The abscissa axis in the diagrams related to families I and II corresponds to explanatory variable X. The ordinate axis at that is not essential and is intended for unfolding the image. The abscissa axis in the diagrams related to families III and IV corresponds to explanatory variable X' and ordinate axis corresponds to explanatory variable X''.

inside region q_i. Optimal partitions are searched inside four partition families that are represented at Fig. 8.3. The best partition will be further referred to as r_b. Let $\mathcal{P}_{pv}(\widetilde{S}) = Q(r_b, \widetilde{S}) \geq Q(r_b, \widetilde{S}_0)$.

Partitions from families I and II correspond to regularities connecting the target variable Y with one single explanatory variable X. Partitions from families III and IV correspond to regularities connecting the target variable Y with pair of variables X' and X''. The simplest one-dimensional family I includes all partitions of single variable range with the help of one boundary point. Besides there are considered one-dimensional family II with two boundary points, two-dimensional family III with two straight boundary lines that are parallel to coordinate axes and two-dimensional family IV with one straight boundary line that is arbitrarily oriented relatively coordinate axes (Fig. 8.3).

8.3.2 Statistical significance

Statistical significance of regularities that are described by optimal partitions may be evaluated with the help of permutation test. Let $\tilde{f}$ is the set of all possible permutations of integers $\{1, ..., m\}$ Let $\widetilde{S}_p(f)$ be the data set that is received from initial data $\widetilde{S}_0$ by random permutation of $\{y_1, ..., y_m\}$:

$$\widetilde{S}_p(f) = \left\{ \left[y_{f(1)}, \mathbf{x}_1\right], ..., \left[y_{f(m)}, \mathbf{x}_m\right]\right\}$$

Let $\tilde{f} = \{f_1, ..., f_{\mathfrak{N}}\}$ is the set of all unique permutations of set of integers $\{1, ..., m\}$

Set $\widetilde{S}_b = \{\widetilde{S}_p(f_1), ..., \widetilde{S}_p(f_{\mathfrak{N}})\}$ includes all possible data sets satisfying conditions:

(a) empirical distribution of Y in $\widetilde{S}$ coincides with empirical distribution of initial data set $\widetilde{S}_0$;
(b) **x**-descriptions in $\widetilde{S}$ completely coincide with **x**-descriptions of $\widetilde{S}_0$.

Statement 4: Ratio

$$\frac{|\{\mathcal{P}_{pv}(\widetilde{S}) = TRUE | \widetilde{S} \in \widetilde{S}_b\}|}{|\widetilde{S}_b|}$$

is equal probability that $\mathcal{P}_{pv}(\widetilde{S}) = TRUE$ if null hypothesis is true: Y is independent on explanatory variables $X_1, ..., X_n$ and observations related to different objects are independent and identically distributed.

Proof: Statement 4 follows from the equal probability of all unique samples from the set sb, as well as from the fact that follows from Statement 2 about the equality of the number of permutations corresponding each unique random sample.

Let $\tilde{f}_g = \left\{f_j | j = 1, ..., N_g\right\}$

be the set of permutations calculated by random numbers generator. Then P-value may be estimated as ratio

$$p = \frac{|\{f_j \in \tilde{f}_g | Q[r_b, \widetilde{S}_p(f_j)] \geq Q(r_b, \widetilde{S}_0)\}|}{N_g} \tag{8.5}$$

The output set of regularities may include only those regularities where P-values calculated by formula (8.5) are less than some threshold p_{thr}. For example, p_{thr} may be equal to 0.05, 0.01, or 0.001. Experiments with artificially simulated data discussed in Ref. [14] showed that such way of

166 Advances of Artificial Intelligence in a Green Energy Environment

regularities selection is effective only in case of simple regularities from family I. At that output set contained great number of partially false two-dimensional regularities not envisaged by the scenario of the experiment. A two-dimensional regularity in output set will be called partially false if only one explanatory variable is tied to target variable Y according to experiments scenario. At that Y is another independent explanatory variable. Occam's razor principle can be used to overcome the problem of redundant partially false two-dimensional regularities. The approach used is based on an attempt to refute the null hypothesis that only one simple one-dimensional model is sufficient to describe the relationship associated with two-dimensional regularity. Suppose that a two-dimensional regularity from family III connects the target Y with the variables X_1 and X_2. Suppose that we have found two one-dimensional regularities r_1 and r_2 from family I connecting Y with the variables X_1 and X_2, respectively. At that boundaries for regularities r_1 and r_2 are correspondingly δ_1 and δ_2. Verification of two-dimensional regularity is based on an attempt to refute both null hypotheses:

- null hypothesis that r_1 exhaustively describes the observed pattern;
- null hypothesis that r_2 exhaustively describes the observed pattern.

It follows from an exhaustive description of the data by regularity r_i that composite null hypothesis must be true:

(1) target Y is independent on variables X_1 and X_2 inside region of feature space satisfying condition $X_1 < \delta_1$;

(2) target Y is independent on variables X_1 and X_2 inside region of feature space satisfying condition $X_1 \geq \delta_1$.

Suppose that the size of subset of $\widetilde{S}_0$ with $X_1 < \delta_1$ and $X_1 \geq \delta_1$ are m_l and m_r, respectively.

Let $\left\{ j_1^l, ..., j_{m_l}^l \right\}$ are numbers of objects from $\widetilde{S}_0$ with $X_1 < \delta_1$ and $\left\{ j_1^r, ..., j_{m_r}^r \right\}$ are numbers of objects from $\widetilde{S}_0$ with $X_1 \geq \delta_1$.

Suppose also that $\widetilde{f}_l = \{ f_1^l, ..., f_{\mathfrak{N}}^l \}$ and $\widetilde{f}_r = \{ f_1^r, ..., f_{\mathfrak{N}}^r \}$ are sets of all random permutations of integers from sets $\left\{ j_1^l, ..., j_{m_l}^l \right\}$ and $\left\{ j_1^r, ..., j_{m_r}^r \right\}$, respectively. Then

$$\widetilde{S}_p \left(f_i^l, f_i^r \right) = \left\{ \left[y_{f_i^l\left(j_1^l\right)}, \mathbf{x}_{j_1^l} \right], ..., \left[y_{f_i^l\left(j_{m_l}^l\right)}, \mathbf{x}_{j_{m_l}^l} \right], \left[y_{f_i^r\left(j_1^r\right)}, \mathbf{x}_{j_1^r} \right], ..., \left[y_{f_i^r\left(j_{m_r}^r\right)}, \mathbf{x}_{j_{m_r}^r} \right] \right\}$$

is a random set received from $\widetilde{S}_0$ by random permutations inside subsamples set by the threshold value δ_1.

Set $\widetilde{S}_b = \{\widetilde{S}_p(f_1^l, f_1^r), \dots, \widetilde{S}_p(f_{\mathfrak{N}}^l, f_{\mathfrak{N}}^r)\}$ includes all possible data sets satisfying conditions:

(a) empirical distribution of Y in $\widetilde{S}$ coincides with empirical distribution of Y in initial data set $\widetilde{S}_0$;

(b) x-descriptions in $\widetilde{S}$ completely coincide with x-descriptions of $\widetilde{S}_0$.

Statement 5: Ratio $\dfrac{\left|\left\{\mathcal{P}_{pv}(\widetilde{S})=TRUE \middle| \widetilde{S} \in \widetilde{S}_b\right\}\right|}{\left|\widetilde{S}_b\right|}$ is an equal probability that $\mathcal{P}_{pv}(\widetilde{S}) = TRUE$ if null hypothesis is true:

(a) Y is independent on explanatory variables X_1, X_2 inside regions of feature space satisfying condition $X_1 < \delta_1$ and $X_1 \geq \delta_1$;

(b) observations related to different objects are independent and identically distributed.

Proof of statement 5 practically coincides with the proof of statement 4.

Let $\widetilde{f}_g = \{(f_j^l, f_j^r) | j = 1, \dots, N_g\}$ be the set of pairs of permutations of integers from sets $\left\{j_1^l, \dots, j_{m_l}^l\right\}$ and $\left\{j_1^r, \dots, j_{m_r}^r\right\}$ calculated by random numbers generator. Then P-value that evaluates contribution of variable X_2 may be estimated as ratio

$$p_2 = \frac{\left|\left\{f_j \in \widetilde{f}_g \middle| Q[r_b, \widetilde{S}_p\left(f_j^l, f_j^r\right)] \geq Q(r_b, \widetilde{S}_0)\right\}\right|}{N_g} \tag{8.6}$$

The P-value p_1 that evaluates contribution of variable X_1 is calculated P-value that evaluates contribution of variable X_2.

But in this case $\left\{j_1^l, \dots, j_{m_l}^l\right\}$ are numbers of objects from $\widetilde{S}_0$ with $X_2 < \delta_2$ and $\left\{j_1^r, \dots, j_{m_r}^r\right\}$ are numbers of objects from $\widetilde{S}_0$ with $X_2 \geq \delta_2$.

Example: Regularity describes relationship between vascular endothelial growth factor (VEGF) and combination of oxygen saturation index (sO2) and S100 proteins level in serum. VEGF is a signal protein involved in process of blood vessels formation. It was shown by many researches that VEGF production is increased in patients with hypoxia. At that hypoxia is strongly related with low values of oxygen saturation factor sO2 in venous blood. Regularity from Fig. 8.4 shows that VEGF activation may be caused by combination of low values of sO2 and high values of S100 (Fig. 8.4).

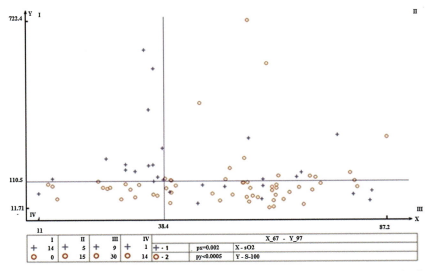

FIGURE 8.4 Association of high (>750 pg/mL) VEGF levels with content of S100 proteins in blood serum and oxygen saturation index sO2 (%). The axes X and Y, respectively, show sO2 (%) and S100 proteins levels (ng/L). The cases with high VEGF are marked as "+." The cases with low VEGF (<750 pg/mL) are marked as "o." Significance is evaluated as $P = .002$ for sO2 and as $P = .0005$ for S100.

8.3.3 Multiple testing

Estimates of P-values calculated by formulas (8.5) and (8.6) evaluate significance of individual one-dimensional patterns without taking into account a total number of evaluated indicators. However, in case of a large number of analyzed parameters, it is possible that significant differences between compared groups can occur incidentally for at least one of them even if these groups are identically distributed. The Bonferroni correction can be used to ensure the significance of the pattern at any given level [17]. Let us suppose that n is the number of analyzed indicators. The pattern with the calculated P-value is considered to be significant at level α if inequality $p < \frac{\alpha}{n}$ obviously equivalent to inequality $pn < \alpha$ holds true.

Note that using two-dimensional models requires even stronger multiple testing correction of statistical significance to compare with correction necessary for one-dimensional models. The latter is associated with the need to calculate P-values for a large number of pairs of indicators. If the verification of one-dimensional models requires testing of n null hypotheses, where n is the number of indicators, then in order to verify all kinds of two-dimensional patterns, it is necessary to test $\frac{n(n-1)}{2}$ hypotheses. For example, in tasks with 50 initial supposed explanatory variables, it is necessary to test 1225 pairs.

In order to achieve the Bonferroni-corrected significance at a level of 0.01, it is necessary that the original significance be no worse than $P < .000005$. Therefore, Bonferroni correction is often not possible because of very strict requirements on initial value. However, Bonferroni correction is actually based on quite rough upper estimates for error probability of the first kind and appears to be very conservative in the sense of unjustified rejection of actually important patterns. The approach based on comparison of a set of significant patterns found on a real sample with a set of patterns found on a random sample seems to be more effective. In this case, random samples are generated using the same procedure of random permutation of positions of target variable as the one used in permutation tests. The use of permutation tests to account for the multiple testing effect makes it possible to take into account the complex interaction of the test statistics [18−23]. Similar approach was used to in work [16] to implement multiple testing correction in OVP analysis. Efficiency of this was evaluated in task where two groups of countries with two different system of basic social institutions.

8.4 Conclusion

Machine learning methods can be successfully used in solving many applied problems. At the same time, a significant drawback of such technologies is the black box problem. The resulting solutions turn out to be incomprehensible to the user. This problem turns out to be very significant when using machine learning in medicine, scientific research, or applied problems with a legal component. One way to overcome the black box problem is to approximate complex machine learning algorithms with simpler models that may be easily assessed and interpreted. A necessary requirement for such models is the ability to describe complex nonlinear effects. Another necessary condition is the statistical significance of the relationships described by the models. However, the fulfillment of only the two mentioned requirements leads to a significant increase in the number of models that largely duplicate each other. The point is that the statistical significance of a complex model may in fact be a consequence of the existence of a simple effect in relation to which the complex model is redundant. The problem of redundancy can be overcome with the concept of completely significant regularities. A regularity is considered completely significant if all elements of associated models are significant. The statistical significance of an element of a complex model can be estimated using the method presented in this article that is based on the Occam's razor principle.

The examples given relate to the problems of verifying piecewise linear regression models and models using two-dimensional partitions with boundaries parallel to the coordinate axes. However, the considered approach can potentially be used for verification and optimization of models of various types.

170 Advances of Artificial Intelligence in a Green Energy Environment

Acknowledgment

Research was supported by Russian Foundation for Basic Research, grant number 20-01-00609, 21-51-53019.

References

[1] M. Mazumdar, J.R. Glassman, Tutorial in Biostatistics. Categorizing a prognostic variable: review of methods, coding for easy implementation and applications to decision making about cancer treatment, Stat. Med. (19) (2000) 113−132.

[2] F. Pesarin, L. Salmaso, Permutation Tests for Complex Data: Theory, Applications and Software, John Wiley and Sons, 2010.

[3] P. Good, Permutation, Parametric and Bootstrap Tests of Hypotheses, third ed., Springer Verlag, New York, 2005.

[4] R. Che, J.R. Jack, A.A. Motsinger-Reif, et al., An adaptive permutation approach for genome-wide association study: evaluation and recommendations for use, BioData Min. (2014) 7−9.

[5] M.J. Anderson, Robinson J Permutation tests for linear models, Aust. N. Z. J. Stat. (43) (2001) 75−88.

[6] M. Abdolell, M. LeBlanc, D. Stephens, R.V. Harrison, Binary partitioning for continuous longitudinal data: categorizing a prognostic variable, Stat. Med. (21) (2002) 3395−3409.

[7] H.J. Kim, M.P. Fay, E.J. Feuer, D.N. Midthune, Permutation tests for joint point regression with applications to cancer rates, Stat. Med. (19) (2000) 335−351.

[8] H. Akaike, A new look at the statistical model identification, IEEE Trans. Automat. Control 19 (6) (1974) 716−723.

[9] G. Schwarz, Estimating the dimension of a model, Ann. Stat. 6 (1978) 461−464.

[10] E. Hannan, B. Quinn, The determination of the order of an autoregression, J. Roy. Stat. Soc. B 41 (1979) 190−195.

[11] J. Rissanen, Modeling by shortest data description, Automatica 14 (5) (1978) 465−658.

[12] P. Domingos, The role of occam's razor in knowledge discovery, Data Min. Knowl. Discov. 3 (4) (1999) 409−425.

[13] O.V. Senko, D.S. Dzyba, E.Y. Pigarova, L.Y. Rozhinskaya, V. Kuznetsova A.V., A method for evaluating validity of piecewise-linear models//Proc. Intern. Conf. Knowledge Discovery and Information Retrieval (KDIR-2014), P.437-443.

[14] O. Senko, A. Kuznetsova, Methods for discovering and analysis of regularities systems - approach based on optimal partitioning of explanatory variables space, Proc. Intern. Conf. Knowledge Discovery and Information Retrieval (KDIR-2011) (2011) 415−418. https://www.scitepress.org/Papers/2011/36391/36391.pdf.

[15] A.V. Kuznetsova, I.V. Kostomarova, O.V. Senko, Modification of the method of optimal valid partitioning for comparison of patterns related to the occurrence of ischemic stroke in two groups of patients//Pattern Recogn. Image Anal. V.24. N.1. P.114-123.

[16] I.L. Kirilyuk, A.V. Kuznetsova, O.V. Sen'ko, A.M. Morozov, Method for detecting significant patterns in panel data analysis, Pattern Recogn. Image Anal. 27 (2017) 94−104.

[17] O.J. Dunn, Multiple comparisons among means, J. Am. Stat. Assoc. 56 (293) (1961) 52−64.

[18] S. Dudoit, J. Popper Shaffer, J.C. Boldrick, Multiple hypothesis testing in microarray experiments, Stat. Sci. (18) (2003) 71−103.

[19] V.G. Tusher, R. Tibshirani, G. Chu, Significance analysis of microarrays applied to the ionizing radiation response, Proc. Natl. Acad. Sci. U. S. A. (98) (2001) 5116−5121.

[20] Y. Ge, S.C. Sealfon, T.P. Speed, Multiple testing and its applications to microarrays, Stat. Methods Med. Res. 18 (2009) 543−563.

[21] J.J. Goeman, A. Solari, Multiple hypothesis testing in genomics, Stat. Med. (33) (2014) 1946−1978.

[22] V. Novik, I. Matveev, I. Litvinchev, Enhancing iris template matching with the optimal path method, Wireless Netw. 7 (2020) 4861−4868, https://doi.org/10.1007/s11276-018-1891-0.

[23] I. Matveev, V. Novik, I. Litvinchev, Influence of degrading factors on the optimal spatial and spectral features of biometric templates, J. Comput. Sci. 25 (2018) 419−424, https://doi.org/10.1016/j.jocs.2017.06.016.

Chapter 9

Artificial intelligence techniques for modeling of wind energy harvesting systems: a comparative analysis

Tigilu Mitiku Dinku[1] and Mukhdeep Singh Manshahia[2]
[1]*Department of Mathematics, Bule Hora University, Bule Hora, Ethiopia;* [2]*Department of Mathematics, Punjabi University, Patiala, Punjab, India*

9.1 Introduction

Pollution, global warming, and environmental degradation problems all over the world are due to intensive use of fossil fuels. Countries with particular geographical locations having major renewable resources (RERs), such as wind, solar, hydropower, etc., are announcing ambitious plans for distributed generation from green resources. Currently, worldwide installed capacity of wind power reached around 744 GW [1].

Artificial Intelligent (AI) methods are significant for modeling and simulation in wind energy harvesting system (WEHS) to extract maximum power and reduce the problem related to quality. To satisfy the increasing energy demand, the worldwide global orientation is to exploit RERs including wind. The main objective of this study is to develop an advanced maximum power point tracking (MPPT) controller for WEHS to ensure the efficiency of the system.

Section 9.2 presents review of works related to the study, Section 9.3 provides explanation about WEHS, Section 9.4 presents about proposed MPPT techniques, and Section 9.5 gives modeling and analysis of the result presented in Section 9.6 before conclusion and future scope.

9.2 Review of related works

Several researchers have applied various MPPT methods to optimize the efficiency of the WEHS. Parween R.Kareem had designed and simulated WEHS in MATLAB/Simulink program which consists of wind turbine strapped with permanent magnet synchronous generator (PMSG) and converters [2,3].

Advances of Artificial Intelligence in a Green Energy Environment
https://doi.org/10.1016/B978-0-323-89785-3.00014-1
Copyright © 2022 Elsevier Inc. All rights reserved.

174 Advances of Artificial Intelligence in a Green Energy Environment

K. Annaraja and S. S. Sundaram [4] have developed an ANN-based MPPT algorithm to produce the reference MPP voltage for the sliding mode controller to produce switching pulses for the positive output Luo converter (POLC) that supplies maximum power output to the RL load. In Ref. [5], ANFIS-based MPPT controllers have been proposed to track irradiance level and operating temperature to produce current at maximum power point to transfer optimal power to the load. Then fuzzy logic controller (FLC) must be tuned to generate an appropriate duty cycle. Omessaad Elbeji et al. [6] have presented a comparative study between two MPPT strategies of PMSG-based WEHS. The boost converter was controlled to produce the maximum power of wind using the tip speed ratio (TSR) and optimal torque strategies. A. Rajavel and N. Rathina Prabha [7] have described the design of three converters, such as buck converter, boost and buck-boost along with the FLC. FLC varies the time for switching ON and OFF of the converter according to the changes in the solar panel power. The FLC was employed in the generation of optimal control pulse for the DC—DC converter under varying climatic conditions. Cuong Hung Tran et al. [8] have presented modeling and simulation of three level boost DC—DC converter with perturb and observe (P&O) algorithm and FLC implemented in WEHS. Fatima Ezzahra Tahiri et al. [9] have designed fuzzy proportional integral (PI) controller to maximize the power transfer of a standalone WEHS. They showed that the fuzzy PI controller improved the system performance better in comparison with the classical PI controller. A.E. Yaakoubi et al. [10] have implemented hill climbing method and FLC-based MPPT for small-scale WTs under various operating conditions to investigate the performances of the two methods.

In this chapter, comparative analysis between PI controller, FLC, ANN, and ANFIS controllers based on MPPT technique is presented to select the best technique that optimizes the power and voltage of the isolated PMSG-based WEHS.

9.3 Modeling of wind energy harvesting system

The proposed WEHS contains different components such 20 kW wind turbine and PMSG, diode bridge rectifier, boost converter with proposed MPPT controllers, bidirectional converter with PI control, and VSI with SPWM control and a resistive load as shown in Fig. 9.1 [11].

9.3.1 Turbine model

Mechanical power available at the output of variable speed wind turbine can be expressed by [13]

$$P_m = \frac{1}{2}C_P(\beta, \lambda)\rho\pi V_w^3 \tag{9.1}$$

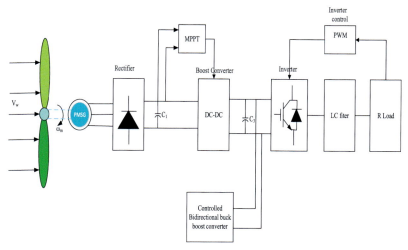

FIGURE 9.1 Overall block diagram of PMSG-based wind energy harvesting system [12].

Here, ρ is the air density, V_w is wind speed, and $C_P(\beta, \lambda)$ is the power coefficient. The power coefficient $C_p(\lambda, \beta)$ is a nonlinear function of the tip speed ration λ and the pitch angle β given by [14,15]

$$C_p(\lambda, \beta) = 0.5716 \left(\frac{116}{\lambda_i} - 0.4\beta - 5 \right) e^{\frac{-21}{\lambda_i}} + 0.0068 \lambda_i \quad (9.2)$$

where

$$\frac{1}{\lambda_i} = \frac{1}{\lambda + 0.08\beta} - \frac{0.035}{\beta^3 + 1} \quad (9.3)$$

The tip speed ratio λ is given by

$$\lambda = \frac{\omega_m V_w}{R} \quad (9.4)$$

where ω_m is the rotational speed of turbine and R is the radius of the rotor. The turbine torque applied to the generator is given by

$$T_m = \frac{P_m}{\omega_m} \quad (9.5)$$

The maximum power output of turbine is given by

$$P_{\max} = \frac{1}{2\lambda_{\text{opt}}^3} \pi \rho R^5 C_{P\max} \omega_m^3 = K_{\text{opt}} \omega_m^3 \quad (9.6)$$

Therefore, from Eq. (9.6), we can get the maximum power by regulating the PMSG speed for different wind speed. The parameter of wind turbine used in the model is presented in Table 9.1.

TABLE 9.1 Parameters of wind turbine.

Rated power	20 kW
R	4.4 m
V_{wrated}	12 m/s
V_{wcutin}	4 m/s
$V_{wcutout}$	25 m/s
ω_m	22 rad/s

9.3.2 Modeling of PMSG

The PMSG generator model is implemented in dq reference frame. Both d and q axes contain a voltage induced by the armature [16]. The d-axis and q-axis current equation of PMSG are expressed as [17,18].

$$\left.\begin{array}{l} \dfrac{di_{sd}}{dt} = -\dfrac{R_s}{L_{sd}}i_{sd} + \omega_e\dfrac{L_{sq}}{L_{sd}}i_{sq} + \dfrac{1}{L_{sd}}u_{sd} \\[3mm] \dfrac{di_{sq}}{dt} = \dfrac{R_s}{L_{sq}}i_{sq} - \omega_e\left(\dfrac{L_{sd}}{L_{sq}}i_{sd} + \dfrac{1}{L_{sq}}\psi_p\right) + \dfrac{1}{L_{sq}}u_{sq} \end{array}\right\} \tag{9.7}$$

Here, i_{sd} and i_{sq} are the dq-axis currents and u_{sd} and u_{sq} are the dq-axis currents; ω_e is the generator electrical angular frequency; L_{sd} and L_{sq} are the inductance of generator; ψ_p is the permanent magnet flux; and R_s is the stator resistance. For surface mounted permanent magnets, L_{sd} and L_{sq} are approximately equal and the electromagnetic torque is expressed as [19]

$$T_e = \frac{3}{2}p\psi_p i_{sq} \tag{9.8}$$

where p is pole pairs. The dynamic equation of mechanical motion of wind turbine system is given by

$$\frac{d\omega_t}{dt} = \frac{1}{J}(T_e - T_t - f\omega_t) \tag{9.9}$$

where J is the total inertia which appears on the shaft of the wind turbine generator and f is the coefficient of viscous friction of the rotor and load that can be neglected in a small-scale wind turbine [20].

The dc output voltage of rectifier can be estimated by

$$V_{rec} = \frac{3\sqrt{3}E_m}{\pi} - \frac{3\omega L_s}{\pi} - 2R_s I_{rec} \tag{9.10}$$

TABLE 9.2 Parameters of wind energy harvesting system.

Parameters	Value
Rated power	20 kW
Rated voltage, peak phase voltage	295.6 V
Stator rated phase current	35.1 A
Stator resistance, Rs	0.1746 Ω
Stator inductance, Ls	4.48 mH
Pole pairs	18
Rated speed	211 rpm
C_{pmax}	0.48
Inertia (J)	1.8 kg m^2/s
Optimal tip speed ratio	8.1

where

$$E_m = K_m \omega_m \qquad (9.11)$$

$$\omega = \frac{P}{2} \frac{2\pi}{60} \omega_m = \frac{P\pi}{60} \omega_m \qquad (9.12)$$

where ω_m is the generator mechanical speed (rpm); E_m is the maximum phase voltage; and L_s is the synchronous inductance of the generator. The generator mechanical speed ω_m can be obtained from Eqs. (9.11−9.13) [21,22].

$$\omega_m = \frac{V_{rec} + 2R_s I_{rec}}{\frac{3\sqrt{3}}{\pi} K_m - \frac{P}{20} L_s I_{rec}} \qquad (9.13)$$

where $K_m = 1.4$ V/rpm is a constant called peak line-to-neutral back emf. The parameter of PMSG used in the model is presented in Table 9.2.

9.4 Maximum power point tracking system

The MPPT-based control strategy is most commonly used to obtain maximum power from WECS. When wind speed is higher, the output voltage of PMSG becomes higher, which will increase the DC voltage output of rectifier [23,24]. To obtain optimum power, generator speed must be set at the optimal speed by setting the duty cycle with the help of different MPPT techniques. Optimal torque control, hill climbing search, P&O, power signal feedback, TSR algorithms, PI, and many others are classified into conventional control methods, whereas FLC, multilayer feed forward neural network (MLFFNN), radial basis

178 Advances of Artificial Intelligence in a Green Energy Environment

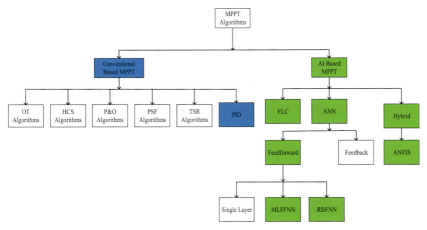

FIGURE 9.2 Maximum power point tracking (MPPT) algorithms.

function neural network (RBFNN), ANFIS, and many others are classified into AI-based algorithms as shown in Fig. 9.2. We selected PI method from conventional approaches and FLC, MLFFNN, RBFNN, and ANFIS controllers from AI method as shown in colors respectively to extract maximum power.

To overcome the problem of conventional MPPT methods like PI controller, FLC was developed by Lotfi Zadeh in 1965 [25]. The basic idea behind FLC is to combine the expert experience of a human operator with the design of the controller that does not need any mathematical model. The inputs for FLC are error signal and change in error signal. The boost converter duty cycle is generated as an output of FLC using control rules. Once the correct rules are formulated, FLC will give better result as compared to PI controller. The proposed FLC-based MPPT diagram is shown in Fig. 9.3.

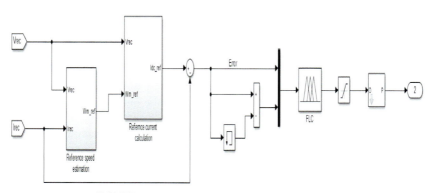

FIGURE 9.3 Block diagram of FLC MPPT controller.

Artificial intelligence techniques for modeling **Chapter | 9 179**

The error between I_{dc-ref} and actual dc current (I_{rec}) and the change in this error are the inputs to the FLC. The output of the controller is the duty cycle of the switch controlling the DC$-$DC boost converter [26,27].

Therefore, inputs to the FLC-based MPPT controller are defined as

$$e(t) = I_{dc\_ref}(t) - I_{rec}(t) \qquad (9.14)$$

$$\Delta e(t) = e(t) - e(t-1) \qquad (9.15)$$

where $e(t)$ and $e(t-1)$ are current and previous error, respectively; I_{dc_ref} is the estimated dc current; and I_{rec} is the actual current produced by rectifier. Output of FLC-based MPPT is duty cycle. FLC changes the duty cycle of the DC$-$DC boost converter according to the error between the reference dc current and the measured dc current resulting in a variation of the load seen by the PMSG.

Both trapezoidal and triangular membership functions are applied. The rule base links input and output with the help of fuzzy conditional statements expressed as the set of IF$-$THEN rules commonly obtained from expert knowledge. Table 9.3 shows 49 fuzzy governing rules when each pair $(E, \Delta E)$ determines the output level of the corresponding variable. The linguistic variables are Negative Big (NB), Negative Medium (NM), Negative Small (NS), Zero (ZE), Positive Small (PS), Positive Medium (PM), and Positive Big (PB).

Mamdani method with minimum-maximum fuzzy inference system is implementation. The flowchart of FLC is shown in Fig. 9.4.

MLFFNN and RBFN the two most commonly used ANNs for MPPT controllers to replace the PI controller to obtain the desired performance by controlling generator speed. The parameters of the network architecture used are given in Table 9.4. The training data set for the network is obtained from the FLC technique. Totally 1001 input data samples of error between the reference dc current and the actual dc current along with the corresponding

TABLE 9.3 Rule matrix for D in fuzzy logic controller.

				D			
E \ CE	NB	NM	NS	ZE	PS	PM	PB
NB	NB	NB	NB	NB	NM	NS	ZE
NM	NB	NB	NM	NM	NS	ZE	PS
NS	NB	NM	NS	NS	ZE	PS	PM
ZE	NB	NM	NS	ZE	PS	PM	PB
PS	NM	NS	ZE	PS	PS	PM	PB
PM	NS	ZE	PS	PM	PM	PM	PB
PB	ZE	PS	PM	PB	PB	PB	PB

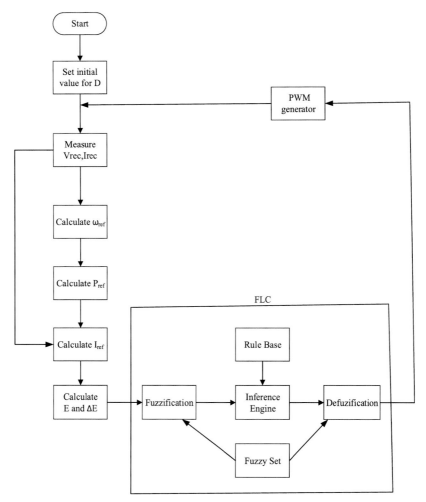

FIGURE 9.4 The flowchart of fuzzy logic controller.

duty cycle data are given to the networks for training. The *tansig* and *purelin* functions are used as activation functions at the input and output layers of MLFFNN, respectively, whereas *gaussian* and *purelin* functions are used as activation functions of RBFNN.

ANFIS developed by Jang in 1993 is a type of hybrid neuro-fuzzy technique which combines fuzzy logic and neural networks in one framework [28]. The general architecture of ANFIS first-order T-S fuzzy system with five layers is shown in Fig. 9.5 [29].

The inputs and output of ANFIS-based MPPT controller are the same as FLC. The training data were taken from FLC. Hybrid learning algorithm of the backpropagation algorithm and least-squares error algorithm is used to train the network to select the proper set of rule base.

Artificial intelligence techniques for modeling **Chapter | 9 181**

TABLE 9.4 Structure of trained networks.

Network layers	MLFFNN	RBFNN
Hidden layer	1	1
Input neurons	10	1
Hidden neurons	1	10
Output neurons	1	1
Learning rate	0.05	0.1
Epochs	5000	45
Goal	1e-6	1e-6

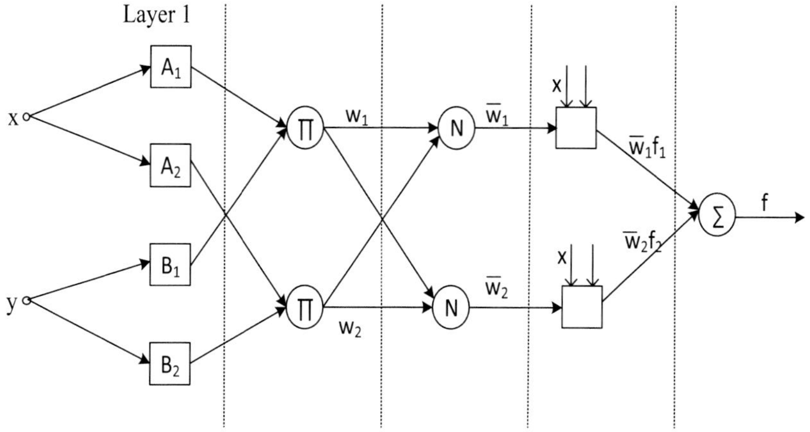

FIGURE 9.5 An ANFIS architecture.

9.5 Load side converter control

The relationship between the DC input voltage and the fundamental output line-to-line voltage is given by [30]

$$V_{ab} = \frac{\sqrt{3}}{2\sqrt{2}} m_a V_{dc} \qquad (9.16)$$

where V_{dc} is the input DC voltage and $m_a = 1$ is the modulation index. The LC filter is connected to the output of inverter to reduce the high-order harmonics produced by the PWM inverters. It delivers a sinusoidal output current with THD less than 5%.

9.6 Results and discussion

The performance of PI, FLC, MLFFNN, RBFNN, and ANFIS-based MPPT is tested under two different conditions based on wind speed and load variations, i.e., (1) step change in wind speed supplying fixed load and (2) continuous change in wind speed supplying variable load to confirm the effectiveness at constant voltage and frequency.

9.6.1 Case 1: step change in wind speed and fixed load

The wind speed steps down from 12 m/s to 8 m/s at t = 1.5 s and from 8 m/s to 10 m/s at t = 3.5 s as shown in Fig. 9.7. The system supplies power to a fixed resistive load of 10 kW at 50 Hz frequency.

The simulation results of power coefficient and TSR are given in Figs. 9.6 and 9.7, respectively. The power coefficient is maintained to its nominal value of 0.48,

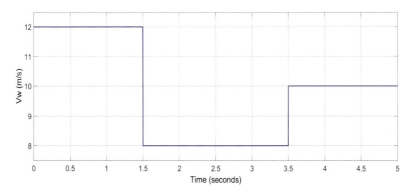

FIGURE 9.6 Wind speed.

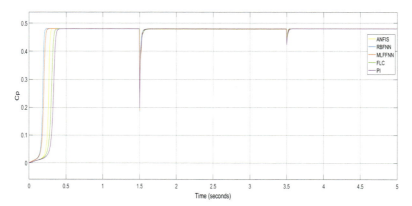

FIGURE 9.7 Power coefficient, Cp.

which shows that the controller works well to track MPP at any wind speed (Fig. 9.12). The TSR of the turbine is maintained approximately to the maximum value of 8.1 as shown in Fig. 9.8 irrespective of change in wind speeds. When wind speed decreased from 12 m/s to 8 m/s, the generator speed is decreasing from its maximum value of around 22 rad/s at base wind speed (12 m/s) to approximately around 15 rad/s to follow the change in wind speed as shown in Fig. 9.9. The simulation results show that the ANFIS has high accuracy and reliability in comparison with the other methods. In PI and FLC, fluctuations in generator speed and torque are very high as compared to ANN and ANFIS network. The mechanical output power of the turbine is given in Fig. 9.10.

The dc-bus voltage must be regulated to stay within a stable region. Accordingly, the dc and battery power are shown in Figs. 9.11 and 9.12, respectively.

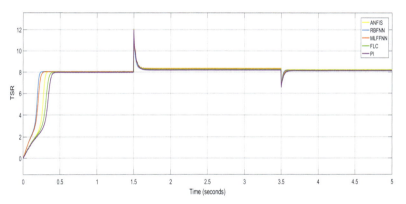

FIGURE 9.8 Tip speed ratio.

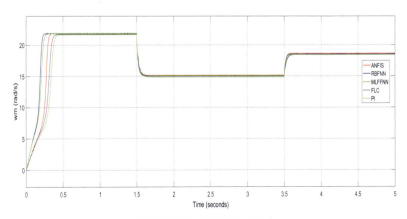

FIGURE 9.9 Generator speed.

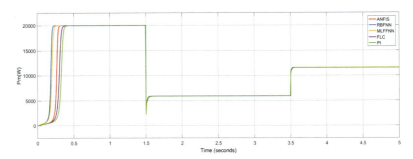

FIGURE 9.10 Mechanical power output of turbine.

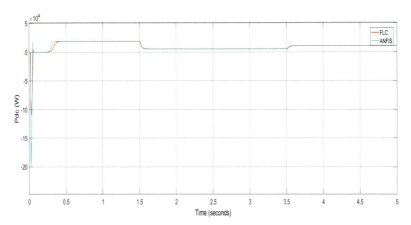

FIGURE 9.11 Boost converter power.

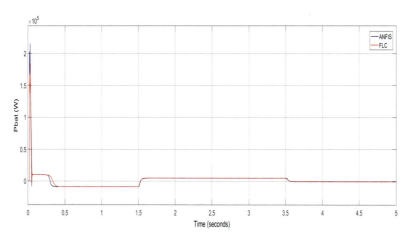

FIGURE 9.12 Battery power.

9.6.2 Case 2: continuous change in wind speed

In this work, the applied wind speed is approximated in to Eq. (9.17) using sine functions from the practical recorded data through trial and error [31].

$$V_w(t) = 9.64 - 0.87 \sin(0.6pi) + 0.75 \sin(pi) - 0.625 \sin(2pi) + 0.5 \sin(6pi) + 0.25 \sin(10pi) + 0.125 \sin(20pi)$$

(9.17)

For the applied wind speed input of Fig. 9.13, the power coefficient, TSR, generator speed, and mechanical power responses are all the methods shown in Figs. 9.14–9.17, respectively. The optimal TSR varies around 8.1 as shown in Fig. 9.15 for this particular wind speed variations. It is seen from the result that the system response is faster for AI-based system than PI and FLC. Moreover, the response of the ANFIS, RBFNN, and MLFFNN tracks the reference generator speed better as shown in Fig. 9.16.

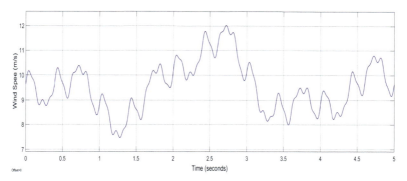

FIGURE 9.13 Wind speed profile.

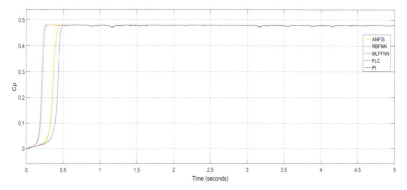

FIGURE 9.14 Power coefficient, Cp.

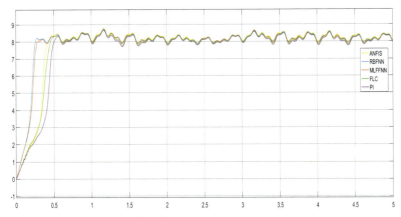

FIGURE 9.15 Tip speed ratio.

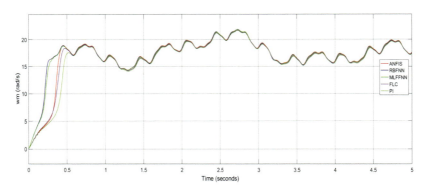

FIGURE 9.16 Generator speed.

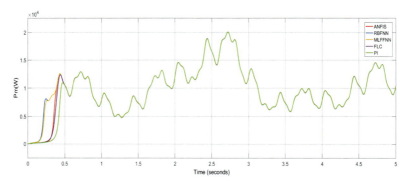

FIGURE 9.17 Mechanical power.

Artificial intelligence techniques for modeling Chapter | 9 **187**

The required load power goes down from 5 kW to 7 kW at time 1.5 sec and also goes up from 7 kW to 10 kW at 3.5 sec when wind speed changes from 12 m/s to 8 m/s. The peak value of the phase voltage is around 311V and the currents change with the variation of the load as shown in Fig. 9.18.

Fig. 9.18 shows the effective value of the load voltage at 400 V and the frequency at 50 Hz. Active and reactive power is given in Fig. 9.19.

The comparison of PI, FLC, ANN, and ANFIS controller based on their complexity, speed of convergence, mechanical power output, and rotor speed are made at nominal wind speed. The obtained results are given in Table 9.5. The results show that ANFIS controller is better than FLC and gives nominal values.

Table 9.6 presents numerical comparisons between different proposed controllers and ANFIS approach based on power coefficient, TSR, rotor speed, and efficiency. According to Table 9.6, it is observed that the proposed approach has higher accuracy in comparison with the presented approaches.

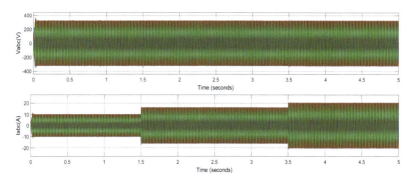

FIGURE 9.18 Peak phase voltage of the load.

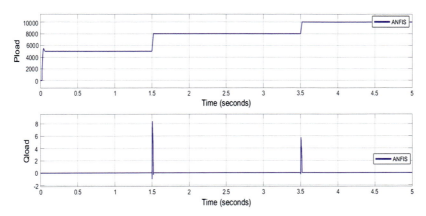

FIGURE 9.19 Active and reactive power of the load.

TABLE 9.5 Comparison of described MPPT algorithm based on different characteristics.

MPPT methods	Implementation complexity	Convergence speed	Memory requirement	Wind speed measurement	Performance under varying wind speeds	Prior training/ knowledge	Robustness
PI	Simple	Slow	No	No	Moderate	No	No
FLC	Simple	Moderate	High	Depends	Good	Required	Yes
MLFFNN	High	Moderate	Yes	Depends	Very good	Required	Yes
RBFNN	High	Moderate	Yes	Depends	Very good	Required	Yes
ANFIS	Simple	Fast	Yes	Depends	Very good	Required	Yes

Artificial intelligence techniques for modeling Chapter | 9 **189**

TABLE 9.6 Output of PI, FLC, MLFFNN, RBFNN, and ANFIS network for MPPT of standalone WECS.

Wind speed (m/s)	Tracking method	C_p	TSR	Wm (rad/s)	Power track (W)
8	PI	0.4798	8.196	14.84	5923
	FLC	0.4792	8.29	15.01	5916
	MLFFNN	0.4789	8.315	15.06	5913
	RBFNN	0.479	8.31	15.05	5913
	ANFIS	0.4785	8.357	15.12	5908
9	PI	0.4799	8.301	16.65	8435
	FLC	0.4795	8.267	16.78	8430
	MLFFNN	0.4794	8.265	16.82	8426
	RBFNN	0.4794	8.267	16.82	8426
	ANFIS	0.4791	8.301	16.9	8421
10	PI	0.48	8.136	18.4	11,570
	FLC	0.4798	8.192	18.54	11,570
	MLFFNN	0.4799	8.179	18.51	11,570
	RBFNN	0.4798	8.202	18.56	115,760
	ANFIS	0.4795	8.244	18.66	11,560
11	PI	0.48	8.083	20.12	15,400
	FLC	0.48	8.107	20.18	15,400
	MLFFNN	0.48	8.129	20.24	15,400
	RBFNN	0.48	8.122	20.22	15,400
	ANFIS	0.4799	8.16	20.32	15,400
12	PI	0.4796	21.67	7.977	19,900
	FLC	0.4796	21.76	8.01	19,900
	MLFFNN	0.48	21.88	8.057	20,000
	RBFNN	0.4799	21.86	8.047	20,000
	ANFIS	0.4799	21.87	8.052	20,000

TABLE 9.7 Comparison among different types of controllers.

Parameters	Control methods				
	PI	FLC	MLFFNN	RBFNN	ANFIS
Rise time (sec)	1.22	1.35	1.52	1.76	1.86
Settling time(sec)	2.46	2.89	3.46	5.67	5.87
Steady state error	0.77	0.5	0	0	0
Peak overshoot (%)	22.3	20.7	17.3	15.4	13.5
Output voltage (V)	740	745	782	798	799.8
THD (%)	6.59	9.56	6.59	3.43	2.5

Comparison of time domain specifications of speed controller at rated wind speed using conventional PI, FLC, MLFFNN, RBFNN, and ANFIS controllers is made in Table 9.7.

9.7 Conclusion

The focus of this research is on comparative analysis of different MPPT techniques for medium-sized PMSG wind turbine system for remote areas. Four MPPT control techniques such as PI controller, FLC, MLFFNN, RBFN, and ANFIS controllers have modeled and simulated using MATLAB/Simulink. The ANFIS, RBFNN, and MLFFNN controllers allow the system to capture maximum power from wind as compared to PI and FLC. Further research includes implementation of hybrid algorithms for MPPT in WECS.

Acknowledgments

The authors wish to extend their great gratitude to Punjabi University, Patiala, and Ministry of Science and Higher Education in Ethiopia.

References

[1] W. W. E. Association, March 24, 2021. [Online]. Available: https://wwindea.org/worldwide-wind-capacity-reaches-744-gigawa.

[2] P.R. Kareem, Modelling and Simulation of Three Phase Inverter Feed from Wind Turbine Based Matlab/Simulink, 2018.

[3] H. Matayoshi, A.M. Howlader, M. Datta, T. Senjyu, Control strategy of PMSG based wind energy conversion system under strong wind conditions, Energy Sustain. Dev. 45 (2018) 211–218.

[4] K. Annaraja, S.S. Sundaram, S. Selvaperumal, G. Prabhakar, ANN-based maximum power point tracking for a large photovoltaic farm through wireless sensor networks, Curr. Signal Transduct. Ther. 14 (1) (2019) 38−48.

[5] A.A. Koochaksaraei, H. Izadfar, High-efficiency MPPT controller using ANFIS-reference model for solar systems, in: 5th Conference on Knowledge Based Engineering and Innovation (KBEI), IEEE, 2019, pp. 770−775.

[6] O. Elbeji, M. Hannachi, M. Benhamed, L. Sbita, Maximum power point tracking control of wind energy conversion system driving a permanent magnet synchronous generator: comparative study, Wind Eng. (2020) 1−10.

[7] A. Rajavel, N. Rathina Prabha, Fuzzy logic controller-based boost and buck-boost converter for maximum power point tracking in solar system, Trans. Inst. Meas. Control 43 (4) (2021) 945−957.

[8] C.H. Tran, F. Nollet, N. Essounbouli, A. Hamzaoui, Maximum power point tracking techniques for wind energy systems using three levels boost converter, IOP Conf. Ser. Earth Environ. Sci. 154 (1) (2018) 012016.

[9] F.E. Tahiri, K. Chikh, M. Khafallah, Designing a fuzzy-PI controller of a stand-alone wind energy conversion system for MPPT, in: The Proceedings of the Third International Conference on Smart City Applications, Spring, 2018, pp. 1093−1106.

[10] A.E. Yaakoubi, L. Amhaimar, K. Attari, M.H. Harrak, M.E. Halaoui, A. Asselman, Nonlinear and intelligent maximum power point tracking strategies for small size wind turbines: performance analysis and comparison, Energy Rep. 5 (2019) 545−554.

[11] M. Baskar, V. Jamuna, Green energy generation using FLC based WECS with lithium ion polymer batteries, Braz. Arch. Biol. Technol. 59 (2) (2016) 1−15.

[12] T. Z., K.L.Z. Khaing, Control analysis of stand-alone wind power supply system with three phase PWM voltage source inverter and boost converter, 2088-8708, Int. J. Electr. Comput. Eng. 5 (4) (2015) 19−25. vol. 3, no. 2.

[13] J.G. N., J.N. N., C.M. M., A.M. M., Power output maximization of a PMSG based stand-alone wind energy conversion system using fuzzy logic, IOSR J. Electr. Electron. Eng. 11 (1) (2016) 58−66.

[14] O. Elbeji, B.h. Mouna, S. Lassaad, Modeling and control of a variable speed wind turbine, in: The Fifth International Renewable Energy Congress IREC, Tunisia, 2014.

[15] H. Slah, D. Mehdi, S. Lassaad, Advanced control of a PMSG wind turbine, Int. J. Mod. Nonlinear Theor. Appl. 5 (2016) 1−10.

[16] R. Tiwari, N.R. Babu, Fuzzy logic based MPPT for permanent magnet synchronous generator in wind energy conversion system, in: IFAC-papers OnLine, India, 2016.

[17] R.I. Putri, M. Rifa'i, M. Pujiantara, A. Priyadi, M.H. Purnomo, Fuzzy MPPT controller for small scale stand alone PMSG wind turbine, ARPN J. Eng. Appl. Sci. 12 (1) (2017) 188−193.

[18] A. Tounsi, H. Abid, M. Kharrat, K. Elleuch, MPPT algorithm for wind energy conversion system based on PMSG, in: 18th International Conference on Sciences and Techniques of Automatic Control and Computer Engineering (STA), 2017.

[19] M. Dursun, A.F. Boz, The analysis of different techniques for speed control of permanent magnet synchronous motor, Teh. Vjesn. 22 (4) (2015) 947−952.

[20] M. Sarvi, S. Abdi, S. Ahmadi, A new method for rapid maximum power point tracking of PMSG wind generator using PSO fuzzy logic, Tech. J. Eng. Appl. Sci. 3 (17) (2013) 1984−1995.

[21] O. Abbaker, Control of wind turbine for variable speed based on fuzzy-PID controller, J. Eng. Comput. Sci. 18 (1) (2017) 40−51.

192 Advances of Artificial Intelligence in a Green Energy Environment

[22] Y.F. Ren, G.Q. Bao, Control strategy of maximum wind energy capture of direct-drive wind turbine generator based on neural-network, in: 2010 Asia-Pacific Power and Energy Engineering Conference, 2010, pp. 1−4.

[23] I. Basheera, M. Hajmeerb, Artificial neural networks: fundamentals, computing, design, and application, J. Microbiol. Methods 43 (2000) 3−31.

[24] H.H. Mousa, A.R. Youssef, E.E. Mohamed, Variable step size P&O MPPT algorithm for optimal power extraction of multi-phase PMSG based wind generation system, Int. J. Electr. Power Energy Syst. 108 (2019) 218−231.

[25] L. Zadeh, Fuzzy sets, Inf. Contr. 8 (3) (1965) 338−353.

[26] A.M. Eltamaly, H.M. Farh, Maximum power extraction from wind energy system based on fuzzy logic control, Elec. Power Syst. Res. 97 (2013) 144−150.

[27] I.a.C.B. Jahmeerbacus, Fuzzy control of a variable-speed wind power generating system, Energize (2008) 41−45.

[28] J.S. Jang, ANFIS: adaptive network based fuzzy inference system, IEEE Trans. Syst. Man Cybern. 23 (1993) 665−685.

[29] M.A. Soliman, H.M. Hasanien, H.Z. Azazi, E.E. El-Kholy, S.A. Mahmoud, Hybrid ANFIS-GA-based control scheme for performance enhancement of a grid-connected wind generator, IET Renew. Power Gener. 12 (7) (2018) 832−843.

[30] N. Chakraborty, M.D. Barma, Modelling of stand −alone wind energy conversion system using fuzzy logic controller, Int. J. Innovative Res. Electr. Electron. Instrum. Contr. Eng. 2 (1) (2014) 861−868.

[31] H. Amine, Wind turbine maximum power point tracking using FLC tuned with GA, Energy Procedia 62 (2014) 364−373.

Further reading

[1] M.A. Shahid, G. Abbas, M.R. Hussain, M.U. Asad, U. Farooq, J. Gu, T. Yazdan, Artificial intelligence-based controller for DC-DC flyback converter, Appl. Sci. 9 (23) (2019) 5108.

[2] R. Esmaili, L. Xu, Sensorless control of permanent magnet generator in wind turbine application, in: Industry Applications Conference Forty-First IAS Annual Meeting, 2006.

Chapter 10

Human paradigm and reliability for aggregate production planning under uncertainty

Selma Gütmen[1], Gerhard-Wilhelm Weber[1,2], Alireza Goli[3] and Erfan Babaee Tirkolaee[4]

[1]*Faculty of Engineering Management, Poznan University of Technology, Poznań, Poland;* [2]*Institute of Applied Mathematics, Middle East Technical University, Ankara, Turkey;* [3]*Department of Industrial Engineering and Future Studies, Faculty of Engineering, University of Isfahan, Isfahan, Iran;* [4]*Department of Industrial Engineering, Istinye University, Istanbul, Turkey*

10.1 Introduction

Production planning has always been one of the inseparable pillars of production. Today, as production conditions become more complex, production planning and control plays an important role in the success of large manufacturing companies. These conditions are:

(1) Increasing the variety of products
(2) Increasing the complexity of demand
(3) Decreasing product life cycle
(4) Rapid changes in market demand and customer tastes
(5) Compression of global competition
(6) Growing need to increase quality and the need to reduce waste costs
(7) Decreasing delivery time to customers
(8) Development of large multinational corporations and increasing supply chain
(9) Uncertainty in the parameters affecting production and supply chain

All these factors have made the production planning and control maintain its special place, especially in large supply chains. Economic and profitable production requires that for all stages of production from the supply of raw materials from suppliers to the production of final products in factories and the distribution of final products to customers, comprehensive and accurate planning is done not only to optimize the use of resources but also to minimize the total cost in the production system.

Nowadays, due to increasing demand fluctuations and strict competition in the market, an organization will achieve its goals that respond to changes more quickly and adapt to environmental changes. Here, aggregate production planning (APP) is a medium-term capacity planning that determines the production and workforce plan with the goal of minimizing the total cost of production to meet customer needs. Simultaneous production planning determines the optimal production, workforce, and inventory level along the planning horizon to meet the total demand for all products and taking into account the organization's resource constraints.

APP is the process of planning and controlling various aspects of production activities related to machinery capacity, inventory levels, workforce levels, etc., in a medium-term planning period such as 3—11 months. Fig. 10.1 illustrated the APP interactions with different components in the production system. According to this figure, uncertainty may occur in each level of planning which needs to be taken into account. The main factor of uncertainty is the demand which can be initially treated by market research and forecast. On the other hand, since production is the main element of supply chains, it plays a key role in minimizing operation costs where the input and output flows to the APP box should be thoroughly examined. Here, product demand and production capacity are two subelements and constant evaluation of their trade-offs over a given planning horizon is the most significant way to minimize the total cost. In the meantime, some options, e.g., subcontracting, are applicable to maintain the optimal trade-off.

APP is to determine the levels of inventory, number of workforce, and optimal production for each period of the planning horizon using available resources and within production constraints. Such planning usually involves a product or a family of similar products with small differences. APP has a significant appeal between both academics and industry activists. Distinctive APP models with different degrees of complexity have been introduced in recent decades. After proposing an approach related to APP by Holt et al. [2], researchers have been developing various models to help solve such problems, each with proponents and opponents.

10.1.1 Human paradigm in APP

Humans can be found active in any engineered environment. The processes of assembly, transportation, installation, use, repair, deconstruction, decision-making, and management, as well as all other planned structures, all require humans. In engineering and management science, "human factors" (HFs) are unexpectedly underrepresented.

Human paradigm and reliability for APP **Chapter** | 10 **195**

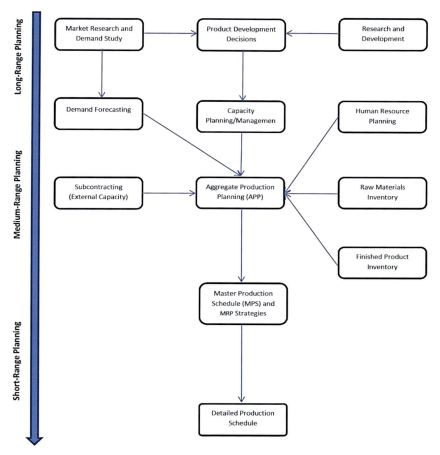

FIGURE 10.1 Aggregate production planning interactions with different components of production planning activities [1].

Although state-of-the-art automation and newest technologies have already helped a lot in our modern industries and in all parts of life, there are still numerous quality deficiencies, and they can affect or endanger outcomes due to a lack of performance of operators, technicians, and workers—of us humans. Indeed, we humans have always been sources and providers of innovation, creation, and problem solution. For all these reasons, HFs can become critical, in fact "bottleneck factors" and "crystallizers," for the reduction of human errors in production, for improving quality outcomes, and for enhancing the overall fulfillment toward top standards on all sides of modern production, manufacturing, and consumption.

196 Advances of Artificial Intelligence in a Green Energy Environment

HFs traditionally have long been overlooked or viewed as somewhat limited to contributions on "safety" standards or "ergonomics," whereas the goals of "quality," "satisfaction," or "fulfillment" remained widely neglected. In fact, when HFs are investigated with respect to quality or satisfaction, customer satisfaction has been the main concern for producers and managers. We should aim at valuing HFs in a much broader and loftier way, so we pay a special and selected attention to "workforce satisfaction."

Satisfaction at work creates happy communities within company. Through this happiness, companies can have better systems and further result in happy customer services. However, job satisfaction is a complicated process including a variety of HFs that combine to achieve a high level of satisfaction. Emotions, experiences, and behaviors are some of the HFs that influence employee cognition and, as a result, their degree of happiness within a business. Herewith, we would take into account the enormous impact which the environment, the economy, and, especially, the financial sector have on our living conditions and thus on our education health, skills, neuronal system, and eventually behavior. That is why, we recognize and explore the need for a "holistic approach" to gain a better understanding of ourselves and to locate the most appropriate, rewarding, and helpful places in the world, such as in current labor markets and industries, and finally within modern companies. In such systems, HFs involving cognitive, emotional, physical, and psychosocial aspects play a vital part in achieving success [3].

10.1.2 Reliability in APP

The definition of HF-related reliability in the context of production system should be clearly provided. Reliability in a general sense is the probability that a system will perform satisfactorily for a given period of time when it is used under certain conditions. Depending on the system under concern, it can be modified. For example, in the case of electrical power system, it is a measure of the ability of the system to generate and supply electrical energy.

If we consider the production system as a single unit system, then that simplest exponential model can be used. On the other hand, if the production system is a system consisting of subsystems, then its reliability should be computed based on reliabilities of its subsystems. For purposes of inner logic, reduced complexity, and comparability of its parts, the entire HF-APP optimization problem should become as "consistent" as possible. Therefore, in the definition of the HF-related reliability goal functions, we should pursue the way how the reliability goal functions are defined for the more classical and "technical" reliability goal functions. Fatigue and learning rates of workers are the two main factors to calculate the reliability of an APP system [4]. Considering different cases in the decision-making process can affect these factors and finally analyzing the trade-off between them leads to reliability maximization.

10.1.3 Uncertainty in APP

Uncertainty means a situation in which there is insufficient information about an event or its consequences so that the status of the event or its future consequences cannot be accurately explained. Many issues, such as production planning, scheduling, placement, transportation, finance, etc., require decisions in the presence of uncertainty, such that APP without uncertainty is not real [5]. Uncertainty optimization has spread rapidly in both theory and algorithm and has been the subject of much research [6].

Galbraith [7] defined uncertainty as the difference between the amount of information needed to do a task and the amount of information already obtained.

In the real world, there are many types of uncertainty in a supply chain that affect both productive and nonproductive processes. Ho [8], for example, categorizes uncertainty into two general categories:

1. Environmental uncertainty
2. System uncertainty

Environmental uncertainty includes uncertainties that exist beyond the production process, such as supply lead-time uncertainty and demand uncertainty.

System uncertainty is related to the uncertainty in the production process such as operation yield uncertainty, lead-time uncertainty, quality uncertainty, production system failure, and changes in product structure.

10.2 Literature review

In this section, the literature of APP is briefly reviewed in terms of optimization methods and uncertainty, and then, human paradigms along with reliability are surveyed.

The uncertainty of APP has been in the center of attention by researchers during recent decades. Fuzzy programming and robust optimization were the two most important uncertainty-handling techniques. For more information, please see Refs. [5,9–11]. They mainly considered the demand as the core uncertain parameter in the problem. On the other hand, exact methods [12], heuristics [13], and metaheuristics [6] have been among the most widely employed solution algorithms to tackle the complexity of the problem. It should be noted that exact solution methods have been at the top of the list. Furthermore, aggregation techniques are also known as applicable and valuable tools to deal with APP [14,15].

As was mentioned, APP establishes a production schedule that includes material, resource plans, deadlines, and constraints. In APP, the decision-makers are human beings. Human actors perform their tasks with the assistance of technologies and decision support systems. During the execution of a plan, several disturbances due to machine failure or human mistakes may occur and it leads to make judgments. In fact, humans play a big role to understand main issue and take the right actions. A solid understanding of the impact of the disruption is required to make strategic decisions for this [16]. Herewith, the importance of the effect of HFs raises. Systems are always in need of having good employment which brings several benefits for both employee's and employers' sides. This leads to satisfied workforce and satisfying systems for business operations.

In today's competitive business world, workforce satisfaction has emerged and become vital for the work system of organizations. Satisfaction at work can have a variety of impact on the company, including increased efficiency, knowledge management, and long-term growth. This has been studied for many years and, finally, it is proven that satisfied employees perform at a high level and contribute properly and efficiently to the success of their companies [17]. More specifically, an analysis by Ostroff [1] found that organizations with satisfied employees were more productive and effective than those with unhappy or unsatisfied employees.

However, there are numerous factors which have positive impact on satisfaction. Emotions, experiences, and behaviors are some of the human elements that influence employee cognition and, as a result, their degree of pleasure inside an organization. For example, the relationship between ethical leadership and workplace happiness was investigated by Tu et al. [17a]. They showed that good experiences with management contribute to job satisfaction. According to Medrano and Trógolo [18], unique experience both outside and inside the workplace facilitates workers' psychological separation from their jobs. Off-work activities that foster psychological detachment, such as leisure activities, according to Medrano and Trógolo [18], can lead to high employee engagement and, as a result, satisfaction.

For many years, the cognitive sides of decision-making have also been studied. For example, Teng and Das [19] offered a study on the meaning of cognitive biases in the context of high-level management's strategic decision-making process. Organizations can improve employee decision-making and drive their behaviors toward happiness by concentrating on their cognitive talents and processes, according to the findings by Cristofaro [20].

Brough et al. [21] looked at job commitment, work satisfaction, and turnover intentions to see how people felt about their jobs. The existing literature in this study found no significant differences in cognitive capacities between young and old workers, indicating that workplace attitude is dependent on a person's cognitive ability. The findings of Brough et al. [21] highlighted the need of organizations focusing on developing systems and models that boost employees' cognitive capacities in order to achieve high levels of job satisfaction.

In addition, Fairlie [22] emphasized the need for models that promote meaningful work, which has a significant impact on employee satisfaction, in order to boost employee satisfaction. Thibault-Landry et al. [22a] also showed that workers' cognitive assessments of work qualities such as task variety, meaningful work, job autonomy, and performance goals are associated to psychological need satisfaction.

HFs are raised to assist practitioners in better understanding human beings and business demands. Work satisfaction has a significant impact on an organization's future performance, according to studies. The studies demonstrated a clear correlation between high employee satisfaction and strong organizational performance [23]. Finally, by analyzing the recently published works, such as by Refs. [24−26], it can be found that the factors of reliability and human paradigm have been totally ignored.

10.3 Discussion and conclusion

This study conducted a survey on the APP in terms of three factors, including human paradigm, reliability, and uncertainty. Accordingly, these factors were investigated in three different sections in order to highlight their importance on the problem individually. It was revealed that each of these factors plays a critical role in the APP which cannot be ignored by companies. In other words, it is necessary to address these factors in the problem through maximizing workforce satisfaction level, minimizing the negative effects of uncertainty, and maximizing the system reliability in order to survive in today's competitive environment. This makes sense for most industries such as automotive, garments, chemicals, and beverages.

Furthermore, a short review was carried out on the most relevant and recently published works in order to find the major gaps, limitations, and challenges. It was revealed that most of the researchers just tried to take the economic aspect into consideration and ignored the other aspects. The following items clarify the necessity of research as future research directions:

I. Although the uncertainty has been addressed in modeling the APP, a lot of potentials is still needed in this field. Due to this, the development of uncertainty techniques, such as fuzzy programming, robust optimization, gray systems, and stochastic optimal control, must be regarded [27−29]. On the other hand, the uncertainty can be investigated for a large group of parameters including demand, supply, capacity, and cost.

II. Consideration of reliability and human resources satisfaction in modeling of the APP has not received enough attention. Therefore, such issues can be addressed in the problem in terms of objectives. In other words, social aspect is shown as one of the most key factors which has been totally neglected in most of the research studies. Accordingly, sustainable APP, as one of the key components of supply chains, leads to sustainable supply chains.

III. Applying mathematical models practically and transferring real successful experiences cannot be regarded as the only appropriate research potential, but also paves the way to accept the qualitative aspects of the human resources management.

References

[1] C. Ostroff, The relationship between satisfaction, attitudes, and performance: an organisational level analysis, J. Appl. Psychol. 77 (6) (1992) 963−974.

[2] C.C. Holt, F. Modigliani, H.A. Simon, A linear decision rule for production and employment scheduling, Manag. Sci. 2 (1) (1955) 1−30.

[3] F. Sgarbossa, E.H. Grosse, W.P. Neumann, D. Battini, C.H. Glock, Human factors in production and logistics systems of the future, Annu. Rev. Control 49 (2020) 295−305.

[4] N. Asadayoobi, M.Y. Jaber, S. Taghipour, A new learning curve with fatigue-dependent learning rate, Appl. Math. Model. 93 (2021) 644−656.

[5] E.B. Tirkolaee, A. Goli, G.W. Weber, Multi-objective aggregate production planning model considering overtime and outsourcing options under fuzzy seasonal demand, in: International Scientific-Technical Conference Manufacturing, Springer, Cham, 2019, pp. 81−96.

[6] A. Goli, E.B. Tirkolaee, B. Malmir, G.B. Bian, A.K. Sangaiah, A multi-objective invasive weed optimization algorithm for robust aggregate production planning under uncertain seasonal demand, Computing 101 (6) (2019) 499−529.

[7] J. Galbraith, Designing Complex Organizations, Mass, Reading, 1973.

[8] C.J. Ho, Evaluating the impact of operating environments on MRP system nervousness, Int. J. Prod. Res. 27 (7) (1989) 1115−1135.

[9] S.M.J.M. Al-e, M.B. Aryanezhad, S.J. Sadjadi, An efficient algorithm to solve a multi-objective robust aggregate production planning in an uncertain environment, Int. J. Adv. Manuf. Technol. 58 (5) (2012) 765−782.

[10] A. Entezaminia, M. Heidari, D. Rahmani, Robust aggregate production planning in a green supply chain under uncertainty considering reverse logistics: a case study, Int. J. Adv. Manuf. Technol. 90 (5−8) (2017) 1507−1528.

[11] N. Gholamian, I. Mahdavi, R. Tavakkoli-Moghaddam, Multi-objective multi-product multi-site aggregate production planning in a supply chain under uncertainty: fuzzy multi-objective optimisation, Int. J. Comput. Integrated Manuf. 29 (2) (2016) 149−165.

[12] S.A.B. Rasmi, C. Kazan, M. Türkay, A multi-criteria decision analysis to include environmental, social, and cultural issues in the sustainable aggregate production plans, Comput. Ind. Eng. 132 (2019) 348−360.

[13] Y. Ning, N. Pang, X. Wang, An Uncertain aggregate production planning model considering investment in vegetable preservation technology, Math. Probl Eng. 2019 (2019) 1−11.

[14] I. Litvinchev, V. Tsurkov, Aggregation in Large-Scale Optimization, vol. 83, Springer Science & Business Media, 2013.

[15] R.T. Rockafellar, R.J.B. Wets, Scenarios and policy aggregation in optimization under uncertainty, Math. Oper. Res. 16 (1) (1991) 119−147.

[16] J. Bendul, Understanding the meaning of human perception and cognitive biases for production planning and control, IFAC-PapersOnLine 52 (13) (2019) 2201−2206.

[17] D. Bakotić, Relationship between job satisfaction and organisational performance, Econ. Res. -Ekonomska istraživanja 29 (1) (2016) 118−130.

[17a] Y. Tu, X. Lu, Y. Yu, Supervisors' ethical leadership and employee job satisfaction: a social cognitive perspective, J. Happiness Stud. 18 (1) (2017) 229−245.

[18] L.A. Medrano, M.A. Trógolo, Employee well-being and life satisfaction in Argentina: the contribution of psychological detachment from work, J. Work Organ. Psychol. 34 (2) (2018) 1−13.

[19] B.-S. Teng, T.K. Das, Cognitive biases and strategic decision processes: an integrative perspective, J. Manag. Stud. 36 (6) (1999) 757−778.

[20] M. Cristofaro, "I feel and think, therefore I am": an affect-cognitive theory of management decisions, Eur. Manag. J. 38 (2) (2020) 344−355.

[21] P. Brough, G. Johnson, S. Drummond, S. Pennisi, C. Timms, Comparisons of cognitive ability and job attitudes of older and younger workers, Equality, Diversity Inclusion Int. J. 30 (2) (2011) 105−126.

[22] P. Fairlie, Meaningful work, employee engagement, and other key employee outcomes: implications for human resource development, Adv. Develop. Hum. Resour. 13 (4) (2011) 508−525.

[22a] A. Thibault-Landry, R. Egan, L. Crevier-Braud, L. Manganelli, J. Forest, An empirical investigation of the employee work passion appraisal model using self-determination theory, Adv. Develop. Hum. Resour. 20 (2) (2018) 148−168.

[23] M.M. Miah, The impact of employee job satisfaction toward organisational performance: a study of private sector employees in Kuching, East Malaysia, Int. J. Sci. Res. Publ. 8 (12) (2018) 270−278.

[24] N. Chiadamrong, N. Sutthibutr, Integrating a weighted additive multiple objective linear model with possibilistic linear programming for fuzzy aggregate production planning problems, Int. J. Fuzzy Syst. Appl. 9 (2) (2020) 1−30.

[25] M.R. Galankashi, N. Madadi, S.A. Helmi, A.R.A. Rahim, F.M. Rafiei, A multiobjective aggregate production planning model for lean manufacturing: insights from three case studies, IEEE Trans. Eng. Manag. (2020), https://doi.org/10.1109/TEM.2020.2995301.

[26] D. Rahmani, A. Zandi, S. Behdad, A. Entezaminia, A light robust model for aggregate production planning with consideration of environmental impacts of machines, Oper. Res. 21 (1) (2021) 273−297.

[27] A. Goli, E.B. Tirkolaee, N.S. Aydin, Fuzzy integrated cell formation and production scheduling considering automated guided vehicles and human factors, IEEE Trans. Fuzzy Syst. 29 (12) (2021) 3686−3695.

[28] S.K. Roy, G. Maity, G.W. Weber, Multi-objective two-stage grey transportation problem using utility function with goals, Cent. Eur. J. Oper. Res. 25 (2) (2017) 417−439.

[29] E.B. Tirkolaee, P. Abbasian, G.W. Weber, Sustainable fuzzy multi-trip location-routing problem for medical waste management during the COVID-19 outbreak, Sci. Total Environ. 756 (2021) 143607.

Chapter 11

Artificial intelligence—based intelligent geospatial analysis in disaster management

R. Subhashini[1], J. Joshua Thomas[2], A. Sivasangari[1], P. Mohana[3], S. Vigneshwari[1] and P. Asha[1]

[1]*School of Computing, Sathyabama Institute of Science and Technology, Chennai, Tamil Nadu, India;* [2]*UOW Malaysia KDU Penang University College, George Town, Pulau Pinang, Malaysia;* [3]*Centre for Remote Sensing and Geo-informatics, Sathyabama Institute of Science and Technology, Chennai, Tamil Nadu, India*

11.1 Introduction

Mankind has to follow the survival of the fittest strategy to fight against the natural disasters over the past few decades. Landslides are one among the prevailing natural hazards which claims lives, livelihoods, sizable infrastructural losses, and environmental damages. By using the remote sensing technology, landslides are identified through topographic surfaces and remote sensing images. Since there is high identification accuracy through visual interpretation, the process requires more man power and time-consuming. Therefore, in recent years, semiautomated and automated methods based on artificial intelligence (AI) have become popular for landslide identification. The role of landslide identification is crucial in case of mitigation and managing risks. Recent researchers are interested in devising a variety of algorithms on AI and machine learning (ML) over geographical information system (GIS). Location information recognition and timely prediction are the two important factors to be considered on landslide research while evaluating the performance.

The identification of landslides using optical images is barely explored and most researchers concentrate only on remote sensing images of rainfall data, terrestrial data, and geological data. The general types of data used for analysis are topographic data, geological data, and rainfall-related data. Land usage, altitude, land cover, contours, slope fluctuation, gradient, wetness index, and incurvation are the possible causes whose features are analyzed as topographical factors. Digital elevation models (DEMs) and satellite images (LISS-

Advances of Artificial Intelligence in a Green Energy Environment
https://doi.org/10.1016/B978-0-323-89785-3.00006-2
Copyright © 2022 Elsevier Inc. All rights reserved.

III) are processed by considering the aforementioned causative factors. The LSZ were analyzed using the weighted overlay analysis in the ARCGIS and mapped the landslide conditioning factors. After the preparation of LSZ map, field verification was carried out in the specific landslide detection locations. This LSZ map was also verified with the historical data and prepared the more detailed hazard analysis regarding the landslide maps.

Recurrent neural networks (RNNs) are widely used in various applications like image captioning, language modeling, etc. It applies the historical information to the current input and further allows feedback in the network. Limitation of RNN is not able to handle problems due to vanishing gradient.

Convolutional neural network (CNN) for recognizing mixed spectral traits is being used as a research technique universally. More remote sensing images could be selected with this mixed band. Time series prediction with common spatial information about the happening of the disaster both before and after could be well predicted using CNN and it is generally used for feature extraction. CNNs have the power to classify high dimensional data. Deep learning algorithms can handle vast amount of data in an excellent manner. To classify remote sensing image scenes, Cheng et al. used SVM. The images may be of high resolution whose features need to be extracted in a well-efficient manner.

Currently, in landslide-related researches, mostly Landslide Susceptibility Zonation (LSZ) has become popular because of hilly region urbanization and in the maintenance of mass wastages. Depending on the causative factors like instability, zonal classifications on stability factors could be performed using LSZ. In order to take up the measures of hazard mitigation, information about the prevailing conditions of slope stability has to be provided by LSM. This study focus on identifying landslide zones using geospatial technology (GST) and AI. The output results will be evaluated and correlating the AI and GST.

11.2 Related work

Alex Krizhevsky et al. [1] employed nonsaturating neurons and a GPU implementation of the convolution process that was exceedingly efficient. We used a recently developed regularization method termed "dropout" to reduce overfitting in fully connected layers, which proved to be quite effective. The goal of this proposed work is to examine and contrast the prediction performance of two AI strategies for GIS: fuzzy expert system (FES), a bivariate statistical technique, and extreme learning machine, a multivariate statistical methodology [2]. Scene classification in high-resolution remote sensing photos is critical for a variety of applications. While much effort has gone into developing alternative scene classification systems, the majority of them rely on handcrafted or shallow learning—based characteristics. The usage of a deep CNN for scene classification is investigated in this research [3]. In Nepal's landslide-prone Rasuwa District, Omid et al. use two training zones and one test zone to independently evaluate the performance of several approaches. Twenty alternative maps are constructed with ANN, SVM, and RF, as well as several CNN instantiations, and then compared against the findings of extensive fieldwork using a mean intersection-over-union method [4]. Using integrated geodatabases, Haojie et al. offer a novel ML and deep learning method for identifying natural terrain landslides. First, landslide-related data, such as topographic, geological, and rainfall-related data, are compiled [5].

A new methodological approach (Mariano Di Napoli et al.) is exhibited, depending on the ensemble of maximal entropy ML algorithms and ANN along with generalized boosting model. Current focus is to employ EM modeling over 100 times, each time with varied train—test mixtures for every three sole selected methods. Ensemble modeling attempts at boosted reliability, asserted by immense scores, accompanied by poorer CV. Ultimately, adopt general consistency over the various ensemble models fixed. Boosted reliability of ensemble modeling assures the utility and aptness of the proposed idea for decision-makers over local and regional land management [6].

The present work (Mehdi Sadighi et al.) attempts at accomplishing a correlative landslide susceptibility assessment over a landslide-prone zone using profuse probabilistic model, specifically multilayer perceptron ANN (artificial neural network) encompassing a BPN (backpropagation algorithm), followed by adaptive neuro-fuzzy inference system (ANFIS), along with the coupled adaptive NFIS-ICA (imperialist competitive algorithm). Although the total performance of the aggregated model is adapted from the assimilation of all the factors, the proposed results show that the adaptive NFIS-ICA model makes use of much of its liability pattern and effectiveness from the remoteness to roads factor [7].

In this research (Pengfei Zhang et al.), the Planet Satellite images by means of spatial resolution of 3 m are used to prepare a seismic landslide deep learning recognition model, so as to automatically extract landslide triggers on 2018 Iburi earthquake, Japan. This chapter gives a scheme to rapidly identify landslides precisely after a major earthquake [8].

Rybarczyk et al. did a systematic review on modeling outdoor air quality index (AQI) utilizing a variety of ML approaches. The review focused on AQI prediction modeling using ML in the geographical whereabouts of North American and Eurasian continents. The study also concentrated on the interaction of a variety of pollutants, their particulate matters, and chemical reaction among them on different seasonal and climatic conditions, and it is due to the ruggedness on geographical landscapes and their complicated geophysical features. Nonlinear models like RNNs, ANNs, and SVMs and ensemble learning methodologies like random forest yielded prediction results accurately. Along with the usual performance evaluation metrics like MSE, root mean square error (RMSE), and mean absolute error (MAE), the coefficient of determination which deals with correlation is also suggested. Detailed descriptive factorial and statistical analysis and regression methodologies need to be followed while synthesizing the results in order to obtain a perfect forecasting model. For analysis, meteorological parameters like pressure, temperature, wind speed, solar radiation, humidity, and levels of gases like oxygen, nitrogen, and carbon need to be validated [9].

Sharma et al. predicted the air quality intensity using long short-term memory (LSTM) and CNN upon experimenting with suspended particulate instances (TSP). Their proposed hybrid model is significant in prognosticating public health prospects. An accumulated ML model comprising random forest, multiple linear regression, M5, and Volterra ML models is designed for comparing the hybrid model and the meteorological data sets that were gathered to get an optimal solution for hourly TSP forecasting. The research was done over the despots of Australia and Queensland. In order to obtain optimal function, ReLu activation function was utilized. Dropout, early stopping, selection, validation and testing, low latency prediction, and low latency are the various phases of hyperparameter selection. Performance evaluation was accomplished using mean absolute percentage error, RMSE, Nash-Sutcliffe efficiency, and other mathematical equations. Thus the authors have bridged the various gaps that existed in the earlier research by designing an efficient predictive system on determining the AQI. The system can be collaborated with IoT to provide handy usage and better prediction [10].

Overfitting and underfitting are the important aspects in training complex models in deep learning. It is advisable to avoid overfitting in many scenarios. In order to ease up the training procedures, big data and cloud computing help in analyzing vast training dataset which may also be a real-time streamlining data. With more enhancements, overfitting problems can be diminished in the training dataset. To gather more attraction, deep learning techniques are widely used in complex models. Multilayer nonlinear mapping employing more hidden layers is the trend followed in deep learning. Compared to conventional ML, deep learning algorithms are faster and yield better performance. In most landslide recognition problems, RGB bands or single bands are used as features. Normalized divergence vegetation index and near-infrared spectroscopy are added up in the process of feature selection especially for enhancing the remote sensing images in multispectral mode (Yu Wang et al.) [11].

Zhao proposes a wavelet transformation and deep autoencoder (DAE)—based algorithm for a content-based image retrieval system. The image is initially wavelet transformed and decomposed into wavelet coefficients for the suggested technique [12].

Haejie Wang et al. proposed a deep learning and ML method to discover the natural terrain landslides by integrating the geodatabases such as RecLD (recent landslide databases), RelLD (relict landslide database), and JLD (joint landslide database). The results have shown that CNN outperforms well compared to other classification ML algorithms with an accuracy of 92.5% on RecLD and shows immense prospective in landslide discovery [13].

The present study has analyzed and compared the predictive productiveness of AI varieties with FES, a bivariate and multivariate statistical methodology for LSZ, and extreme learning. In the present research work (Bipin et al.), two LSMs were generated and validated in terms of FR assessment and ROC assessment. Although the spatial distribution of susceptibility zones was originating satisfactory in both maps, the statistical validation shows more accuracy and reliability [2].

Adil et al. analyzed the global air quality and the health concerns especially in urban locality. The authors have undergone their observations using LSTM and deep autoencoder techniques and predicted the rate of optimal learning of particulate matter. The performance was incredible when their proposed methodology was applied to Korean meteorological dataset. With a recognition of their LSTM methodology which utilizes recursive RNN, more number of hidden states were introduced delegating learning capacity of the model, thereby improving the air quality prediction appraisal. Reinstatement of the output data is being performed by the typical neural network model of autoencoders. There has been a big challenge in implementing the autoencoders due to the huge real-time meteorological dataset size depicting the air quality index value with every now and then change in values. To overcome this difficulty, sparse autoencoder model was designed modifying the existing autoencoder design and the sparse autoencoder model was utilized which was promising in reducing the reconstruction errors. The sparse encoder model can further be recast into DAE or stacked autoencoder, where layer-wise training was solicited which utilized a greedy approach. The hyperparameters around the network have been adjusted with the help of backpropagation and gradient descent optimization methods. The performance rate between the actual values and predicted air pollution values was measured using RMSE calculation technique, which is minimal, thereby showing improved performance of the model. The dynamic values are captured and analyzed every 10 days for 1 month and the LSTM model and DAE models were compared. The results with the LSTM model were more accurate than DAE. This research could be further improved by taking into consideration the GIS-based spatial data and by applying revolutionary deep learning algorithms [14].

Hossain et al. explored modern deep learning techniques to predict air quality index in Bangladesh cities of Dhaka and Chattogram. The authors have combined together both the sturdy variants of RNNs which are LSTM and gated recurrent unit (GRU) models. In this methodology, four dense hidden layers have been utilized where the second layer is LSTM with 256 neurons and the first and foremost hidden layer as GRU with 64 neurons. The third layer consisted of 64 neurons and the last layer with 1 neuron. Training and validation of the dataset was gathered from Bangladesh Ministry of Environment, Forest and Climate Change website from June 1, 2017 to July 25, 2020 with a total number of 1151 observations. The missing values were made up with the mean value of the previous 20 days observation. The normalized

AQI values range from 0 to 1. Mean squared error, mean absolute error, and RMSE were considered as the cost functions and their GRU-LSTM combined model manifested minimal errors when compared against two existing baseline models [15].

Ping et al. utilized admissible spatial-temporal affinity by employing adaptive deep learning. Being scared of the health concerns and other pulmonary and cardiovascular diseases and their impacts, the authors designed a hybrid neural network comprising LSTM, CNN, and ANN. The research was conducted on rural as well as urban outskirts of Beijing and Taiwan meteorological data sets. Spatial and temporal affinity from the existing models was taken into account. Synergy among locations was found on the basis of the historical patterns which are temporal in nature and this was employed to analyze more sequence delays. Two variants of KNN such as ED (Euclidean distance) and DTWD (distance between a pair of sequence with minimized errors) algorithms. Taiwan and Beijing data sets were used for performance evaluation with the metrics of MAE which was being curtailed. To create an acquainted report, the authors also flourished a chatbot application associated with social media like Facebook. This web service was handy and is accessible on laptops, tablets, and mobile phones which will emphasize people about the current outdoor AQI [16].

11.3 Proposed work

The Sirumalai Hills are situated in the Dindigul Taluk and Dindigul District of Tamil Nadu, India, between $10°$ $07'$ N and $10°$ $18'$ N longitude and $77°$ $55'$E and $78°$ $12'$ E longitude (Fig. 11.1). They are a small, isolated ridge about 6.5 km south of Madurai City. The hills are rectangular in shape, with a north-south length of 19.3 km and an east-west width of 12.8 km, covering a total area of 288.4 square kilometers. May has the highest temperatures ($29.5°C$), while January has the lowest temperatures ($18.5°C$).

In the present study, geocoded IRS P6 satellite images of LISS-III sensor were used. The thematic layers for LSZ of drainage, DEM, slope, lithology, lineament, and geomorphology were prepared.

The workflow of this study is as follows (Fig. 11.2):

1. For LSZ map preparation, various thematic layers were prepared using remote sensing images, and weightage was assigned to each parameter based on their contributions to landslides, and overlay analysis was performed.
2. Using training data sets and validating the output of the study region, a multilayer feedforward CNN is used to detect landslides.
3. The most effective CNN settings are determined by structuring the CNN with different layers of depths according to the range of input window size CNN.
4. Perform the testing and validation of method by using multiple parameters.

Intelligent geospatial analysis in disaster management Chapter | 11 **209**

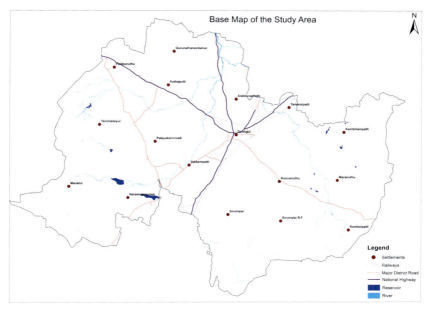

FIGURE 11.1 Basemap of the study area.

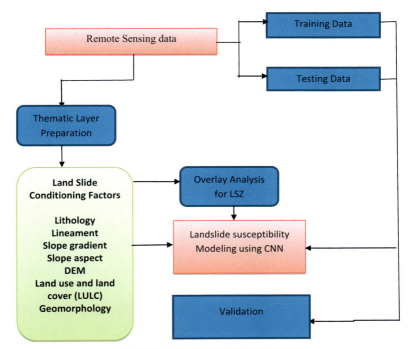

FIGURE 11.2 Proposed flow diagram.

FIGURE 11.3 Landslide incidences marked in geoeye data.

11.3.1 Preparation of various thematic maps

1. **Database creation**: Database creation of previous ground landslide incidents were reported based on the landslide observations and information from the locals and government agencies. In addition, the spatial coordinates of landslide incidence were registered in the GIS environment (Fig. 11.3).

11.3.2 Thematic maps

11.3.2.1 Lithology

A rock unit's lithology is a summary of its physical characteristics, such as color, texture, grain size, and composition, that can be seen in outcrop, hand or core samples, or with low magnification microscopy. It may be a detailed explanation of these characteristics or a rundown of a rock's overall physical characteristics. Through the field verification and satellite data image processing, the surface geology has been studied in the study area. This area encountered with hard rock of Archaean (Precambrian) age. There are five major classifications, such as charnockite, granite, a rock type from Archean formations, fissile hornblende biotite gneisses, migmatite gneisses, and quartzite (Fig. 11.4). The hilly portion of the study area comprises of charnockite type of rock with lot of structural characteristics which is the reason for possible zones of the landslides.

11.3.3 Land use and land cover

Land use and land cover also influences some surface changes through soil reinforcement and slope gradient. Depending on the land cover type, the root strengthening of the crop/trees, barren lands, waste lands, drainage

Intelligent geospatial analysis in disaster management **Chapter | 11 211**

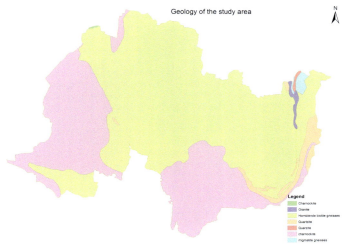

FIGURE 11.4 Geology map of the study area.

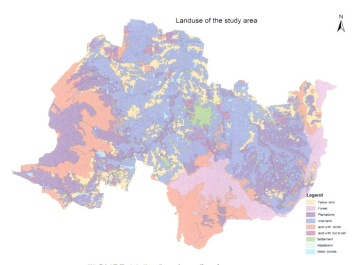

FIGURE 11.5 Land use/land cover map.

pattern, slope gradient, and its stability are the main criteria for the assessment of the landslide occurrence possible zones. In this study, IRS P6 satellite images of 2017 were used to prepare the land use and land cover map of the study area (Fig. 11.5). The land use categories of third level were delineated from the remote sensing data showing the detailed land use classification like forest, scrub forest, forest plantation, crop lands, water bodies and settlements, etc.

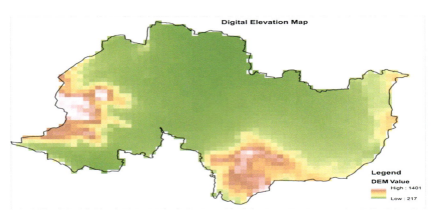

FIGURE 11.6 Digital elevation map.

11.3.4 Digital elevation map

DEM is used for the representation of the earth's topographic surface. DEM is used to classify the elevation or relief of certain area at minimum and maximum level from the MSL. In this study, SRTM DEM with 30m resolution is used to class the relief and to assess the slop gradient (Fig. 11.6). Increases in the slope gradient and the area with high density of lineament are chances for the LSZ.

11.3.5 Slope

In any landslide hazard zonation mapping, slope is a critical factor. On the basis of 1:50,000 scale topographic maps, different slope categories are developed. The slope map was created with the aid of a spot height map. If the slope is steeper, there is a greater risk of a landslide. The slope varies between 0% and 70% in the research field. The entire slope chart is divided into three categories: steeply sloping (35%−>70%), moderately sloping (3%−35%), and gently sloping (0%−3%). The majority of the research area falls into the moderately steep to steep slope group (Fig. 11.7).

11.3.6 Slope gradient

Slope gradient is measured as an angle in degree or a percentage of the slope and the aspect can be defined as a direction of slope. At the western and southern part of the study area with high gradient, the streams and rivers are driven through the path.

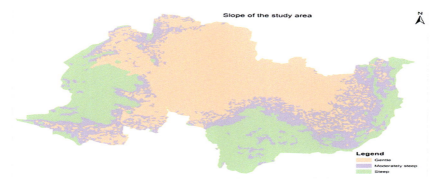

FIGURE 11.7 Slope map.

11.3.7 Lineaments

Lineaments are broad linear structural features that reflect faults, fractures, or joints and influence the geology's structure and permeability. They control the movements of surface and groundwater, drainages, and streams in the hard rock terrain. Mostly the direction of the lineaments depends upon the tectonic movement of the surface and subsurface geology. Lineaments in the study area observed from the satellite imagery through the structural interpretations were observed mostly in the hilly zones of study area (Fig. 11.8). The direction of the lineaments mostly occurred through the west to east direction toward the streams and river direction. The NW−SE trending lineaments are spread over the area. The lineaments are traversing several kilometers across granite and biotite gneisses geological formations mostly in weaker zones of the slope gradient. The landslide zonation is used to classify the density of lineaments.

FIGURE 11.8 Lineaments map.

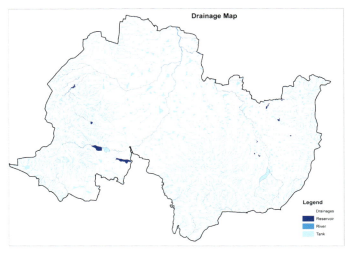

FIGURE 11.9 Drainage map.

11.3.8 Drainage

Deep drainage is a hydrologic process where water moves downward from surface water to groundwater. Drainage system, an indivisible part of the land, forms an important component in groundwater analysis by GIS. A number of factors influence the drainage of a land and include topography, soil type, bedrock type, climate, and vegetation cover, influence input, output, and transport of sediment and water. Larger areas of drainage tanks are found in the northern half of the study area. Various small-sized tanks are scattered all over the area (Fig. 11.9).

Drainage density is grouped into three classes as high, medium, and low. The suitability of landslides hazard zonation is indirectly related to drainage density because of its relation with surface runoff and permeability.

11.3.9 Landslide susceptibility zone map preparation

In the current research, landslide susceptibility zonation in hilly zones of Dindigul taluk of Dindigul District, Tamil Nadu, involves in generating complete spatial and nonspatial data collected from different sources. The multicriteria evaluation method of weighted overlay model is used in GIS techniques for allocating or dividing the each thematic layer characteristics into three parameters like high, medium, and low category depending upon locations favorable for LSZ as shown in Fig. 11.10. In this model, each and every individual raster cell in thematic layer was reclassified into the units of suitability by giving weightage for each characteristic. The algorithm behind this model is

$$S = \sum W_i X_i \text{(Eastman, 2001)} \qquad (11.1)$$

where W_i = The weight of ith factor
X_i = Category score of class of factor i
S = Suitability index for each pixel in the layer

Landslides are the slow to rapid downhill slope movements of earth materials caused by a wide range of natural processes, as well as by human activities causing land surface disturbances. Landslides are responsible for a lot of economic harm and loss of life each year, whether caused naturally or by human activities. Policy and decision-makers will use these maps to reduce social and economic losses from landslides. Our proposed work should be to provide guidance into the creation of more reliable and accurate predictive landslide models to assist decision-makers.

The first step in this work is to define historical locations of landslides and to define a collection of variables for conditioning landslides. Landslide-related study, especially LSZ (land slide susceptibility zonation), has received more attention as people become more aware of the effects of mass landslides and the need for urbanization in hilly areas. The Land Surface Stability Zone (LSZ) is a method of categorizing the land surface into different stability zones based on the estimated value of causative factors in causing instability. By providing firsthand knowledge of slope stability condition in the sense of susceptibility zones, an LSM will assist planners and field engineers in implementing hazard mitigation plans, developing sustainable development practices, and geoenvironmental planning.

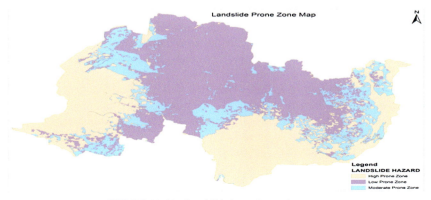

FIGURE 11.10 Landslide hazard zonation map.

216 Advances of Artificial Intelligence in a Green Energy Environment

After preprocessing with geometric rectification, projection transformation, and image fusion, remote sensing images with a resolution of 2 m and four bands were obtained. Since landslide data require changes in remote sensing pictures before and after disasters, the picture pixels should be as close together as possible. We are using in the extremely landslide-prone zones of Palani Hills, two training zones and one test zone to independently assess the efficiency of various methods.

11.3.10 Convolution neural networks

Presently, CNN provides state-of-the-art precise results in these domains. CNNs are a type of feedforward neural network that has a deep structure and uses convolutional or correlation computation. It is one of the most commonly used deep learning models for multidimensional array data.

It has distinct advantages in speech recognition and image processing thanks to its unique method of local weight distribution. The network would accept multidimensional vector input images directly, removing the need for complex feature extraction and classification processes. The workflow of this study is as follows:

- Spectral information and topographic variables are two separate data sets that need to be designed.
- Using training and validating data sets, a multilayer feedforward CNN is used to detect landslides in the study area.
- The most effective CNN settings are determined by structuring the CNN with different layers of depths according to the variety of key in window size CNN.
- Using various parameters, perform method checking and validation.

CNN's multilayer feedforward neural networks can extract an image's active feature representations, allowing them to recognize visual policies in the image without the use of complex rules created by humans. CNNs have a particular architecture in which convolutional and pooling layers are usually included in each so-called hidden layer, whereby the convolution is considered to be the main key component of any CNN.

In CNN, the primary parameter (function x) is the input, the next parameter (function w) is the kernel function, and the outputs are eigenmaps. A number of feature maps are stored in each convolutional layer. A convolutional filter extracts a function of the input and it is a plane composed of neurons which creates a feature map. The learning algorithm optimizes the kernel as a multidigit array and the input is multidimensional array. The original input image converges with a group of kernels that can be trained to scan the entire input image, yielding a set of feature maps. Each feature map is the product of the kernel's convolution, with each local area corresponding to the original input picture. Various input window sizes are used for landslide detection.

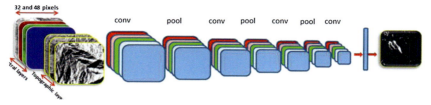

FIGURE 11.11 Convolutional neural network architecture.

The kernel is convolutional with the limited area on the unique input image that corresponds to each function map to produce it. In addition, an elementwise nonlinear activation feature is removed from the effects of a convolutional layer. The pooling layer is used to generate a condensed set of feature maps. The most popular and commonly used pooling layer is max pooling, which allows you to hold only the function maps' maximal values. The main activity of every CNN is known to be max pooling. In this research, a deeper CNN with a seven-layer depth was prepared to apply only for large input window sizes. It is given away in Fig. 11.11.

The angle of the slope and the geometry of the slope are important factors in slope instability. Slope angle patterns can influence the moisture content, pore pressure, and hydraulic behavior of a slope. As a result, estimating the slope angle is an essential aspect of hazard assessment. DEM was used to calculate the slope angle. The effects of prevailing winds, different weather conditions, and incident solar radiation are depicted by slope aspect, which determines the direction of slopes. Depending on infiltration capacity, which is affected by slope angle, soil type, and vegetation cover, the soil becomes saturated more quickly in the orientations that receive more quantity or heavy rainfall.

The classification accuracy of this land cover chart needs to be assessed. To test the classification precision, thousands of categorized image items are randomly selected as reference samples. True positive (TP), false negative (FN), and false positive (FP) are the three main types of graded pixels used to compare accuracy (FN). Pixels that have been correctly identified as landslide areas are referred to as TPs. FPs are pixels that were classified as landslide region based on the definition but are not landslides based on the ground truth values. FNs indicate landslide record areas that the machine does not recognize.

11.4 Performance analysis

To evaluate the performance measures, the real data and predicted data were compared to attain the following evaluation measures. Accuracy, F-measure, precision, and recall are measured by using the true positive (X), false positive (Y), true negative (M), and false negative (N) values.

Accuracy measure is used to classify the balance dataset in a very precise manner compared to the unbalanced dataset.

$$\text{Accuracy} = (X + M)/(X + Y + M + N) \quad (11.2)$$

Precision is used to calculate the positive class predictions from the entire collection.

$$\text{Precision} = X/(Y + X) \quad (11.3)$$

Recall is the ratio of positive class predictions from positive samples in the dataset.

$$\text{Recall} = X/(N + X) \quad (11.4)$$

F-measure is a weighted harmonic mean value.

$$F - \text{Measure} = (2 \times P \times R)/(P + R) \quad (11.5)$$

Fig. 11.12 shows the landslide feature analysis, i.e., the visualization of height from sea level and landslide prediction with the ground truth value.

Fig. 11.13 represents the RMSE comparison with the proposed CNN and RNN. The error rate is comparatively lesser for CNN. Further, Fig. 11.14 shows the accuracy comparison between CNN and RNN. Accuracy of CNN is precisely better than the RNN.

The evaluation metrics such as F-measure, MAE, accuracy, recall, and precision are calculated for CNN and RNN under various time slots from D1 to D10 and are plotted in Figs. 11.15 and 11.16.

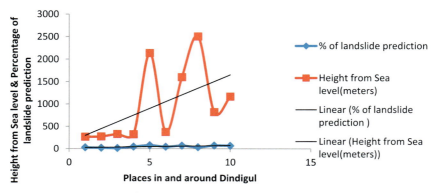

FIGURE 11.12 Landslide feature analysis.

Intelligent geospatial analysis in disaster management **Chapter | 11** **219**

RMSE comparison

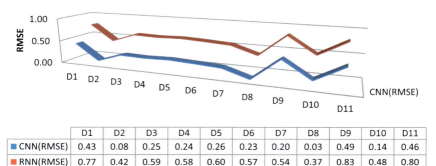

	D1	D2	D3	D4	D5	D6	D7	D8	D9	D10	D11
CNN(RMSE)	0.43	0.08	0.25	0.24	0.26	0.23	0.20	0.03	0.49	0.14	0.46
RNN(RMSE)	0.77	0.42	0.59	0.58	0.60	0.57	0.54	0.37	0.83	0.48	0.80

FIGURE 11.13 Root mean square error comparison.

Accuracy comparison

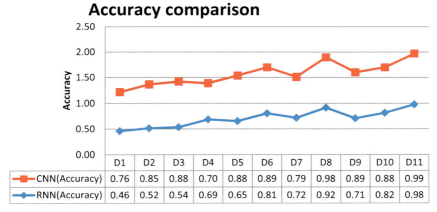

	D1	D2	D3	D4	D5	D6	D7	D8	D9	D10	D11
CNN(Accuracy)	0.76	0.85	0.88	0.70	0.88	0.89	0.79	0.98	0.89	0.88	0.99
RNN(Accuracy)	0.46	0.52	0.54	0.69	0.65	0.81	0.72	0.92	0.71	0.82	0.98

FIGURE 11.14 Accuracy comparison.

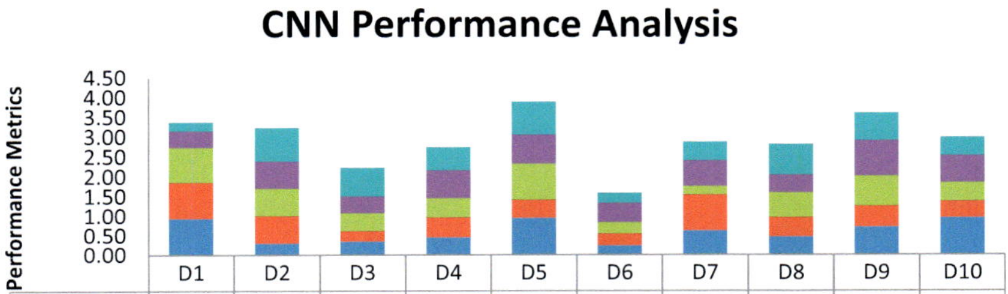

CNN Performance Analysis

	D1	D2	D3	D4	D5	D6	D7	D8	D9	D10
MAE	0.22	0.85	0.72	0.58	0.83	0.26	0.47	0.77	0.70	0.46
Accuracy	0.41	0.68	0.43	0.70	0.73	0.50	0.64	0.45	0.89	0.68
F-Measure	0.88	0.70	0.46	0.50	0.91	0.29	0.22	0.64	0.76	0.48
Recall	0.92	0.71	0.27	0.52	0.47	0.31	0.93	0.50	0.55	0.43
Precision	0.94	0.30	0.35	0.45	0.95	0.24	0.62	0.47	0.72	0.95

FIGURE 11.15 Convolutional neural network performance analysis.

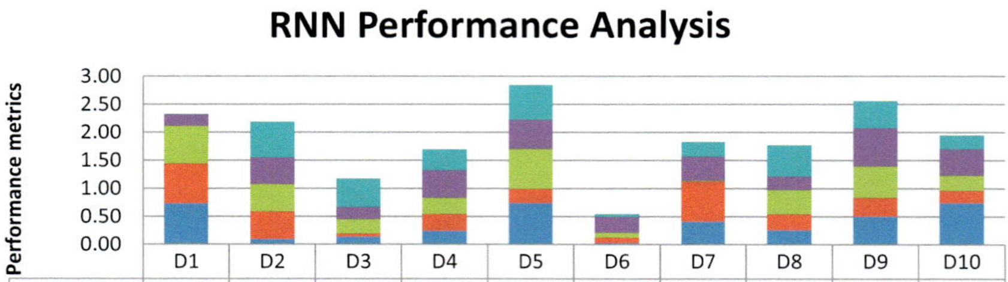

RNN Performance Analysis

	D1	D2	D3	D4	D5	D6	D7	D8	D9	D10
■ MAE	0.01	0.64	0.51	0.37	0.62	0.05	0.26	0.56	0.49	0.25
■ Accuracy	0.20	0.47	0.22	0.49	0.52	0.29	0.43	0.24	0.68	0.47
■ F-Measure	0.67	0.49	0.25	0.29	0.70	0.08	0.01	0.43	0.55	0.27
■ Recall	0.71	0.50	0.06	0.31	0.26	0.10	0.72	0.29	0.34	0.22
■ Precision	0.73	0.09	0.14	0.24	0.74	0.03	0.41	0.26	0.51	0.74

FIGURE 11.16   Recurrent neural network performance analysis.

11.5 Conclusion

Landslide-prone zone identification is obtained through human interference, and various statistical models are less accurate and there are still lot of challenges in the traditional methods. Such investigation is time-consuming, and further due to the complex factors such as size, shape, elevation, and form of the landslides, landslide affected zone mapping is becoming challenging. This could be resolved by integrating the methodology with AI technologies for automatic extraction of landslide detection details from the satellite images such as IRS P6 satellite imagery and GIS-based integration of various thematic layers in weightage overlay analysis method. Various AI techniques such as ML and deep learning perform better in object detection from satellite images. In the automatic extraction of landslide recognition, compared to the traditional algorithms like ML and also with RNN, the performance of the CNN is better in terms of precision, recall, accuracy, and F-measure. The effectiveness and feasibility of the proposed method is also further fine-tuned by adjusting the hyperparameters and layers in the CNN architecture.

The validation of both the model results can be concluded that the limitations of the AI model are more challenged for the low landslide density area and give high accuracy level of high landslide density area in the study area, but, in GIS-based analysis, it gives more accuracy level in the low and high intensity of landslide-prone zones. Further, this approach is not area-specific and it can be applied to any geographical zone.

References

[1] A. Krizhevsky, Sutskever, G. Hinton, ImageNet classification with deep convolutional neural networks, Commun. ACM 60 (6) (2017) 84–90.

[2] B. Peethambaran, R. Anbalagan, K.V. Shihabudheen, A. Goswami, Robustness evaluation of fuzzy expert system and extreme learning machine for geographic information system-

based landslide susceptibility zonation: a case study from Indian Himalaya, Environ. Earth Sci. 78 (2019) 231.

[3] G. Cheng, C. Ma, P. Zhou, X. Yao, J. Han, Scene classification of high resolution remote sensing images using convolutional neural networks, in: Proceedings of the 2016 IEEE International Geoscience and Remote Sensing Symposium (IGARSS), IEEE, Beijing, China, July 2016.

[4] O. Ghorbanzadeh, T. Blaschke, K. Gholamnia, S.R. Meena, D. Tiede, J. Aryal, Evaluation of different machine learning methods and deep-learning convolutional neural networks for landslide detection, Rem. Sens. 11 (2019) 196, https://doi.org/10.3390/rs11020196.

[5] H. Wang, L. Zhang, K. Yin, H. Luo, J. Li, Landslide identification using machine learning, Geosci. Front. 12 (2021) (2021) 351−364.

[6] Di; N. Mariano, I.F. Carotenuto, A. Cevasco, P. Confuorto, D. Calcaterra, Machine learning ensemble modelling as a tool to improve landslide susceptibility mapping reliability, Landslides 17 (2020) 1897−1914.

[7] M. Sadighi, B. Motamedvaziri, H. Ahmadi, A. Moeini, Assessing landslide susceptibility using machine learning models: a comparison between ANN, ANFIS, and ANFIS-ICA, Environ. Earth Sci. 79 (2020) 536.

[8] P. Zhang, C. Xu, S. Ma, X. Shao, Y. Tian, B. Wen, Automatic extraction of seismic landslides in large areas with complex environments based on deep learning: an example of the 2018 Iburi Earthquake, Japan, Rem. Sens. 12 (2020) 3992, https://doi.org/10.3390/rs12233992.

[9] Y. Rybarczyk, R. Zalakeviciute, Machine learning approaches for outdoor air quality modelling: a systematic review, Appl. Sci. 8 (12) (2018) 2570.

[10] E. Sharma, R.C. Deo, R. Prasad, A.V. Parisi, N. Raj, Deep air quality forecasts: suspended particulate matter modeling with convolutional neural and Long short-term memory networks, IEEE Access 8 (2020) (2020) 209503−209516.

[11] Y. Wang, X. Wang, J. Jian, Remote sensing landslide recognition based on convolutional neural network, Hindawi Math. Probl. Eng. 2019 (2019) 12, https://doi.org/10.1155/2019/8389368. Article ID 8389368.

[12] X. Zhao, B. Nutter, Content based image retrieval system using Wavelet transformation and multiple input multiple task deep autoencoder, in: Proceedings of the 2016 IEEE Southwest Symposium on Image Analysis and Interpretation (SSIAI), Santa Fe, NM, USA, 6−8, 2016, pp. 97−100.

[13] X. Haejie Wang, B. Nutter, Content based image retrieval system using Wavelet transformation and multiple input multiple task deep autoencoder, in: Proceedings of the 2016 IEEE Southwest Symposium on Image Analysis and Interpretation (SSIAI), Santa Fe, NM, USA, 6−8, 2016, pp. 97−100.

[14] A. Masood, K. Ahmad, A model for particulate matter (PM2.5) prediction for Delhi based on machine learning approaches, Procedia Comput. Sci. 167 (2020) 2101−2110, https://doi.org/10.1016/j.procs.2020.03.258. ISSN 1877-0509.

[15] E. Hossain, M.A.U. Shariff, M.S. Hossain, K. Andersson, A novel deep learning approach to predict air quality index, in: Proceedings of International Conference on Trends in Computational and Cognitive Engineering, Springer, Singapore, 2020, pp. 367−381.

[16] P.-W. Soh, J.-W. Chang, J.-W. Huang, Adaptive deep learning-based air quality prediction model using the most relevant spatial-temporal relations, IEEE Access 6 (2018) 38186−38199.

Chapter 12

Optimizing the daily use of limited solar panels in closely located rural schools in Zimbabwe

Elias Munapo
Business Statistics and Operations Research Department, Economic Sciences, FEMS, NWU, Mafikeng Campus

12.1 Introduction

Most of the rural schools in Zimbabwe do not have access to electricity and even if they do have it, it is not always available. The quality of life in Zimbabwe has been seriously affected by the unreliable supply of electricity to both urban and rural areas [1]. The study in Ref. [1] covered the period 1992−2018 and what emerged from the study is that access to electricity for the people of Zimbabwe has a significant positive impact on the country's economy.

To alleviate the serious challenge of electricity supply faced by some Zimbabwean rural schools, some donor agencies operating in Zimbabwe have donated solar panels so that they can be used by the disadvantaged schools. Unfortunately, these solar panels are not enough for all the rural schools. Even if the schools have access to electricity, it is always not available. Electrical power is necessary for computers and the science experiments. There is a need to move the limited solar panels from one school to the next in such a way that all the rural schools can use the solar panels and that the overall distance traveled is minimal. In this chapter the challenge of moving solar panels from one school to the next is modeled as a traveling salesman problem (TSP).

The TSP is believed to be difficult to solve and has so many practical applications. A lot has been done on TSP without a breakthrough. Some of the works done on this difficult model are in Applegate et al. [2], Baniasadi et al. [3], Bektas and Gouveia [4], Berman and Karpinski [5], Gutin and Punnen [6],

---

Advances of Artificial Intelligence in a Green Energy Environment
https://doi.org/10.1016/B978-0-323-89785-3.00012-8
Copyright © 2022 Elsevier Inc. All rights reserved.

Mitchell [7], Nadef [8], Padberg [9], Papadimitriou [10], and Winston [11]. Since the TSP model has so many important practical applications, research is still ongoing as given in Campuzano [12], Ebadinezhad [13], Hougardy [14], and Munapo [15].

12.2 Modeling the solar panel problem

The problem of taking the limited solar panels to all the disadvantaged rural schools in such a way that all these schools will have their allocation time and the total distance traveled to all to these schools is minimized can be formulated as a TSP.

Assumptions: To make things work, we make the following assumptions:

- It is possible to group the closely located schools into a single set.
- The allocation of the solar panel equipment can be made per unit of time (e.g., hourly, daily, or weekly), and this depends on the size of the set of schools. For smaller sets with smaller number of schools, allocation can be daily, but for bigger sets with large number of closely located schools, it can be weekly.
- Since there is a limited number of solar equipment, there is one unit of solar equipment for each set of schools. In addition, one unit of solar equipment is sufficient for the power needs of the school for the allocated time.
- The allocation of solar panel equipment is repeated throughout the academic period.
- There are at least two roads or routes leading to any of the schools in the set. Modification can be done to any road network so that this is satisfied.
- The proposed optimizing technique has to work for the general TSP.

12.2.1 The TSP model

The objective in a TSP is to start from a point of origin (starting school) and then move to the next node in such a way that every school is visited once and then return to the starting schools. The total distance that must be traveled from the starting school to all other schools and back to the starting school must be minimized and that there must be at least two routes or roads reaching any given school [2,6] in such a way that every node (or school) is visited once and that the total distance traveled is minimized (Fig. 12.1) [2,6].

The nodes S_1 and S_n represent the first and last node in the set, respectively.

12.2.1.1 Available methods

For solving TSP models, there are exact and heuristic approaches.

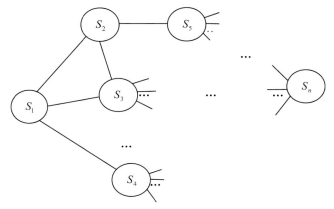

FIGURE 12.1 Traveling salesman problem model.

Methods that give exact solutions

Methods that give exact solutions to the TSP model that are known are the following:

(a) Dynamic programming–based methods are given in Ref. [11]. These methods have the problem that they become inefficient as the number of nodes of the TSP increases.
(b) Assignment-based branch and bound methods are given in Ref. [11]. The assignment model generated in every subproblem is easier to solve, but the number of subproblems increases exponentially as the number of nodes of the TSP increases.
(c) Hybrids of linear programming (LP), cutting planes, and branch and bound methods. These methods combine LP and cutting planes to yield a branch and cut method or combining LP, cutting planes, pricing, and branch and bound methods to obtain a more powerful hybrid called branch cut and price method given in Refs. [7,9].
(d) Exhaustive enumeration method: In this method, a possible route that can be used to visit all schools is explored and the one giving the minimal cost is selected as the optimal solution. This method is good for small TSPs and not feasible for large TSPs.

An efficient general purpose algorithm for the TSP is still elusive, even though a lot has been done in terms of research.

Heuristics

Heuristics are the efficient methods that give near-optimal solutions to the TSP. These methods are necessary in those situations where very quick decisions are required such as military operations or rescue missions. The main challenge of these heuristics is that there is no guarantee the solution is close to the optimal solution.

12.2.1.2 Linear programming formulation and its weakness

The TSP model can be modeled as an LP which can be solved efficiently by the simplex method or interior point algorithms. The main weakness of the LP formulation is that there is existence of subtours. Even though the LP formulations of TSPs are easy to solve, detecting and removing all the subtours is a very big challenge.

12.3 TSP network features

12.3.1 Network feature 1

Every school or node has at least two routes or arc leading to it. Suppose there are ℓ routes or roads leading to a School k (Fig. 12.2).

Since there are at least two routes or roads leading to a node or School k, then Eq. (12.1) holds.

$$x_{k1} + x_{k2} + \ldots + x_{k\ell} = 2. \qquad (12.1)$$

12.3.2 Network feature 2

Any LP formulation for n schools that is made up of constraints of the type (2) has an optimal solution that is always an integer. In other words, the constraint formulation from all the n schools is given in (2). In this chapter the equality constraints in (2) are called standard constraints.

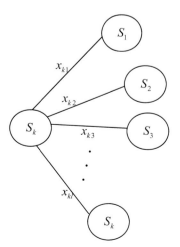

FIGURE 12.2 Routes or roads leading to School k.

$$
\left.\begin{array}{l}
x_{12} + x_{13} + \ldots + x_{1r} = 2; \\
x_{12} + x_{23} + \ldots + x_{2l} = 2; \\
\ldots \\
x_{in} + \ldots + x_{jn} + \ldots + x_{kn} = 2.
\end{array}\right\} \tag{12.2}
$$

Suppose A is the matrix coefficients of (2), and if this matrix of coefficients satisfies all the five conditions from (i) to (v) given below, then the matrix A is totally unimodular.

(i) In every column of A, there are no more than two nonzero elements.

(ii) All elements of matrix A are -1, 0, or 1.

(iii) The two rows of matrix A can be partitioned into P_1 and P_2 which are two disjoint sets.

(iv) If two nonzero elements of the same sign are contained in any column of A, then one nonzero element is in a row of P_1 and the other nonzero element is in a row of P_2.

(v) If two nonzero elements of the opposite same sign are contained in any column of A, then both are contained in rows of P_1 or in rows of P_2.

More on unimodular matrices can be found in Rebman [16]. We can use these equality constraints given in (2) on their own to find the optimal solution as explained in Section 12.3.3.

12.3.3 TSP network feature 3

In TSP network diagram, there is existence of subtours. These subtours make it very difficult to solve a TSP to optimality. An illustration is given in Fig. 12.3. The three subtours that can be identified in Fig. 12.3 are given in (12.3).

$$
\begin{array}{l}
s_1 \rightarrow s_2 \rightarrow s_6 \rightarrow s_4 \rightarrow s_1 \\
s_3 \rightarrow s_8 \rightarrow s_{15} \rightarrow s_{20} \rightarrow s_{19} \rightarrow s_{18} \rightarrow s_{12} \rightarrow s_{13} \rightarrow s_7 \rightarrow s_3 \\
s_9 \rightarrow s_{10} \rightarrow s_{11} \rightarrow s_{17} \rightarrow s_{16} \rightarrow s_3
\end{array} \tag{12.3}
$$

12.3.4 TSP network feature 4

If a constant λ is subtracted or added to all roads emanating from the same school, this process does not affect the optimal solution of given TSP.

12.3.4.1 Justification

Suppose the roads emanating from School S_k are as given in Fig. 12.3. In a TSP model, all schools are visited once in a given period. This means that two routes are required to reach the school. In other words, two roads are required

228 Advances of Artificial Intelligence in a Green Energy Environment

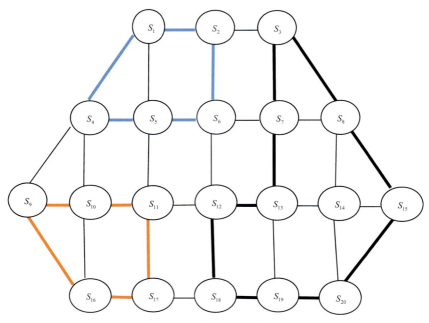

FIGURE 12.3 Subtours in TSP model.

to visit this school. One road is for getting into school and the other road is for leaving the school. The sums of distances that are possible in this operation before introducing the constant λ are presented in (12.4).

$$\left.\begin{array}{l} c_{k1} + c_{k2} \\ c_{k1} + c_{k3} \\ \ldots \\ c_{k1} + c_{k\ell} \\ c_{k2} + c_{k3} \\ \ldots \\ c_{k2} + c_{k\ell} \\ \ldots \\ c_{k3} + c_{k\ell} \end{array}\right\} \tag{12.4}$$

After adding a constant (λ), the routes leading to School k become as given in Fig. 12.4.

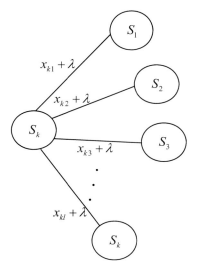

FIGURE 12.4 Adding a constant λ to all routes or roads leading to School k.

Given a TSP model of schools, every school is supposed to be visited which means that there must be at least one road into the school and another leaving the school. All the possible roads emanating from the school are given in (12.5).

$$\left.\begin{aligned} c_{k1} + c_{k2} + 2\lambda \\ c_{k1} + c_{k3} + 2\lambda \\ \ldots \\ c_{k1} + c_{k\ell} + 2\lambda \\ c_{k2} + c_{k3} + 2\lambda \\ \ldots \\ c_{k2} + c_{k\ell} + 2\lambda \\ \ldots \\ c_{k3} + c_{k\ell} + 2\lambda \end{aligned}\right\} \quad (12.5)$$

All the possible sums increase by the same constant 2λ. Similarly if we are subtracting a constant λ from all routes leading to School k, the possible sums decrease by the same constant 2λ.

12.3.5 TSP network feature 5

In any TSP network structure, we have a tree, leaf, spanning tree, minimal spanning tree, tour, subtour, and optimal tour (Fig. 12.5). These terms are defined as follows.

Tree—In a TSP of schools, a tree is a set of schools connected forming a single structure.

Leaf—In a TSP of networks, this is a school that is connected to a tree and there are no schools that are connected to this school.

Node S_1 represents a leaf since it has no other nodes (children) connected to it.

12.3.5.1 Spanning tree

Given a TSP network of schools, a spanning tree is a single structure of roads that connects all schools and does not form a loop.

12.3.5.2 Minimum spanning tree

If a spanning tree gives the shortest total distance as compared to all the other possible spanning trees that can be formed from a TSP, then it is called the minimum spanning tree (MST).

12.3.5.3 Tour

Tour is that set of roads in the TSP network diagram that forms a loop. The loop may be made up of n school or less.

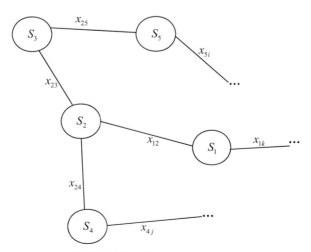

FIGURE 12.5 An example of a tree.

Optimizing the daily use of limited solar panels **Chapter | 12** **231**

12.3.5.4 Optimal tour

Optimal tour is a single loop that is made up of all the n schools.

12.3.5.5 Minimum spanning tree algorithm

For any given TSP network model, an algorithm that is used to find an MST is called an MST algorithm. This algorithm is made up of the following steps.

Step 1: Select any school i from the TSP network as starting point and join this school to the closest school j. A connected set of schools $C = \{i,j\}$ and road (i,j) are formed by the schools i and j. The connected set of schools is C and the remaining set of unconnected schools is $\overrightarrow{C}$.

Step 2: Now we select a school in $\overline{C}$ that is closest to any of the schools in C. Suppose n is a school in $\overline{C}$ and m represents the school in C that is closest to school n. In this case the additional road to the MST is (m,n). The new set of schools in C and $\overline{C}$ are updated accordingly.

Step 3: Return to Step 2 until all schools are connected.

Preference in this chapter is given to that road that has the highest chance of forming a leaf.

The MST algorithm is not a new approach as given in Ref. [6]. The strategy of the algorithm proposed in this chapter lies in solving the TSP as an MST and then use branching to remove leaves. The following proof shows that the MST algorithm will always give an MST.

Using contradiction to prove the MST algorithm, define

- the MST as S.
- after iteration k of MST algorithm, the schools that are connected as C_k.
- after iteration k of MST algorithm, the schools that are not connected as $\overrightarrow{C_k}$.
- after the k iterations of the MST algorithm, the set of roads that have been connected as A_k.

We assume that the MST algorithm is not giving an MST.

If this is true then one of the roads (a_k) is not in S, i.e., $a_k \notin S$. This road is selected at iteration k. From this we can show that $A_{k-1} \in S \Rightarrow \overline{a}_k \in S$ and $\overline{a}_k$ leads from a school in C_{k-1} to a school in $\overline{C}_{k-1}$. This means that a shorter spanning tree can be obtained by replacing $\overline{a}_k$ by a_k. This contradicts the fact that all roads selected by the MST algorithm lie in S. The MST algorithm finds the MST.

12.3.5.6 TSP tree

If an MST has no leaves, then it is called a TSP tree.

12.4 Dummies and their use in elimination of subtours

12.4.1 Dummy schools

This is an additional school that is added to a TSP so as to eliminate a subtour. This additional school is ignored in the optimal solution.

12.4.2 Dummy point

Dummy point is the point in the TSP network diagram where the dummy school is added in order to eliminate a subtour.

The existence of a subtour, dummy schools, and dummy point are illustrated in Figs. 12.6 and 12.7.

In Fig. 12.6, the dummy point is any point along the dotted line. The school S_t^i is the first subtour and S_j^t is a school in the second subtour. The dummy school D can be created as given in Fig. 12.7. After adding dummies, there is no way the schools that are enclosed in dotted lines can form a subtour.

12.4.3 Identifying dummy points

In this chapter a dummy point can be determined as that point which is three schools from the previous dummy point. This depends on the angle that is being considered. In a traveling salesman network model, a salesman must enter a school and leave. Identifying dummy points is illustrated in Fig. 12.8.

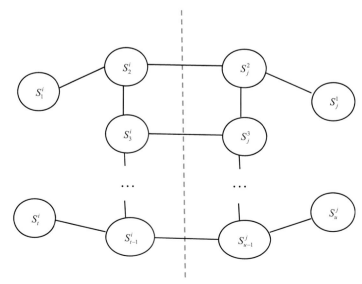

FIGURE 12.6 Existence of a subtour and dummy point.

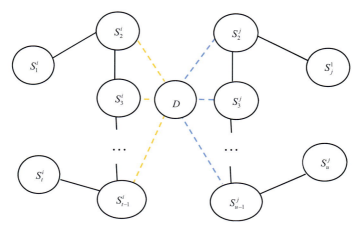

FIGURE 12.7 Creating a dummy node.

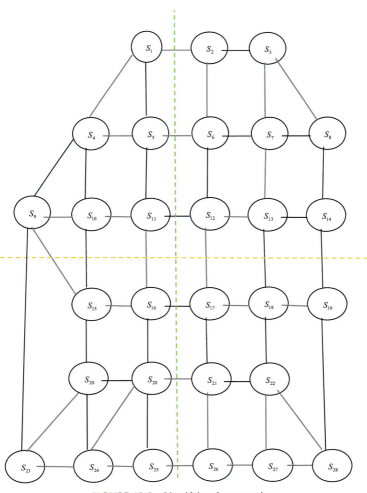

FIGURE 12.8 Identifying dummy points.

234 Advances of Artificial Intelligence in a Green Energy Environment

The two dotted lines show the positions of the dummy points.

In other words, for Fig. 12.7 at the dummy school D, we have Eq. (12.6).

$$x_{s_2^i D} + x_{s_3^i D} + \dots + x_{Ds_{t-1}^i} + x_{Ds_2^j D} + x_{s_3^i D} + \dots + x_{Ds_{t-1}^j} = 2. \qquad (12.6)$$

In this case the road entering the dummy school must be the same as the road leaving the dummy.

$$x_{s_2^i D} = x_{Ds_2^j}, x_{s_3^i D} = x_{s_3^i D}, \dots, x_{Ds_{t-1}^i} = x_{Ds_{t-1}^j}. \qquad (12.7)$$

This is done to ensure that a road leaving a given school and going to a specific school will still go to the specific school after being diverted to a dummy. Thus, the equation given in (12.6) simplifies to (12.8).

$$x_{s_2^i D} + x_{Ds_2^j} + x_{s_3^i D} + x_{s_3^i D} + \dots + x_{Ds_{t-1}^i} + x_{Ds_{t-1}^j} = 2,$$

$$2x_{s_2^i D} + 2x_{s_3^i D} + \dots + 2x_{Ds_{t-1}^i} = 2,$$

$$x_{s_2^i D} + x_{s_3^i D} + \dots + x_{Ds_{t-1}^i} = 1. \qquad (12.8)$$

The equation given in (12.8) is not enough. A subtour with the dummy as one of the schools can still form.

12.4.4 Subtour eliminators

There is a need to generate subtour eliminators and these eliminators depend on the number of roads entering the dummy school D. For Fig. 12.7, the following subtour eliminators can be generated as given in (12.9).

$$x_{s_2^i D} + x_{S_2^i S_2^j} \leq 1,$$

$$x_{s_3^i D} + x_{S_3^i S_3^j} \leq 1,$$

$$\dots \qquad (12.9)$$

$$x_{s_{t-1}^i D} + x_{S_{t-1}^i S_{t-1}^j} \leq 1$$

With these subtour eliminators, subtours involving the dummy school cannot form.

12.5 Proposed algorithm for TSP

The proposed algorithm is summarized as given in Section 12.5.1.

12.5.1 Proposed algorithm

Step 1: From the given TSP network model, identify all the possible dummy points.

Step 2: Use the dummy points to construct the dummy school to form a new TSP network model.

Step 3: Use the new TSP network model to formulate an LP model. The LP model must contain an objective function, standard constraints, and the subtour eliminators.

Step 4: Solve the formulated LP model.

Step 5: Adjust the optimal solution obtained in Step 4 by eliminating the dummies so that it is now optimal in terms of the original problem.

12.5.2 Numerical illustration

Consider the TSP network model for 10 Zimbabwean rural schools given in Fig. 12.9. The distances between the 10 schools are given in kilometers. Use the proposed approach to determine the route that gives the shortest distance in visiting all the 10 schools.

The TSP network model can be solved as given in Section 12.5.2.1.

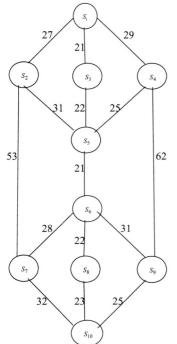

FIGURE 12.9 TSP network for numerical illustration.

12.5.2.1 Identifying dummy points

From Fig. 12.9, the dummy points can be identified as given in Fig. 12.10. The dummy points are represented by dotted lines S_2S_8.

The dummy points can be used to select a suitable position for the dummy node as given in Fig. 12.11.

12.5.2.2 Introducing dummy node to the network

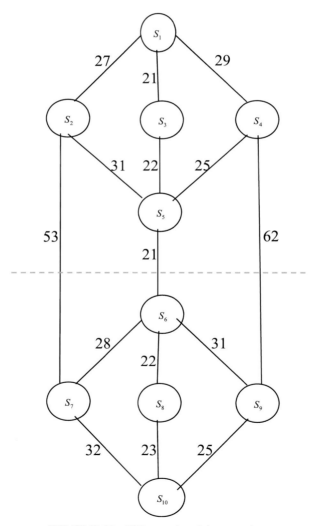

FIGURE 12.10 TSP network and dummy points.

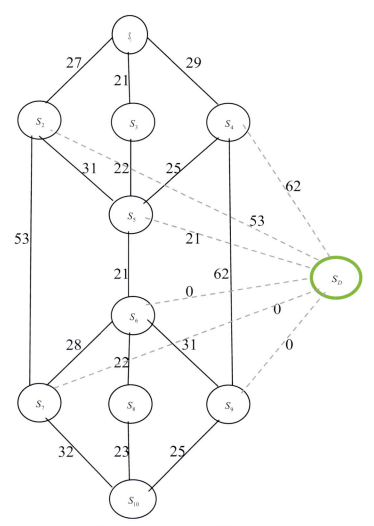

FIGURE 12.11 Dummy added TSP network diagram.

12.5.2.3 Linear programming formulation

$$\text{Min} \begin{bmatrix} Z_o = 27x_{12} + 21x_{13} + 29x_{14} + 31x_{25} + 53x_{27} + 22x_{35} \\ +25x_{45} + 62x_{49} + 21x_{56} + 28x_{67} + 22x_{68} + 31x_{69} + \\ 32x_{7(10)} + 23x_{8(10)} + 25x_{9(10)} + 53x_{2D} + 62x_{4D} + 21x_{5D} \end{bmatrix}$$

238 Advances of Artificial Intelligence in a Green Energy Environment

such that

Node $S_1: x_{12} + x_{13} + x_{14} = 2,$

Node $S_2: x_{12} + x_{25} + x_{27} + x_{2D} = 2,$

Node $S_3: x_{13} + x_{35} = 2,$

Node $S_4: x_{14} + x_{45} + x_{49} + x_{4D} = 2,$

Node $S_5: x_{25} + x_{35} + x_{45} + x_{56} + x_{5D} = 2,$ (12.10)

Node $S_6: x_{56} + x_{67} + x_{68} + x_{69} + x_{5D} = 2,$

Node $S_7: x_{27} + x_{67} + x_{7(10)} + x_{2D} = 2,$

Node $S_8: x_{68} + x_{8(10)} = 2,$

Node $S_9: x_{49} + x_{69} + x_{9(10)} + x_{4D} = 2,$

Node $S_{10}: x_{7(10)} + x_{8(10)} + x_{9(10)} = 2,$

Dummy Node $D: x_{2D} + x_{4D} + x_{5D} = 1,$

Subtour eliminators:

$$x_{27} + x_{2D} \leq 1,$$
$$x_{56} + x_{5D} \leq 1,$$
$$x_{49} + x_{4D} \leq 1.$$

where $x_{ij} \geq 0$ and binary $\forall ij$ and $i \neq j$.

12.5.2.4 Optimal solution of the formulated linear program

$$Z_o = 308,$$
$$x_{12} = x_{13} = x_{27} = x_{35} = x_{45} = x_{67} = x_{68} = x_{8(10)} = x_{9(10)} = x_{4D} = 1,$$
$$x_{14} = x_{25} = x_{49} = x_{56} = x_{69} = x_{7(10)} = x_{2D} = x_{5D} = 0.$$

(12.11)

The optimal solution in Fig. 12.12 reduces to Fig. 12.13. In this case we now discard the dummy node and the associated artificial arcs as they do not have a meaning to the original problem.

Optimizing the daily use of limited solar panels **Chapter | 12** **239**

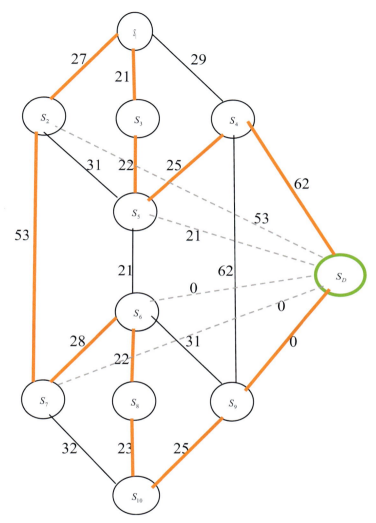

FIGURE 12.12 Optimal solution.

12.5.2.5 Adjusted optimal solution of the formulated linear program

$Z_o = 308,$
$x_{12} = x_{13} = x_{27} = x_{35} = x_{45} = x_{49} = x_{67} = x_{68} = x_{8(10)} = x_{9(10)} = 1,$
$x_{14} = x_{25} = x_{56} = x_{69} = x_{7(10)} = 0.$

(12.12)

FIGURE 12.13 Optimal solution in terms of the original problem.

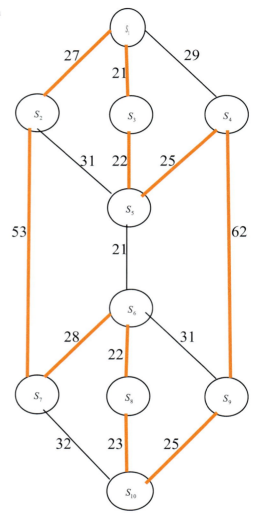

12.6 Other applications of the traveling salesman

In addition to minimal allocation or optimization of energy resources, the traveling salesman model has other very important practical applications.

12.6.1 Wiring problem

The total amount of wire must be kept minimal when wiring a computer backboard. The wire that is used to wire computer backboard is expensive and hence there is a need to optimize the total amount of wire to be used. Besides the length being minimal, there is also a need to minimize the computing time. The wiring problem can be formulated as a QAP and its optimal solution is the minimal amount of wire that can be used so as to save costs and also to improve the computing time.

12.6.2 Hospital layout

When designing and planning a hospital layout, there is a need to consider some important factors. The important factors that can be considered include the patients, X-ray room, clinics, hospital staff, drug store, emergency room, etc. The objective in designing the hospital layout is to determine the optimal distance traveled by patients and medical staff in treating patience. The problem of hospital layout can be formulated as a QAP.

12.6.3 Dartboard design

A dart game is a game whereby two or more competitors aim sharp missiles at a round target or dartboard. The game of darts is interesting and the points to be earned by players decrease as we move out of the center of the dartboard. The idea is to aim at the center of the board where there are higher points. Dartboard design is a QAP when mathematically modeled.

12.6.4 Designing a typewriter keyboard

Smart phones and tablets are now in fashion these modern days. This improvement in technology has resulted in most users losing interest and abandoning the conventional keyboards. Most people are now opting for the convenient and advanced virtual keyboards to capture data or text. The only challenge with this modern technology is how to select the virtual keyboard which is best suited for their devices. This problem of modern technology and virtual technology matching can be formulated as a quadratic assignment problem.

12.6.5 Production

The objective of a production process is to optimize the product costs, time-constraint penalties, and sum of product-dependent changeover costs, production costs, and time-constraint penalties. This problem can be modeled as a quadratic assignment problem.

12.6.6 Scheduling

Timetabling of classes at colleges, universities, and other institutions of higher learning is done in a way to minimize clashes and the necessary convenience to the students. After attending one class, students require ample time to prepare for the next class. Even lecturers need space to prepare another lecture after delivering one. At hospitals, airports, railway organizations, bus companies, and other organizations, scheduling is a very important part of their planning. These scheduling problems can be formulated as QAP.

12.7 Conclusions

The TSP is very difficult to solve but has so many applications which include the allocation of solar energy resources. The chapter proposed an efficient approach to solve the TSP model which is formulated when allocating energy resources. In solving the TSP model, dummies are introduced to eliminate subtours. The proposed approach and formulation for the traveling salesman presented in this chapter will give other researchers in this area a new avenue to explore the difficult TSP model.

References

[1] S. Mhaka, R. Runganga, D.T. Nyagweta, N. Kaseke, S. Mishi, Impact of rural and urban electricity access on economic growth in Zimbabwe, Int. J. Energy Econ. Pol. 10 (6) (2020) 427—434.

[2] D.L. Applegate, R.E. Bixby, V. Chvatal, W.L. Cook, The Traveling Salesman Problem: A Computational Study, Princeton University Press, 2006.

[3] P. Baniasadi, M. Foumani, K. Smith-Miles, V. Ejov, A transformation technique for the clustered generalized traveling salesman problem with applications to logistics, Eur. J. Oper. Res. 285 (2) (2020) 444—457.

[4] T. Bektaş, L. Gouveia, Requiem for the Miller—Tucker—Zemlin subtour elimination constraints? Eur. J. Oper. Res. (2014) 820—832.

[5] P. Berman, M. Karpinski, 8/7 - approximation algorithm for (1,2) - TSP, in: Proc. 17th ACM-SIAM SODA conference, 2006, pp. 641—648.

[6] G. Gutin, A.P. Punnen, The Traveling Salesman Problem and Its Variants, Springer, 2006.

[7] J.F. Mitchell, Branch and cut algorithms for integer programming, in: A.F. Christodous, P.M. Pardalos (Eds.), Encyclopedia of Optimization, Kluwer Academic Publishers, 2001.

[8] D. Nadef, Polyhedral theory and branch and cut algorithms for the symmetric TSP, in: G. Gutin, A. Punnen (Eds.), The Traveling Salesman Problem and Its Variations, Kluwer, Dorcdrecht, 2002, pp. 29—116.

[9] M. Padberg, G. Rinaldi, A branch and cut algorithm for the resolution of large-scale symmetric traveling salesman problems, SIAM Rev. 33 (1991) 60—100.

[10] C.H. Papadimitriou, The Euclidean traveling salesman problem is NP-complete, Theor. Comput. Sci. 4 (1977) 237—244.

[11] W.L. Winston, Operations Research: Applications and Algorithms, fourth ed., Thomson Brooks/Cole, 2004.

[12] G. Campuzano, C. Obreque, M. Aguayo, Accelerating the Miller-Tucker-Zemlin model for the asymmetric traveling salesman problem, Expert Syst. Appl. 148 (2020) 113229.

[13] S. Ebadinezhad, DEACO: adopting dynamic evaporation strategy to enhance ACO algorithm for the traveling salesman problem, Eng. Appl. Artif. Intell. 92 (2020) 103649.

[14] S. Hougardy, F. Zaiser, X. Zhong, The approximation ratio of the 2-opt heuristic for the metric traveling salesman problem, Oper. Res. Lett. 48 (4) (2020) 401—404.

[15] E. Munapo, Network Reconstruction — A New Approach to the Traveling Salesman Problem and Complexity, Intelligent Computing and Optimization Proceedings of the 2nd International Conference on Intelligent Computing and Optimization 2019 (ICO 2019) 123, 2020.

[16] K.R. Rebman, Total unimodularity and the transportation problem: a generalization, Lin. Algebra Appl. 8 (1974) 11—24.

Further reading

[1] L.A. Wolsey, Heuristics analysis, linear programming and branch and bound, Math. Program. Study 13 (1980) 121—134.

Chapter 13

Review on recent implementations of multiobjective and multilevel optimization in sustainable energy economics

Timothy Ganesan[1], Igor Litvinchev[2], Jose Antonio Marmolejo-Saucedo[3], J. Joshua Thomas[4] and Pandian Vasant[5]

[1]*University of Calgary, Alberta, Canada;* [2]*Neuvo Leon State University (UANL), Neuvo Leon, Mexico;* [3]*Universidad Panamericana, Facultad de Ingeniería, Mexico City, Mexico;* [4]*UOW Malaysia KDU Penang University College, George Town, Pulau Pinang, Malaysia;* [5]*MERLIN Research Centre, TDTU, Ho Chi Minh City, Vietnam*

13.1 Introduction

Advanced techniques in computational intelligence are increasingly being utilized to solve industrial-scale optimization problems. The primary driver of these developments is the spike in complexity of optimization problems—especially those encountered in the real world. The idea of "complexity" used here encompasses the following notions: nonlinearity, nonconvexity, multiple objective functions, large-scale (multivariate), and formulations with multiple levels of objectives. Faced with such challenging scenarios, various sophisticated computational innovations have been utilized to effectively and efficiently resolve these problems. Recent review of such problems and their solution methods could be seen in Refs. [1−5]. A general definition of an MO optimization problem (considered as minimization) is as follows:

$$\text{Min}_i \rightarrow f_i\left(x_j, \ldots, x_J\right), f_{i+1}\left(x_j, \ldots, x_J\right), \ldots, f_I\left(x_j, \ldots, x_J\right)$$

such that

Advances of Artificial Intelligence in a Green Energy Environment
https://doi.org/10.1016/B978-0-323-89785-3.00013-X
Copyright © 2022 Elsevier Inc. All rights reserved.

245

246 Advances of Artificial Intelligence in a Green Energy Environment

$$g_k\left(x_j, .., x_J\right) \le 0$$
$$g_k\left(x_{j+1}, .., x_{J+1}\right) \le 0$$
$$g_{k+1}\left(x_j, .., x_J\right) \le 0 \qquad\qquad (13.1)$$
$$\dots$$
$$g_K\left(x_j, .., x_J\right) \le 0$$
$$i \in [1, I], j \in [1, J], \ k \in [1, K]$$

where f denotes the objective function, x represents the decision variables, and g is the constraints, i is the index for the objectives function, j is the index for the decision variables, and k is the index of the number of constraints. The maximum number of objective functions, decision variables, and constraints are represented by I, J, and K, respectively.

In multiobjective (MO) optimization, real-world problems in energy management systems have been recently encountered. For instance, in Ref. [6], a novel (multiobjective) real-time control strategy was developed for battery/ultracapacitor hybrid energy management systems. The control strategy was targeted toward minimizing power loss, enhancing battery life cycle, enhancing the stability of the ultracapacitor's stable terminal voltage, and improving its ability to speedily charge/discharge. Transforming the problem into a uni-objective convex framework, they applied the Karush-Kuhn-Tucker conditions to solve it. Another interesting work in energy systems could be seen in Ref. [7]; a hybrid modular framework was employed for electrical power system forecasting. The authors used an MO dragonfly algorithm to optimize the Elman neural network model. Upon validation with empirical data, it was found that the developed approach was an effective tool for planning and managing power grids. Another MO application to energy systems is given in Ref. [8]. In that work, two scenarios were considered: energy arbitrage-peak shaving and energy arbitrage. To solve these problems: a mixed-integer linear programming (MILP) optimization model was developed to minimize operational costs and carbon dioxide emissions of a community situated in Cernier, Switzerland. Rigorous analyses were conducted to evaluate the performance of both scenarios in terms of the optimization outcome. In Ref. [9], a wind power application is considered where wind speed forecasting models are developed. Due to the inadequacy of traditional forecasting methods, where key features arising from randomness and intermittency are not captured, the authors develop a novel hybrid forecasting system that modularly combines preprocessing, optimization, and forecasting. To accomplish this, they used a fuzzy time series method which was optimized by an MO differential evolution (DE) method. This approach balances the conflicts between stability and forecasting accuracy—which was validated using experimental results taken at various time intervals (where the mean absolute error was below 4%). Other works involving MO applications in energy systems are given in Refs. [10−19].

Besides energy systems, problems in communications and networks have also become a critical platform for implementing MO techniques. In Ref. [20], a novel congestion control algorithm was proposed for wireless sensor networks. Due to limitations in conventional methods (Poisson process), the authors employ an MO optimization hybrid swarm algorithm for rate optimization and regulating the arrival rate of data from child node to the parent node. The proposed algorithm was compared against cuckoo search (CS) and adaptive cuckoo search methods, where the proposed algorithm produced superior results. In Ref. [21], fog computing technology was considered for MO optimization—with this technology, cloud services can be extended to the edge of the network to decrease congestion of the network. In that work, a two-level resource scheduling scheme was proposed based on the theory of the improved nondominated sorting genetic algorithm II (NSGA-II). Numerical simulations showed that the proposed scheme could reduce service latency and improve the stability of task execution. Similarly, in Ref. [22], the spectrum allocation problem is addressed specifically for the connectivity demands of the Internet of Things (IoT). The optimization problem was targeted toward maximizing the end-to-end throughput while maximizing transmission to maximize the spectrum utilization. The proposed solution method to solve the mentioned MO allocation problem was NSGA-II. Simulation results were used to validate the efficiency of the proposed method in terms of efficiency in searching for optimal solutions as well as satisfying the requirements of the spectrum allocation for diverse cases.

In material and structural engineering, MO techniques have been used to optimize various processes. This can be seen in the work of Ref. [23]. In that work, the authors integrated the Taguchi method, grey relational analysis (GRA), and NSGA-II to optimize the milling process. This was done by reducing the energy consumption (under nanofluid assisted machining) while simultaneously considering material removal quantity as well as the machining product quality. The primary goal of this work was to achieve a sustainable manufacturing process. Another example of MO applied in material and structural engineering is seen in Ref. [24]. In that research work, the authors considered a novel optimization framework combining MO optimization and multicriteria decision-making. The novel solution method combining multiobjective artificial bee colony (MOABC), best worst (BW) method, GRA, and visekriterijumsko kompromisno rangiranje (VIKOR) was developed. The primary goal of the work was to optimize the energy-absorbing structures in train collisions. To empirically verify the solution method, a multicell thin-walled aluminum energy-absorbing structure was developed and experimentally tested. Another MO implementation in material engineering is given in Ref. [25]. In that work, the authors investigate the turning operations using a metamodel framework. Utilizing seven metamodels, the work focused on maximizing the material removal rate and minimizing the interface temperature as well as tool wear depth (obtained from FE simulations using

248 Advances of Artificial Intelligence in a Green Energy Environment

DEFORM-2D software). Recent developments on MO optimization in material engineering could be seen in Refs. [2,25−30]. MO optimization has also been applied broadly across various industries such as transportation [31−33], bioprocess engineering [34,35], water engineering [36−38], portfolio management [39,40], supply chains [41], scheduling [42,43], path planning [44,45], and risk and reliability [46,47].

In addition to MO optimizations, industrial problems have also taken the shape of multilevel (or bilevel [BL]) systems—where the optimization problem(s) are embedded within another [48]. The mathematical representation of the problem formulation would be as follows:

$$\text{Min}_i \rightarrow f_i\left(x_j, \ldots, x_J, y_m\ldots, y_M\right), f_{i+1}\left(x_j, \ldots, x_J, y_m\ldots, y_M\right), \ldots,$$
$$f_I\left(x_j, \ldots, x_J, y_m\ldots, y_M\right)$$

such that

$$g_k\left(x_j, \ldots, x_J, y_m\ldots, y_M\right) \leq 0$$
$$g_{k+1}\left(x_j, \ldots, x_J\right) \leq 0$$
$$\ldots$$
$$g_{K-1}\left(x_j, \ldots, x_J, y_m\ldots, y_M\right) \leq 0$$
$$g_K\left(x_j, \ldots, x_J\right) \leq 0$$

such that

$$\text{Min}_n \rightarrow f'_n(y_m\ldots, y_M), f'_{n+1}(y_m\ldots, y_M), \ldots, f'_N(y_m\ldots, y_M)$$
$$g'_p(y_m\ldots, y_M) \leq 0$$
$$g'_{p+1}(y_m\ldots, y_M) \leq 0$$
$$\ldots$$
$$g'_P(y_m\ldots, y_M) \leq 0$$

$$i \in [1, I], \ j \in [1, J], \ k \in [1, K], \ p \in [1, P], m \in [1, M], n \in [1, N]$$

(13.2)

where f' denotes the objective function for the lower level problem, y represents the decision variables, and g' is the constraints for the lower level problem. n is the index for the number of objectives, m is the index for the number of decision variables, and p is the index of the number of constraints for the lower level problem. The maximum number of objective functions, decision variables, and constraints for the lower level problem are represented by N, M, and P, respectively. Since BL optimization problems are very frequently viewed in the context of a Stackelberg game (a form of game theory), the upper level problem is often referred to as the leader problem, while the lower level subproblem is called the follower problem.

The field of energy systems has recently been a focus of many BL optimization implementations. For instance, in Ref. [49], the restoration problem for the operation of power systems is considered. This problem was formulated as a BL optimization problem where the optimal start-up sequence of non-black start units was determined for the upper level problem. The lower level problem was solved by determining the optimal transmission path with minimal number of switching and maximum reliability between any two necessary buses. The upper level problem was solved using the teaching-learning—based optimization algorithm while the lower level problem was tackled using a path searching graph-based algorithm. The proposed approach was successfully implemented on IEEE 24 and IEEE I88-bus systems. Another example could be seen in the work of Ref. [50]. In that research work, the authors considered the hybrid renewable energy systems. The BL optimization problem was formulated by integrating component sizing (upper level) with the energy management strategy (lower level)—which is based on an economic model predictive control approach. The upper and lower level problems were solved using genetic algorithms (GAs) and MILP, respectively. The results obtained from the proposed methodology showed reduction in operations costs. Another interesting BL optimization implementation in an energy system is given in Ref. [51]. In that work, a BL optimization model was developed for integrated energy systems with application to building thermal systems, where the upper level problem maximizes the total benefit of the integrated energy system while the lower level problem was set to minimize the heating bills of residents. The authors employed a MILP approach in tandem with the Karush-Kuhn-Tucker optimality condition to solve the problem. In energy allocation, the recent work of Ref. [52,53] provides some insights on solving such problems using BL optimization. In their formulation, the upper level problem targets flexibility and maximizes net revenue while the lower level minimizes the deviation between the level 1 target and the achievable solution ensuring no violation in physical and operational constraints. The solution methods employed were sequential quadratic programming (SQP) and GA with continuous decision variables. Rigorous comparative analysis was carried out on the computational results.

In the field of communication and networks, a BL framework was employed in the work of Cheng et al. [54] for optimizing the utilization of Multi-Mobile Edge Computing (MEC) server system. Their aim was to achieve energy-efficient offloading and an optimized resource allocation strategy. The proposed technique was validated via simulations where its performance was compared with conventional schemes. In Ref. [55], a BL programming model was developed for Cloud Manufacturing Services based on Extension Theory. The model was formulated such that the upper level problem focused on the service demanders (consumers) while the lower level problem considered the service providers. The problem was then solved using linear programming (and extension evaluation was applied) where the outcome of the model was validated with experimental results.

250 Advances of Artificial Intelligence in a Green Energy Environment

In civil engineering, a BL MO framework was employed to solve the evacuation location assignment problem [56]. The aim of that work was to determine the best location for placement of shelters such that the traffic assignment of the evacuees on the existing network is optimized. The authors of that work used a mixed solution method to solve the problem—where the Karush-Kuhn-Tucker was used to reduce the problem to a single level. Then the MILP model was solved using an off-the-shelf solver. Similarly, in Ref. [57], a MO BL programming model was formulated to optimize traffic signal timings. In that work, the objective functions were to (1) maximize the reserve capacity of a road network and (2) minimize vehicle emissions (to achieve environmentally friendly signal timings). In Ref. [57], the upper level problem was tackled using DE and the weighted sum approach, while the lower level formulation was defined as a stochastic traffic assignment problem. Another excellent implementation of BL optimization in civil engineering could be seen in the work of Ref. [58]. There the authors employed a network-wide BL optimization simulation approach for variable speed limit systems—aiming to improve travel time reliability. The central idea was to maximize travel time reliability on certain critical paths on a network. In that work, the upper level problem focused on travel time optimization while the lower level problem was designed to assign traffic to the network using a dynamic traffic assignment simulation tool. The computational technique used to solve this problem was simulated annealing (SA) [59]; the proposed method was applied to a real roadway network.

BL optimization methods have also made their way to industrial chemical process and design applications. This can be seen in the work of Ref. [60], where a BL optimization framework was used to address the overproduction of biochemicals (succinate in *Escherichia coli*) problem. The developed formulation was in a MILP form. The work of Ref. [60] provides insights on the computational hurdles faced during strain design and other tested methodologies. Another application in chemical engineering is given in the research work of Ref. [61]. In that work, the authors combined ideas from BL optimization, direct dynamic optimization, and penalization schemes for the production of recombinant erythropoietin in *Pichia pastoris* growing on glucose—biopharmaceutical engineering. In Ref. [62], bioprocess productivity was optimized using metabolic-genetic network models with BL dynamic programming. Utilizing BL programming, the authors of that work obtained an optimal strategy for dynamic genetic and process level manipulations for enhancing bioprocess productivity. The developed computational methodology was then applied to maximize ethanol productivity.

Another interesting application of BL programming could be observed in the field of manufacturing. For instance, in Ref. [63], BL optimization was applied to production planning for multiple products with short life cycles. In that work, decision-making models were developed for a manufacturer for obtaining optimal production quantities based on the most relevant scenarios.

Besides swarm-based metaheuristics, GAs have also been utilized for MO optimization of economic dispatch systems. One example of this could be seen in the work of Ref. [79]. In that research, the MO environmental economic dispatch problem was solved using nondominated sorting genetic algorithm (NSGA-II). The application problem involved a 30-bus network with wind power sources where the fuel cost and pollutant emission objectives were considered. The NSGA-II was improved by adding a parameter-free self-tuning reinforcement learning mechanism. The proposed method was shown to be more effective as compared to the classical NSGA-II technique in terms of solution quality and saved time (spent on parameter tuning). Another application of GA to the MO EED problem is given in the work of Ref. [80]. In that work, the authors used a NSGA-II technique in combination with a modified version of the marginal histogram model (estimation of distribution algorithm). The implementation on the MO economic emission power dispatch problem focused on improving the convergence behavior while maintaining the diversity properties of the obtained set of solutions. The nondomination and elitism-based mechanisms were included in the marginal histogram model to handle the multiple target objectives. Computational experiments showed that the proposed method improved convergence (and had good solution diversity) as compared to other conventional MO evolutionary algorithms.

In addition to GA variants, DE has also been recently employed to solve economic dispatch-related problems. This is observed in the work of Ref. [81], where an MO DE was applied to an EED problem in a Chinese manufacturing environment. The problem contained conflicting fuel costs and pollution emission as the target objectives. In that work, two mutation strategies were implemented with the aim of improving the performance of the conventional DE technique. The performance of the proposed method was benchmarked on six test functions and compared with other MO evolutionary algorithms. In Ref. [82], the MO neural network trained with the DE method was used to solve the dynamic economic dispatch problem. Rigorous comparative analysis was carried out in that work, where the proposed method was compared with other current techniques with respect to 24-h cost and emissions. The work also compared the performance of a fully and partially connected networks—showing that dynamic optimization of the topology of neural network improved its performance as compared to optimizing the network weights. In Ref. [83], the MO DE was employed to solve the dynamic economic dispatch problem which incorporates a wind power plant. The problem formulation was highly constrained and nonlinear with two objectives: fuel costs and pollution emission. The optimization formulation accounted for valve point effect, spinning reserve, real power loss, and the ramping rate. The authors used real-time output adjustment and penalty factor methods for constraint handling. Summation-based DE and decomposition-based MO evolutionary algorithms were used in the work of Refs. [52,53] to solve an

254 Advances of Artificial Intelligence in a Green Energy Environment

economic-environmental power dispatch problem (which was dual objective and nonlinear in nature). The problem was solved while accounting for network security constraints (such as transmission line capacities and bus voltage limits), generator capabilities, and prohibited operating zones for the thermal units. In that work, the simulation results obtained using both techniques were analyzed and compared. Other current applications of evolution-based techniques in MO emission dispatch problems are seen in Refs. [84−87]. A more comprehensive review of MO optimization in EED applications is given in Refs. [88,89]. Table 13.1 provides a summary of the application, technique, and the respective reference of some of the recent works in MO optimization in economic load dispatch:

MO optimization has been applied extensively to load dispatch problems for the past 15−20 years. The same could not be said for BL optimization in such applications. BL optimization has seen many applications in the field of environmental economic load dispatch in the recent 5 years. For instance, in Ref. [90], the MO BL environmental economic power generation and dispatch problem was solved using GA and fuzzy goal programming. In that work, the GA was executed in two stages where the objectives were directly optimized, and the goal achievement function was evaluated, respectively, to reach an optimal power generating decision. The proposed methodology was tested in a IEEE 30-bus system. Similarly, in Ref. [91], a BL MO environmental economic dispatch model with wind farms was developed based on the decomposition coordination principle and users' side resources. The MO model was developed to determine the fitting load of the system and the time-sharing price—to attain the energy generation plan. To achieve this, the authors employed a hybrid universal gravitation algorithm and the taboo search technique. The computational outcome showed that the following criteria were achieved:

- Improved load level of the system operation.
- Satisfaction of using electricity and absorptive capacity of wind power.
- Reduce generation costs and pollution emission.

Another wind power application utilizing BL optimization is seen in the work of Ref. [92]. In that research, a BL dynamic EED model was proposed, which takes into consideration wind power integration. In that BL model, the upper level objective functions minimized fuel cost and emission simultaneously while the lower level objective function seeks to minimize the interval reduction of wind power output. The solution method utilized in this work was nested approach of the teaching-learning—based optimization and linear programming. The computational methodology was then applied to three cases with different ratios of wind power and their efficiency and feasibility were evaluated. In the work of Ref. [93], a BL energy-saving dispatch in a smart grid was considered to enhance the energy consumption and carbon dioxide emission reduction. The objective function of the upper level consists of the

Recent implementations of MO optimization Chapter | 13 **255**

TABLE 13.1 Recent multiobjective (MO) optimization applications in economic load dispatch.

References	Application	Technique
[67]	Combined economic emission dispatch (EED) for distributed energy resources sizing	Whale optimization algorithm
[68]	Combined EED	Interior search algorithm
[69]	Combined EED	Chaotic self-adaptive interior search algorithm
[71]	Optimal power flow problem for wind power-enhanced power system	Grey wolf optimizer
[72]	EED	Fast nondominated time-varying acceleration coefficient PSO
[80]	EED	Hybrid NSGA-II—estimation of distribution algorithm
[79]	EED	NSGA-II with self-tuning mechanism
[73]	Economic load dispatch	Mine blast algorithm
[81]	EED	MO differential evolution
[82]	Dynamic economic dispatch	MO neural network—differential evolution
[83]	Dynamic EED	MO differential evolution
[52]	EED	Summation-based DE and decomposition-based evolutionary algorithm
[78]	Economic load dispatch	MO cuckoo search
[118]	EED	MO collective decision optimization
[74]	Economic load dispatch	MO bee colony optimization
[77]	Economic load dispatch	Cuckoo search algorithm
[76]	Environmental economic load dispatch	Grey wolf optimization
[86]	MO economic dispatch in cascaded hydropower systems	Improved Pareto GA
[75]	MO dynamic EED	MOPSO
[85]	Combined heat and power EED	Theta-dominance—based evolutionary algorithm
[84]	Wind-thermal dynamic EED	Hybrid MO DE-PSO

256 Advances of Artificial Intelligence in a Green Energy Environment

electric power generation and the carbon emission costs of thermal units. The objective function of the lower level problem includes the compensation and incentive costs of electricity consumers according to the influences of their regulation-based demand response to the power grid. The authors employed an iterative algorithm as well as the NSGA-II technique as the solution methods.

An interesting example is given in the work of Ref. [94] where an energy optimal dispatching model of an integrated energy system is constructed based on uncertain BL programming. In that work, the upper level objective takes the transformation matrix of the energy hub as the decision-maker (minimizing operating cost) while the lower level objective accounts for operational economy of each subnetwork with the operations as necessary constraints. The authors used a firefly algorithm with chaotic search as a solution method. The computational results showed the effectiveness of the energy optimal dispatching model and the economy of the system operation under multiple decision-maker hierarchy. In that spirit, the work of Ref. [95] presents a low-carbon BL optimization framework for an industrial park while accounting for multi-energy price incentives. In their research, the upper level problem consists of the optimal net income of the integrated energy service agency (along with the carbon emission constraints of the real-time unit integrated energy supply). On the other hand, the lower level problem takes the minimum integrated energy cost to the multi-energy users as the target objective. The coordinated interplay between the upper and lower level objectives optimizes the price of energy sold and the energy utilized by the integrated demand response. The interior point method was employed to solve the multidimensional, nonlinear, and BL optimization model—and the obtained results were observed to improve the mentioned net income while ensuring the environmental protection. A similar application tackled in Ref. [95] was solved by Ref. [96]: BL optimal dispatch strategy for a multi-energy system of industrial parks. In Ref. [96], the proposed BL optimal dispatch strategy was proposed to decrease the difficulty in coordinated control and interaction between lower level factories and upper level energy operators in industrial parks. The central thesis of this work was to obtain a strategy that maximizes the benefits to users and multi-energy operators. The authors used the binary MILP as the solution method. The computational experiments showed that the optimal solutions could be achieved under peak shifting constraint and there was a reduction in the overall peak power value of the industrial park.

In Ref. [97], a BL optimization framework was applied for economic dispatch of a power generation system, with multiple energy sources. In that work, an optimal economic dispatch model for balancing power and heat demand was proposed. The power system in question incorporated combined heat and power units: wind energy, solar energy, and a battery energy storage system. The BL optimization model accounted for the uncertain nature of the wind and solar (PV) energy outputs and the storage constraints. The solution technique utilized was the hybrid PSO and SQP approach, which was used to determine the optimal power supplied by the individual energy sources. The proposed model was simulated and tested on the modified IEEE 24 and IEEE 118-bus systems. The experiments as well as the comparative studies showed that the methodology provided a powerful solution for obtaining the optimal power dispatch. A different take on applying BL optimization for sustainable energy dispatch is seen in the work of Ref. [98]. In that work, an electric vehicle (EV) charging demand with wind energy was represented in a BL energy dispatch framework. The primary hypothesis of that work was to reduce greenhouse gases (GHG) and driving cost by strategic coordination between EV charging demand and wind supply. Toward this goal, the authors proposed BL optimization method where the Markov decision process (MDP) was utilized as a working methodology. The authors also leveraged in the aggregation relationship to develop a BL simulation-based policy improvement method for a large number of EVs. The proposed methods were validated using numerical experimentation. A combination of BL optimization and risk assessment was carried out in Ref. [99]. In that work, a BL optimization dispatch model was proposed to obtain coordinated dispatch between the upper and lower level objective functions. The upper level objective was to minimize the system's operational cost while the lower level objective was to maximize the virtual power plant's net profit. The BL optimization model consists of a virtual power plant and an independent system operator. Uncertainties in the system were unit outage, load forecast deviation, and renewable energy forecast error. These uncertainties contribute to the overall risk, where they are evaluated using the metric: expected energy not supplied. The solution method used in that work was the chance-constrained programming. The BL model was aimed to maintain a balance between the potential risk and economic profit. A similar application problem was tackled in work of Ref. [100]. In that work, the BL paradigm was utilized to formulate a balance between the two target objectives: independent system operator (upper level) and load aggregator (lower level). As done in Ref. [99], the chance-constrained programming approach was used to tackle the proposed BL problem. Using this approach, the authors were able to make risk-benefit coordinated decisions while considering uncertainties on the bidding information (basic) as well as power supply-demand prediction. Simulations were conducted validate the BL optimized strategy—the results proved that the optimized strategy was effective for attaining satisfactory decision schemes. A tabular representation of BL optimization in economic load dispatch applications discussed in this section is given in Table 13.2:

13.3 Bioenergy and biofuel supply chains

Bioenergy supply chains have also been seen to have recent implementations within an MO framework. In Ref. [101], the MO optimization of a biomass-based renewable jet fuel supply chain system was carried out. In that work,

TABLE 13.2 Recent bilevel (BL) optimization applications in economic load dispatch.

References	Application	Technique
[90]	Environmental economic power generation	GA and fuzzy goal programming
[91]	Environmental economic dispatch with wind farm	Hybrid universal gravitation algorithm and taboo search
[92]	Dynamic EED	TLBO and LP
[93]	Energy-saving dispatch on a smart grid	Iterative algorithm and NSGA-II
[94]	Energy dispatch of integrated energy system	Firefly algorithm with chaotic search
[95]	Low-carbon energy dispatch	Interior point method
[96]	Dispatch strategy for multi-energy system	Mixed-integer linear programming
[97]	Optimal economic dispatch with multiple energy sources	Hybrid PSO and SQP algorithm
[98]	BL energy dispatch for electric vehicle charging	BL Markov decision process
[99]	Dispatch of virtual plants	Chance-constrained programming
[100]	Dispatch of electric supply demand	Chance-constrained programming
[143]	Load supply capability in an active distribution network	

a MILP model was employed to account for the spatial, agricultural, techno-economic, and environmental data for the supply chain system. The model was applied to evaluate the sustainability performance of three pathways: alcohol-to-jet, Fischer—Tropsch [102], and hydrothermal liquefaction processes. The research showed that the Fischer—Tropsch supply chain pathway was the most sustainable in terms of cost and GHG emissions.

In Ref. [103], an MO bioenergy supply chain was considered while accounting for social benefits. The following aspects were included in the case study: job creation, biomass purchase and impacts of economic activities in different locations. In that work, the MO MILP model was formulated where the epsilon-constraint technique was employed as a solution method for the biobjective problem. The computational results showed that the bioethanol supply chain could foster economic and social benefits in a given area—the location of the case study was the Liaoning Province in Northeast China.

Similarly, in Ref. [104], the MO mixed-integer model was proposed for the optimal design and planning of a lignocellulosic bioethanol supply chain. The model accounted for economic, environmental, and social considerations—represented as target objective functions. The utility of the proposed model was to enable strategic decisions, biomass sourcing and allocation, locations, capacity levels, and determining technology types for biorefinery facilities. Additionally, the model could also be employed to aid in making tactical decisions such as inventory levels, production amounts, and shipments among the network. For efficient handling of uncertainty, a robust possibilistic programming approach was developed and applied to a real-world case study. The solutions obtained were shown gauged using performance measures and compared against solutions achieved using deterministic techniques.

In Ref. [105], an MO optimization model was formulated to redesign a global bioenergy supply network. In that work, the authors formulated an MO model which incorporates the economic, environmental, and social objectives. The model was solved using the epsilon-constrained method—where the solutions were used to construct the Pareto frontier. The model was validated using real-world data and the geographical maps of the countries considered (Iran and Armenia) were studied closely using the geographic information system (GIS) software. Another interesting application of MO optimization in sustainable supply chains is observed in the work of Ref. [106]. In that work, the MO optimization of a biogas supply network for electricity generation was considered. The MO model was formulated using MILP utilizing a four-layer biogas supply chain. Three objectives were taken into account in that work: (1) maximization of economic profit, (2) maximization of economic profit while considering GHG emissions, and (3) maximization of sustainability profit. The authors went further to consider various scenarios in their model, e.g.:

- Scenario where carbon prices were incremented (steadily) to the prevalent ecological costs/benefits of global warming.
- Scenario where all the electricity auction prices were multiplied by certain factors to search for the profitability break-even factor.
- Scenario where shorter time periods were implemented, and investment cost of biogas storage was reduced.
- Scenario where the capacity of the biogas plant was varied from 1 to 5 MW.

The MO models were applied to the real-world case study of the Slovenian biogas plants and the effects of the correlations between the economic, economic (with GHG), and sustainability profits on the supply chain and its associated effects to decision-making were obtained. In Ref. [107], the MO optimization of a food waste to bioethanol supply chain was carried out. The MO model in that work considered the site locations and the amount of food wastes collected as well as the transportation plans of bioethanol to demand points. The model also factored in the number, sizes, and locations of the

260 Advances of Artificial Intelligence in a Green Energy Environment

biorefineries. Considering the economic, social, and environmental target objectives, the real-world implementation of the proposed MO framework was in Egypt. In Ref. [108], a spatiotemporal MO optimization model was developed for biomass value chains. The objectives were to maximize synergies and minimize conflicts within the food-energy-water-environment nexus. Utilizing the MILP approach, the model considered planning horizons (temporal), facilities, and transport locations as well as land use models (spatial). The model incorporated various types of biomass—food crops, energy crops, and forestry resources. The work discusses various strategic sustainable decision-making advantages which could be attained by using the proposed MO model. The work of Ref. [109] was focused on the MO optimization model for a biofuel supply chain. In that work, the biobjective problem was considered: maximization of resource utility and the minimization of environmental pollution (caused by straw burning). Unlike many other works in MO optimization of sustainable supply chains, this work used the two-stage fuzzy approach to transform the MO problem into a single objective before solving it. The optimization results showed that the solutions that balance the trade-off between both target objectives were attained, and the correct utilization of the model could significantly support managers and decision-makers in the design of biofuel supply chains.

MILP has also been used hand in hand with the hub-and-spoke structure to design and manage biofuel supply chains [110]. In Ref. [110], the proposed MO model aimed to optimize carbon dioxide emissions and social impacts (job creation potential). Upon application of the optimization techniques, a set of Pareto optimal solutions were obtained. The case study considered in that work was the Midwest region of the United States—where the computational estimates of the quantity and cost of cellulosic ethanol delivered under various scenarios were produced. In Ref. [111], an MO biofuel supply chain (which utilizes forest and wood residues to produce bioenergy and biofuels) was optimized. These biofuel/bioenergy resources generate additional revenue streams for forestry companies and generates new development opportunity for forest-dependent communities. The MO biofuel supply chain model was developed and implemented in that work using MILP; which maximizes social benefit, net present value, and GHG emission savings potential. The case study considered in that work was based in the province of British Columbia, Canada—where different utilization paths for available forest and wood residues are investigated. Using the Pareto curve constructed using the generated solutions to the MO problem, a positive correlation between the social benefit (job creation) and savings from GHG was uncovered. A very interesting solution method was implemented in a research thesis investigating biorefinery supply chains; see Ref. [112]. In that work, a model encompassing of multiple layers was constructed—economic, political, technological, social, and environmental. The modeling work utilized a mixed-integer programming framework and the solution method used was the evolutionary algorithm:

NSGA-II. The optimization framework discussed in that thesis was applied to a case study involving the deployment of a biorefinery in Columbia. Various scenarios were proposed by the methodology and sustainability trade-off solutions were evaluated by integrating stakeholder preferences. Besides, biorefinery supply chains and straw-to-electricity supply chains [113] have also been studied on an MO perspective, as seen in the work of Ref. [114]. In that research, the MO straw-to-electricity supply chain which uses agriculture waste products was optimized. The work considered two aspects of the supply chain problem: MO optimization and system dynamics simulation. Using the normalized normal constraint method to tackle the MO part of the integer programming model, the Pareto frontier was constructed. In that work the case study considered was from a key bioenergy plant located in the Cangxi County, Guangyuan City, Sichuan Province, southwest of China. Detailed analysis in terms of the sustainability policies as well as the environmental and economic aspects of the mentioned supply chain was performed by the authors. In Ref. [115], the MO optimization of a bioethanol and bioelectricity supply chain was carried out. Utilizing a MILP framework, the multi-echelon supply chain was simultaneously optimized with respect to economic performance (net present value) and environmental considerations (GHG emissions). The bioethanol and bioelectricity supply chains are assessed accounting for multiple feedstocks—Miscanthus, stover, Arundo donax, corn, poplar, and wood residues. The authors of that work also performed sensitivity analysis on the economic and environmental parameters in addition to gauging the effectiveness of the model, in providing decision-makers with optimal supply chain design configurations. A thorough review of recent biomass supply chain models and optimization which also contain MO scenarios is given in Ref. [116]. Table 13.3 provides a brief version of the previously discussed literature studies on the recent developments in MO optimization in bioenergy and biofuel supply chains.

On the other hand, biomass-related supply chain optimization has also been recently explored in the light of multilevel optimization. This could be observed in the work of Ref. [117]. In that work, a BL biomass-coal co-firing method was applied for carbon emissions control, where the hierarchical relationship between the government and multiple power plants was considered. The authors in Ref. [117] transformed the BL problem into a single level using the Karush-Kuhn-Tucker approach. A case study on the Heilongjiang Province was used in that work where the proposed optimization method was verified and tested. The results showed that the power plants having lower biomass utilization were more sensitive to carbon emissions constraints. On the other hand, the authors also found that the relaxed carbon constraints promoted the use of biomass while strict carbon constrains hampered economic development.

Another similar study involving BL optimization on biofuel/bioenergy supply chains could be observed in the work of Ref. [118]. In that study, the

262 Advances of Artificial Intelligence in a Green Energy Environment

TABLE 13.3 Recent multiobjective (MO) optimization applications in bio-energy and biofuel supply chains.

References	Application	Modeling and optimization techniques
[101]	Biomass-based renewable jet fuel supply chain system	Mixed-integer linear programming technique
[103]	Bioenergy supply chain	Mixed-integer linear programming and epsilon constraint methods
[104]	Sustainable bioethanol supply chain design	MO robust possibilistic programming
[105]	Global bioenergy supply network design	Epsilon-constrained method
[106]	Four-layer biogas supply chain	Mixed-integer linear programming
[107]	Food waste to bioethanol supply chain	Mixed-integer linear programming
[108]	Biomass supply chains	Mixed-integer linear programming
[109]	Biofuel supply chain	Mixed-integer linear programming and two-stage fuzzy method
[110]	Biomass supply chain	Mixed-integer linear programming and epsilon-constraint method
[111]	Forest-based biofuel/bioenergy supply chain	Mixed-integer linear programming and augmented epsilon-constraint method
[112]	Biorefinery supply chain design	Mixed-integer programming and NSGA-II
[114]	Straw-to-electricity supply chain	Integer programming method and normalized normal constraint method
[115]	Bioethanol and bioelectricity supply chain	Mixed-integer linear programming

authors employed a BL programming strategy to build a cooperative relationship between the governing authority and the coal-fired power plants. The MO optimization model (dynamic programming) was formulated and utilized to handle the biomass available time conflict. The MO model was applied to find a balance between economic development and environmental sustainability, similar to the work of Ref. [117]. Additionally, uncertainty in certain variable (e.g., straw prices) was taken into account in Ref. [118] using a fuzzy formulation. The authors then applied and verified the proposed methodology on a case study situated in Jiangsu, China. The linear BL programming model was then solved using a hierarchical (leader—follower) random search optimization framework. The computational results showed the practicality and the efficiency of the proposed methodology.

In Ref. [119], a biofuel supply chain was modeled and optimized using a Stackelberg game theoretic approach. To optimize the BL biofuel supply chain model, the authors utilized a mixed-integer BL programming methodology. Using this method the authors obtained the optimal supply chain design and operational parameters, where the decision-maker has the freedom to make discreet decisions, e.g., technology selection, facility location, and opening/shutting down of product lines. The solution method also comprises of a reformulation and decomposition algorithm [120]. The authors proved that the solution modeling framework as well as the solution methodology successfully optimized the non-cooperative supply chain (in the spirit of Stackelberg game theory). The BL problem was efficiently transformed to a single-level problem and solved using the reformulation and decomposition algorithm.

In the work of Ref. [121], a waste-to-energy incentive policy design was modeled and optimized using Stackelberg game theory. The policy design was targeted toward the dairy sector. The model was developed using the Stackelberg game-based BL mixed-integer fractional programming framework. In the formulation of this scenario, the government takes the role of the leader while the dairy farms play the role of the followers, effectively creating a single-leader-multiple-follower Stackelberg game. The problem was then efficiently solved a tailored global optimization algorithm, which integrates a parametric algorithm and projection-based reformulation and decomposition algorithm. The proposed methodology was implemented in a real-world case study in farms located in New York State, US. The authors of that work showed that the numerical results demonstrated that the incentive policies can effectively promote bioelectricity generation, which could potentially refund the capital investment to a farm much higher than the subsidy on bioelectricity generation.

13.4 Sustainable capacity planning and optimization

Sustainable efforts have also been recently directed toward capacity planning of energy systems in the context of MO optimization. This could be observed in the works of Ref. [122], where a rule-based energy management scheme for long-term optimal capacity planning of a grid-independent microgrid was optimized using nature-inspired approach. The authors of that work employed an MO grasshopper optimization algorithm to solve the MO optimization problem by maximizing the system reliability while minimizing the energy cost. The proposed MO grasshopper optimization approach was then compared with the PSO and cuckoo search algorithm. The authors noted that the rule-based energy management scheme contributed to a cleaner energy production system; reducing fuel consumption (92.4%), emissions (92.3%), and cost of energy (79.8%). In addition, the authors of Ref. [122] also carried out sensitivity analysis to underscore the effect of future uncertainties on the system inputs.

264 Advances of Artificial Intelligence in a Green Energy Environment

Another interesting capacity planning application could be seen in the work of Ref. [123]. In that research work, potential prospects of connecting short-term flexibility value into long-term capacity planning for achieving high renewable energy fraction microgrid was investigated. The authors considered a system involving photovoltaic (PV), wind, and energy storage components as well as several optimal component-sizing scenarios. In consideration of this, an MO problem was constructed, with the aim to minimize the loss of power supply probability index while simultaneously minimizing the total life cycle costs under each scenario. The problem was then solved using the multi-objective particle swarm optimization (MOPSO) algorithm while the demand response program's forecast was done using the scikit-learn package in Python. The computational results showed that the mentioned forecast and optimization resulted in significant cost saving due to reduction in component sizing.

In the work of Ref. [124], the authors employed an MO stochastic expansion planning for a regional integrated energy system with renewable energy sources. The objective of their work was to improve energy efficiency using multi-energy complementarity. Thus the MO stochastic expansion planning was based on multidimensional correlation scenario set generation method. The MO stochastic model for the energy network minimized the investment cost as well as the energy pipeline risk. The authors applied the proposed methodology on a real-world case located in Yangzhong City, China. In addition to comparative analysis, the research work generated Pareto fronts of the optimized capacity planning schemes, successfully obtaining a reference for balancing the energy network planning scheme economy as well as the energy pipeline risk.

The MO optimization of an autonomous hybrid microgrid was carried out in Ref. [125]. In that work, the authors proposed an optimal strategy for performance evaluation of various hybrid microgrid configurations for the Long San Village in Sarawak, Malaysia. The mathematical model for the microgrid was developed for component sizing to meet the maximum load demand under changing weather conditions and at the lowest possible cost. The authors then used the MOPSO technique to seek the exact dynamic energy price of the selected optimal configuration in the context of system uncertainties. The computational results presented insights with respect to microgrid operational feasibility, economic feasibility, and environmental impacts.

MO-based strategies have also been employed for capacity planning for net zero energy communities in Ref. [126]. In that work, an MO energy model was developed using goal programming. The MO model was formulated to determine the optimal energy mix by

(a) Minimizing life cycle impacts and costs
(b) Maximizing contributions of renewables and operational energy savings

The variation in terms of the changing decision priorities and stakeholders' interest was captured by changing the weights of the goal programming strategy. The computational results showed that the proposed approach found the best possible strategies with the lowest total undesirable deviations from the desired levels of the goals.

On the other hand, in Ref. [127] the capacity optimization of a distributed generation system was considered. In that work, an MO nonlinear optimization model was developed for modeling device capacity while keeping in mind the following target objectives: economic, environmental, and energy. The MO optimization problem was resolved using an integrated solution procedure: (1) NSGA-II technique for order preference by similarity to an ideal solution and (2) Shannon entropy approach. The proposed solution procedure was applied to a case study of an indoor swimming pool in Changsha city of China. The results showed that the capacity of the distributed energy system was optimized, and the results provided significant decision-making support. Table 13.4 presents an overview of recent works in sustainable capacity planning and optimization (as previously discussed), as well as other recent research efforts focused in a similar direction.

BL optimization has also been recently implemented to application problems in capacity planning. This can be seen in Ref. [128], where a wind-photovoltaic-battery hybrid generation system was optimized in terms of planning and operations. The central goal was to plan equipment capacity while maximizing the annual net income as well as the return on investment. In that work, the authors effectively optimized the real-world system using the adaptive weighted PSO technique and the configuration results were analyzed in detail.

A similar line of research was carried out by Ref. [129]. In that research, an isolated microgrid with load demand management was optimized with respect to its capacity. The problem was formulated in a BL form, where the model considered load demand management as well as load and renewable generation uncertainties. Additionally, the problem also considered energy and power balance constraints. The BL optimization problem was solved using the proposed MILP method. The results showed the effectiveness of the proposed model as well as the optimization technique in reducing the planning costs and prolonging the service life of the microgrid. Another interesting BL implementation in microgrids could be observed in the work of Ref. [130]. In that research, the authors employed a BL formulation for microgrid design planning, where the upper level problem is related to the planning of microgrids (minimizing the initial investment, utility's demand and emission costs) while the lower level problem minimizes the operation and maintenance costs (via initializing an energy management system). The nonlinear BL problem was then transformed to a linear single-level problem using the Karush-Kuhn-Tucker conditions. The proposed approach was then compared against the weighted-sum MO approach to gauge its efficiency.

266 Advances of Artificial Intelligence in a Green Energy Environment

TABLE 13.4 Recent MO optimization applications in sustainable capacity planning.

References	Application	Modeling and optimization techniques
[122]	Long-term optimal capacity planning of a grid-independent microgrid	MO grasshopper optimization algorithm
[123]	Capacity planning of renewable energy microgrid	MOPSO
[124]	Capacity expansion planning for integrated energy system with renewable energy sources	Novel multidimensional correlation scenario set generation method
[125]	Hybrid microgrid configurations	MOPSO
[126]	Capacity planning for net zero energy communities	Goal programming
[127]	Capacity optimization of a distributed generation system	NSGA-II and Shannon entropy
[101]	MO techno-economical optimization for smart/ microgrid	Hybrid GA, PSO and backward-forward sweep algorithms
[52]	Distributed energy systems	Weight distributed TOPSIS multiobjective decision-making method
[8]	Design passive filters for single-phase distributed energy grid integration systems	PSO
[62]	Networked microgrid energy management	Compromised program (CP) method
[57]	Off-grid microgrids	Goal programming
[90]	Microgrid energy management	Mixed-integer linear programming and fuzzy logic

In Ref. [131], a BL optimization model was developed for capacity planning by identifying the optimal non–storage-based small hydropower projects. The BL model consists of penstock diameter and length, intake location, and turbine number and capacity. BL optimization was carried out on two real-world scenarios (Mamquam River in Canada and the Guder River in Ethiopia) using the greedy GA. The optimization was implemented such that the greedy GA maximizes the net annual benefit. Comparative analysis was then done against the previous model applied to the same application cases and then it was found that the proposed method outperformed the previous optimization with respect to the annual net benefit.

The MO and BL optimization of a hybrid electric propulsion system was done in the work of Ref. [132]. In that research work, the hybrid electric propulsion system was optimized in terms of its sizing (upper level) as well as with respect to energy management (lower level). The optimization strategy was a combination of an MOPSO technique as well as an adaptive equivalent consumption minimization strategy. The BL optimization results were then compared to two single-level optimization models, and this exercise showed that the proposed BL approach was superior to the single-level optimization frameworks. The work of Ref. [132] optimized fuel consumption, GHG emission, and net present cost.

In Ref. [133], a BL robust planning model for active management of a distributed generation network (with renewable energy sources) was proposed. The central idea of that work was to employ active management elements such as electrical storage system, capacitor bank, on-load tap changer, and static VAR compensation (SVC) to accommodate uncertainty arising from wind and photovoltaic power. The BL problem consists of (1) investment and (2) operational levels. In that work the problem was formulated using mixed-integer second-order conic programming and solved using the column and constraint algorithm. Some recent BL implementations in sustainable capacity planning could be seen in Refs. [134–139]. A summary of some of the recent implementations in BL optimization in sustainable capacity planning is given in Table 13.5.

13.5 Outlook

In the previous sections, various optimization strategies have been seen to be utilized to tackle MO and multilevel problems in ELED, bioenergy supply chains, and sustainable capacity planning. In each of these key areas of sustainable energy economics, the following strategies have been applied for effective optimization: hybrid optimization, problem reduction, nonlinear programming, fuzzy methodology, adaptive metaheuristics, MO programming, Stackelberg game theory, Shanon entropy, chaotic search, chance-constrained programming, and MDP. The enhancements and strategic employment of existing optimization methodologies by the literature mentioned in this chapter have effectively optimized the industrial problems involving sustainable energy economics. Future review efforts could focus on the recent overlap between the field of optimization and blockchain-enabled real-world systems, e.g., smart grids [140], healthcare [141], supply chains [142] and IoT [143].

268 Advances of Artificial Intelligence in a Green Energy Environment

TABLE 13.5 Recent MO optimization applications in sustainable capacity planning.

References	Application	Modeling and optimization techniques
[128]	Wind-photovoltaic-battery hybrid generation	Adaptive weighted PSO
[129]	Isolated microgrid with load demand management	Mixed-integer linear programming
[130]	Microgrid design planning and operations	Karush-Kuhn-Tucker—based hybrid technique
[131]	Non-storage-based small hydropower projectss	Greedy GA
[132]	Hybrid electric propulsion system	Hybrid MOPSO and adaptive equivalent consumption minimization strategy
[133]	Active management of distributed generation network (with renewable energy sources)	Mixed-integer second-order conic programming and column and constraint algorithm
[134]	Operational scheduling of smart distribution networks	Hybrid Karush-Kuhn-Tucker and game theory approach.
[138]	Wind-solar capacity allocation	Mixed-integer linear programming and branch and bound method
[135]	Allocation of distributed generation and electric vehicle charging stations	Improved harmonic particle swarm optimization algorithm
[136]	Design, operation, and subsidies for standalone solar/diesel multi-generation energy systems	Hybrid mixed-integer linear programming and grid search
[139]	Virtual power plant—distributed energy resource planning	Hybrid simulated annealing-genetic algorithm with self-adaptive parameters
[137]	Renewable energy networks	Mixed-integer linear programming

References

[1] Y. Cui, Z. Geng, Q. Zhu, Y. Han, Multi-objective optimization methods and application in energy saving, Energy 125 (2017) 681–704.

[2] N. Gunantara, A review of multi-objective optimization: methods and its applications, Cogent Eng. 5 (1) (2018) 1502242.

[3] V.V. Kalashnikov, S. Dempe, G.A. Pérez-Valdés, N.I. Kalashnykova, J.F. Camacho-Vallejo, BL programming and applications, Math. Probl Eng. 2015 (2015).

[4] J. Liu, Y. Fan, Z. Chen, Y. Zheng, Pessimistic BL optimization: a survey, Int. J. Comput. Intell. Syst. 11 (1) (2018) 725–736.

[5] A. Sinha, P. Malo, K. Deb, A review on BL optimization: from classical to evolutionary approaches and applications, IEEE Trans. Evol. Comput. 22 (2) (2017) 276.

[6] X. Lu, Y. Chen, M. Fu, H. Wang, Multi-objective optimization-based real-time control strategy for battery/ultracapacitor hybrid energy management systems, IEEE Access 7 (2019) 11640–11650.

[7] J. Wang, W. Yang, P. Du, Y. Li, Research and application of a hybrid forecasting framework based on multi-objective optimization for electrical power system, Energy 148 (2018) 59–78.

[8] T. Terlouw, T. AlSkaif, C. Bauer, W. van Sark, Multi-objective optimization of energy arbitrage in community energy storage systems using different battery technologies, Appl. Energy 239 (2019) 356–372.

[9] P. Jiang, H. Yang, J. Heng, A hybrid forecasting system based on fuzzy time series and multi-objective optimization for wind speed forecasting, Appl. Energy 235 (2019) 786–801.

[10] T. Ganesan, M.S. Aris, I. Elamvazuthi, Multiobjective strategy for an industrial gas turbine: absorption chiller system, in: Handbook of Research on Emergent Applications of Optimization Algorithms, IGI Global, 2018, pp. 531–556.

[11] T. Ganesan, Multi-objective optimization of industrial power generation systems: emerging research and opportunities, IGI Glob. (2020).

[12] T. Ganesan, P. Vasant, P. Sanghvi, J. Thomas, I. Litvinchev, Random matrix generators for optimizing a fuzzy biofuel supply chain system, J. Adv. Eng. Comput. 4 (1) (2020) 33–50.

[13] T. Ganesan, P. Vasant, I. Litvinchev, M.S. Aris, Extreme value metaheuristics and coupled mapped lattice approaches for gas turbine-absorption chiller optimization, in: Research Advancements in Smart Technology, Optimization, and Renewable Energy, IGI Global, 2021, pp. 283–312.

[14] T. Ganesan, P. Vasant, I. Litvinchev, Chaotic simulator for bilevel optimization of virtual machine placements in cloud computing, J. Oper. Res. Soc. China (2021), https://doi.org/10.1007/s40305-020-00326-5.

[15] T. Ganesan, M.S. Aris, P. Vasant, Extreme value metaheuristics for optimizing a many-objective gas turbine system, Int. J. Energy Optim. Eng. 7 (2) (2018) 76–96.

[16] S. Saib, A. Gherbi, R. Bayindir, A. Kaabeche, Multi-objective optimization of a hybrid renewable energy system with a gas micro-turbine and a storage battery, Arabian J. Sci. Eng. (2019) 1–14.

[17] Z. Tian, B. Si, Y. Wu, X. Zhou, X. Shi, Multi-objective optimization model predictive dispatch precooling and ceiling fans in office buildings under different summer weather conditions, in: Building Simulation, vol. 12 (6), Tsinghua University Press, 2019, pp. 999–1012.

[18] X. Xiong, W. Wu, N. Li, L. Yang, J. Zhang, Z. Wei, Risk-based multi-objective optimization of distributed generation based on GPSO-BFA algorithm, IEEE Access 7 (2019) 30563−30572.

[19] Y. Yang, S. Peng, L. Zhu, D. Zhang, Z. Qiu, H. Yuan, L. Xian, A modified multiobjective self-adaptive differential evolution algorithm and its application on optimization design of the nuclear power system, Sci. Technol. Nuclear Install. 2019 (2019).

[20] K. Singh, K. Singh, A. Aziz, Congestion control in wireless sensor networks by hybrid multi-objective optimization algorithm, Comput. Netw. 138 (2018) 90−107.

[21] Y. Sun, F. Lin, H. Xu, Multi-objective optimization of resource scheduling in fog computing using an improved NSGA-II, Wireless Pers. Commun. 102 (2) (2018) 1369−1385.

[22] R. Han, Y. Gao, C. Wu, D. Lu, An effective multi-objective optimization algorithm for spectrum allocations in the cognitive-radio-based Internet of Things, IEEE Access 6 (2018) 12858−12867.

[23] A.M. Khan, M. Jamil, K. Salonitis, S. Sarfraz, W. Zhao, N. He, M. Mia, G. Zhao, Multiobjective optimization of energy consumption and surface quality in nanofluid SQCL assisted face milling, Energies 12 (4) (2019) 710.

[24] H. Zhang, Y. Peng, L. Hou, G. Tian, Z. Li, A hybrid multi-objective optimization approach for energy-absorbing structures in train collisions, Inf. Sci. 481 (2019) 491−506.

[25] K. Amouzgar, S. Bandaru, T. Andersson, A.H. Ng, Metamodel-based multi-objective optimization of a turning process by using finite element simulation, Eng. Optim. (2019) 1−18.

[26] A.T. Abbas, K. Hamza, M.F. Aly, E.A. Al-Bahkali, Multiobjective optimization of turning cutting parameters for J-steel material, Adv. Mater. Sci. Eng. 2016 (2016).

[27] D. Gao, H. Liang, G. Shi, L. Cao, Multiobjective optimization of carbon fiber-reinforced plastic composite bumper based on adaptive genetic algorithm, Math. Probl Eng. 2019 (2019).

[28] A.M. Gopakumar, P.V. Balachandran, D. Xue, J.E. Gubernatis, T. Lookman, Multiobjective optimization for materials discovery via adaptive design, Sci. Rep. 8 (1) (2018) 1−12.

[29] S. Klancnik, M. Hrelja, J. Balic, M. Brezocnik, Multi-objective optimization of the turning process using a gravitational search algorithm (GSA) and NSGA-II approach, Adv. Prod. Eng. Manag. 11 (4) (2016) 366.

[30] A. Solomou, G. Zhao, S. Boluki, J.K. Joy, X. Qian, I. Karaman, R. Arróyave, D.C. Lagoudas, Multi-objective Bayesian materials discovery: application on the discovery of precipitation strengthened NiTi shape memory alloys through micromechanical modeling, Mater. Des. 160 (2018) 810−827.

[31] L. Lemarchand, D. Massé, P. Rebreyend, J. Håkansson, Multiobjective optimization for multimode transportation problems, Ad. Operat. Res. 2018 (2018).

[32] D. Schmaranzer, R. Braune, K.F. Doerner, Multi-objective simulation optimization for complex urban mass rapid transit systems, Ann. Oper. Res. (2019) 1−38.

[33] J. Szlapczynska, R. Szlapczynski, Preference-based evolutionary multi-objective optimization in ship weather routing, Appl. Soft Comput. 84 (2019) 105742.

[34] S. Garcia, C.T. Trinh, Comparison of multi-objective evolutionary algorithms to solve the modular cell design problem for novel biocatalysis, Processes 7 (6) (2019) 361.

[35] S. Garcia, C.T. Trinh, Multiobjective strain design: a framework for modular cell engineering, Metab. Eng. 51 (2019) 110−120.

[36] S. Beygi, M. Tabesh, S. Liu, Multi-objective optimization model for design and operation of water transmission systems using a power resilience index for assessing hydraulic reliability, Water Resour. Manag. 33 (10) (2019) 3433−3447.

[37] S. Carpitella, B. Brentan, I. Montalvo, J. Izquierdo, A. Certa, Multi-criteria analysis applied to multi-objective optimal pump scheduling in water systems, Water Supply 19 (8) (2019) 2338−2346.

[38] T.T. Tanyimboh, A.M. Czajkowska, Joint entropy based multi-objective evolutionary optimization of water distribution networks, Water Resour. Manag. 32 (8) (2018) 2569−2584.

[39] Y. Hu, H. Chen, M. He, L. Sun, R. Liu, H. Shen, Multi-swarm multi-objective optimizer based on-optimality criteria for multi-objective portfolio management, Math. Probl Eng. 2019 (2019).

[40] K. Liagkouras, A new three-dimensional encoding multiobjective evolutionary algorithm with application to the portfolio optimization problem, Knowl. Base Syst. 163 (2019) 186−203.

[41] H.G. Resat, B. Unsal, A novel multi-objective optimization approach for sustainable supply chain: a case study in packaging industry, Sustain. Prod. Consum. 20 (2019) 29−39.

[42] Y. Fu, H. Wang, M. Huang, J. Wang, A decomposition based multiobjective genetic algorithm with adaptive multipopulation strategy for flowshop scheduling problem, Nat. Comput. 18 (4) (2019) 757−768.

[43] H. Lu, R. Zhou, Z. Fei, J. Shi, A multi-objective evolutionary algorithm based on Pareto prediction for automatic test task scheduling problems, Appl. Soft Comput. 66 (2018) 394−412.

[44] Y. Ma, M. Hu, X. Yan, Multi-objective path planning for unmanned surface vehicle with currents effects, ISA Trans. 75 (2018) 137−156.

[45] B. Saicharan, R. Tiwari, N. Roberts, Multi objective optimization based path planning in robotics using nature inspired algorithms: a survey, in: 2016 IEEE 1st International Conference on Power Electronics, Intelligent Control and Energy Systems (ICPEICES), IEEE, 2016, pp. 1−6.

[46] B.N. Chebouba, M.A. Mellal, S. Adjerid, Multi-objective system reliability optimization in a power plant, in: 2018 International Conference on Electrical Sciences and Technologies in Maghreb (CISTEM), IEEE, 2018, pp. 1−4.

[47] J. Zhao, S. Si, Z. Cai, M. Su, W. Wang, Multiobjective optimization of reliability−redundancy allocation problems for serial parallel-series systems based on importance measure, Proc. Inst. Mech. Eng. O J. Risk Reliab. 233 (5) (2019) 881−897.

[48] S. Dempe, BL Optimization: Theory, Algorithms and Applications, TU Bergakademie Freiberg, 2018 (Fakultät für Mathematik und Informatik).

[49] A. Ketabi, A. Karimizadeh, M. Shahidehpour, Optimal generation units start-up sequence during restoration of power system considering network reliability using bi-level optimization, Int. J. Electr. Power Energy Syst. 104 (2019) 772−783.

[50] P. Rullo, L. Braccia, P. Luppi, D. Zumoffen, D. Feroldi, Integration of sizing and energy management based on economic predictive control for standalone hybrid renewable energy systems, Renew. Energy 140 (2019) 436−451.

[51] C. Wu, W. Gu, Y. Xu, P. Jiang, S. Lu, B. Zhao, Bi-level optimization model for integrated energy system considering the thermal comfort of heat customers, Appl. Energy 232 (2018) 607−616.

272 Advances of Artificial Intelligence in a Green Energy Environment

[52] P.P. Biswas, P.N. Suganthan, B.Y. Qu, G.A. Amaratunga, Multiobjective economic-environmental power dispatch with stochastic wind-solar-small hydro power, Energy 150 (2018) 1039−1057.

[53] A. Biswas, Y. Chen, C. Hoyle, A bi-level optimization approach for energy allocation problems, in: ASME 2018 International Design Engineering Technical Conferences and Computers and Information in Engineering Conference, American Society of Mechanical Engineers Digital Collection, 2018.

[54] K. Cheng, Y. Teng, W. Sun, A. Liu, X. Wang, Energy-efficient joint offloading and wireless resource allocation strategy in multi-MEC server systems, in: 2018 IEEE International Conference on Communications (ICC), IEEE, 2018, pp. 1−6.

[55] J. Zhao, Y. Zhou, Bi-level programming model of cloud manufacturing services based on extension theory, Math. Probl. Eng. 2018 (2018).

[56] A.W. Hammad, A BL multiobjective optimisation approach for solving the evacuation location assignment problem, Adv. Civ. Eng. 2019 (2019).

[57] O. Baskan, A multiobjective BL programming model for environmentally friendly traffic signal timings, Adv. Civ. Eng. 2019 (2019).

[58] S.G. Machiani, A. Jahangiri, A. Ahmadi, A network-wide BL optimization-simulation approach for variable speed limit systems to improve travel time reliability, ASCE-ASME J. Risk Uncertainty Eng. Syst. A: Civ. Eng. 3 (3) (2017) 04016017.

[59] W. Yong, J. Zhou, D. Jahed Armaghani, M.M. Tahir, R. Tarinejad, B.T. Pham, V. Van Huynh, A new hybrid simulated annealing-based genetic programming technique to predict the ultimate bearing capacity of piles, Eng. Comput. 37 (3) (2021) 2111−2127.

[60] A. Chowdhury, A.R. Zomorrodi, C.D. Maranas, BL optimization techniques in computational strain design, Comput. Chem. Eng. 72 (2015) 363−372.

[61] V.N. Emenike, R. Schenkendorf, U. Krewer, Model-based optimization of biopharmaceutical manufacturing in Pichia pastoris based on dynamic flux balance analysis, Comput. Chem. Eng. 118 (2018) 1−13.

[62] B. Jabarivelisdeh, S. Waldherr, Optimization of bioprocess productivity based on metabolic-genetic network models with BL dynamic programming, Biotechnol. Bioeng. 115 (7) (2018) 1829−1841.

[63] X. Zhu, P. Guo, BL programming approaches to production planning for multiple products with short life cycles, 4OR (2019) 1−25.

[64] B. Behnia, I. Mahdavi, B. Shirazi, M.M. Paydar, A bi-objective bi-level mathematical model for cellular manufacturing system applying evolutionary algorithms, Scientia Iranica. Trans. E Ind. Eng. 26 (4) (2019) 2541−2560.

[65] A. Migdalas, P.M. Pardalos, P. Värbrand (Eds.), Multilevel Optimization: Algorithms and Applications, vol. 20, Springer Science & Business Media, 2013.

[66] J.F. Bard, Practical BL Optimization: Algorithms and Applications, vol. 30, Springer Science & Business Media, 2013.

[67] B. Dey, S.K. Roy, B. Bhattacharyya, Solving multi-objective economic emission dispatch of a renewable integrated microgrid using latest bio-inspired algorithms, Eng. Sci. Technol. Int. J. 22 (1) (2019) 55−66.

[68] N. Karthik, A.K. Parvathy, R. Arul, Multi-objective economic emission dispatch using interior search algorithm, Int. Trans. Electr. Energy Syst. 29 (1) (2019) e2683.

[69] A. Rajagopalan, P. Kasinathan, K. Nagarajan, V.K. Ramachandaramurthy, V. Sengoden, S. Alavandar, Chaotic self-adaptive interior search algorithm to solve combined economic emission dispatch problems with security constraints, Int. Trans. Electr. Energy Syst. 29 (8) (2019) e12026.

Recent implementations of MO optimization **Chapter | 13 273**

[70] X. Xu, Z. Hu, Q. Su, Z. Xiong, Multiobjective collective decision optimization algorithm for economic emission dispatch problem, Complexity 2018 (2018).

[71] S. Haddi, O. Bouketir, T. Bouktir, Improved Optimal Power Flow for a Power System Incorporating Wind Power Generation by Using Grey Wolf Optimizer Algorithm, 2018.

[72] H. Nourianfar, H. Abdi, Solving the multi-objective economic emission dispatch problems using fast non-dominated sorting TVAC-PSO combined with EMA, Appl. Soft Comput. 85 (2019) 105770.

[73] E.S. Ali, S.A. Elazim, Mine blast algorithm for environmental economic load dispatch with valve loading effect, Neural Comput. Appl. 30 (1) (2018) 261−270.

[74] S.K. Gachhayat, S.K. Dash, P. Ray, Multi objective directed bee colony optimization for economic load dispatch with enhanced power demand and valve point loading, Int. J. Electr. Comput. Eng. 7 (5) (2017) 2088−8708.

[75] B. Lokeshgupta, S. Sivasubramani, Multi-objective dynamic economic and emission dispatch with demand side management, Int. J. Electr. Power Energy Syst. 97 (2018) 334−343.

[76] Y.V.K. Reddy, M.D. Reddy, Solution of multi objective environmental economic dispatch by grey wolf optimization algorithm, Int. J. Intell. Syst. Appl. Eng. 7 (1) (2019) 34−41.

[77] J. Valder, P.P. AJ, Deregulated multi-objective economic load dispatch using cuckoo search algorithm, Int. J. Adv. Sci. Res. Manag. 4 (1) (2019) 15−21.

[78] Z.M. Yasin, N.F.A. Aziz, N.A. Salim, N.A. Wahab, N.A. Rahmat, Optimal economic load dispatch using multiobjective cuckoo search algorithm, Indonesian J. Electr. Eng. Comput. Sci. 12 (1) (2018) 168−174.

[79] T.C. Bora, V.C. Mariani, L. dos Santos Coelho, Multi-objective optimization of the environmental-economic dispatch with reinforcement learning based on non-dominated sorting genetic algorithm, Appl. Therm. Eng. 146 (2019) 688−700.

[80] K.O. Alawode, G.A. Adegboyega, J. Abimbola Muhideen, NSGA-II/EDA hybrid evolutionary algorithm for solving multi-objective economic/emission dispatch problem, Elec. Power Compon. Syst. 46 (10) (2018) 1160−1172.

[81] X. Yu, X. Yu, Y. Lu, J. Sheng, Economic and emission dispatch using ensemble multi-objective differential evolution algorithm, Sustainability 10 (2) (2018) 418.

[82] K. Mason, J. Duggan, E. Howley, A multi-objective neural network trained with differential evolution for dynamic economic emission dispatch, Int. J. Electr. Power Energy Syst. 100 (2018) 201−221.

[83] B.Y. Qu, J.J. Liang, Y.S. Zhu, P.N. Suganthan, Solving dynamic economic emission dispatch problem considering wind power by multi-objective differential evolution with ensemble of selection method, Nat. Comput. 18 (4) (2019) 695−703.

[84] G. Liu, Y.L. Zhu, W. Jiang, Wind-thermal dynamic economic emission dispatch with a hybrid multi-objective algorithm based on wind speed statistical analysis, IET Gener., Transm. Distrib. 12 (17) (2018) 3972−3984.

[85] Y. Li, J. Wang, D. Zhao, G. Li, C. Chen, A two-stage approach for combined heat and power economic emission dispatch: combining multi-objective optimization with integrated decision making, Energy 162 (2018) 237−254.

[86] J. Wang, W. Huang, G. Ma, S. Chen, An improved partheno genetic algorithm for multi-objective economic dispatch in cascaded hydropower systems, Int. J. Electr. Power Energy Syst. 67 (2015) 591−597.

[87] L. Yang, D. He, B. Li, A selection hyper-heuristic algorithm for multiobjective dynamic economic and environmental load dispatch, Complexity 2020 (2020).

274   Advances of Artificial Intelligence in a Green Energy Environment

[88]   B.Y. Qu, Y.S. Zhu, Y.C. Jiao, M.Y. Wu, P.N. Suganthan, J.J. Liang, A survey on multi-objective evolutionary algorithms for the solution of the environmental/economic dispatch problems, Swarm Evolut. Comput. 38 (2018) 1−11.

[89]   F.P. Mahdi, P. Vasant, V. Kallimani, J. Watada, P.Y.S. Fai, M. Abdullah-Al-Wadud, A holistic review on optimization strategies for combined economic emission dispatch problem, Renew. Sustain. Energy Rev. 81 (2018) 3006−3020.

[90]   D. Chakraborti, P. Biswas, B.B. Pal, March. Modelling multiobjective BL programming for environmental-economic power generation and dispatch using genetic algorithm, in: International Conference on Computational Intelligence, Communications, and Business Analytics, Springer, Singapore, 2017, pp. 423−439.

[91]   D.C. Xing, P.A.N. Chao, Q.Y. Lv, B.S. Qin, Bi-level multi-objective environmental economic dispatch with wind farm, in: DEStech Transactions on Engineering and Technology Research, (ICAMM), 2016.

[92]   Z. Hu, M. Zhang, X. Wang, C. Li, M. Hu, Bi-level robust dynamic economic emission dispatch considering wind power uncertainty, Electr. Power Syst. Res. 135 (2016) 35−47.

[93]   J. Liu, J. Li, A bi-level energy-saving dispatch in smart grid considering interaction between generation and load, IEEE Trans. Smart Grid 6 (3) (2015) 1443−1452.

[94]   X. Song, H. Lin, G. De, H. Li, X. Fu, Z. Tan, An energy optimal dispatching model of an integrated energy system based on uncertain BL programming, Energies 13 (2) (2020) 477.

[95]   H. Gu, Y. Li, J. Yu, C. Wu, T. Song, J. Xu, Bi-level optimal low-carbon economic dispatch for an industrial park with consideration of multi-energy price incentives, Appl. Energy 262 (2020) 114276.

[96]   Y. Zhao, K. Peng, B. Xu, H. Li, Y. Liu, X. Zhang, BL optimal dispatch strategy for a multi-energy system of industrial parks by considering integrated demand response, Energies 11 (8) (2018) 1942.

[97]   A.A. Eladl, A.A. ElDesouky, Optimal economic dispatch for multi heat-electric energy source power system, Int. J. Electr. Power Energy Syst. 110 (2019) 21−35.

[98]   Q. Huang, Q.S. Jia, X. Guan, Coordinating EV charging demand with wind supply in a bi-level energy dispatch framework, in: 2016 American Control Conference (ACC), IEEE, 2016, pp. 6233−6238.

[99]   G. Zhang, C. Jiang, X. Wang, B. Li, Risk assessment and bi-level optimization dispatch of virtual power plants considering renewable energy uncertainty, IEEJ Trans. Electr. Electron. Eng. 12 (4) (2017) 510−518.

[100]   Q. Xu, Y. Ji, Q. Huang, Y. Sheng, Bi-level optimised dispatch strategy of electric supply−demand balance considering risk−benefit coordination, IET Smart Grid 1 (4) (2018) 169−176.

[101]   E. Huang, X. Zhang, L. Rodriguez, M. Khanna, S. de Jong, K.C. Ting, Y. Ying, T. Lin, Multi-objective optimization for sustainable renewable jet fuel production: a case study of corn stover based supply chain system in Midwestern US, Renew. Sustain. Energy Rev. 115 (2019) 109403.

[102]   F. Lu, X. Chen, Z. Lei, L. Wen, Y. Zhang, Revealing the activity of different iron carbides for Fischer-Tropsch synthesis, Appl. Catal. B Environ. 281 (2021) 119521.

[103]   C. Gao, D. Qu, Y. Yang, Optimal design of bioenergy supply chains considering social benefits: a case study in Northeast China, Processes 7 (7) (2019) 437.

[104]   S. Bairamzadeh, M.S. Pishvaee, M. Saidi-Mehrabad, Multiobjective robust possibilistic programming approach to sustainable bioethanol supply chain design under multiple uncertainties, Ind. Eng. Chem. Res. 55 (1) (2016) 237−256.

Recent implementations of MO optimization **Chapter | 13 275**

[105] S. Razm, S. Nickel, H. Sahebi, A multi-objective mathematical model to redesign of global sustainable bioenergy supply network, Comput. Chem. Eng. 128 (2019) 1–20.

[106] J.M. Egieya, L. Čuček, K. Zirngast, A.J. Isafiade, Z. Kravanja, Optimization of biogas supply networks considering multiple objectives and auction trading prices of electricity, BMC Chem. Eng. 2 (1) (2020) 1–23.

[107] F.A. Al-Noweam, I.A. El-Khouly, K.S. El-Kilany, Multi-objective optimization of biomass supply chain networks, in: Proceedings of the International Conference on Industrial Engineering and Operations Management, Paris, France, 2018.

[108] S. Samsatli, A multi-objective MILP model for planning, design and operation of biomass supply chains: capturing the trade-offs within the food-energy-water-environment Nexus, in: Proceedings of the 2018 AIChE Annual Meeting. AIChE, 2018 AIChE Annual Meeting, Pittsburgh, USA United States, 2018.

[109] Y. Zhang, R. Zhang, A multi-objective optimization model considering inventory strategy for biofuel supply chain design, in: 2017 6th International Conference on Energy and Environmental Protection (ICEEP 2017), Atlantis Press, 2017.

[110] M.S. Roni, S.D. Eksioglu, K.G. Cafferty, J.J. Jacobson, A multi-objective, hub-and-spoke model to design and manage biofuel supply chains, Ann. Oper. Res. 249 (1–2) (2017) 351–380.

[111] C. Cambero, T. Sowlati, Incorporating social benefits in multi-objective optimization of forest-based bioenergy and biofuel supply chains, Appl. Energy 178 (2016) 721–735.

[112] A.T.E. Perez, Biorefinery Supply Chain Design Optimization under Sustainability Dimensions (Doctoral dissertation), 2017.

[113] W. Wen, Q. Zhang, A design of straw acquisition mode for China's straw power plant based on supply chain coordination, Renew. Energy 76 (2015) 369–374.

[114] Y. Liu, R. Zhao, K.J. Wu, T. Huang, A.S. Chiu, C. Cai, A hybrid of multi-objective optimization and system dynamics simulation for straw-to-electricity supply chain management under the belt and road initiatives, Sustainability 10 (3) (2018) 868.

[115] L. Ascenso, F. d'Amore, A. Carvalho, F. Bezzo, Assessing multiple biomass-feedstock in the optimization of power and fuel supply chains for sustainable mobility, Chem. Eng. Res. Des. 131 (2018) 127–143.

[116] N. Zandi Atashbar, N. Labadie, C. Prins, Modelling and optimisation of biomass supply chains: a review, Int. J. Prod. Res. 56 (10) (2018) 3482–3506.

[117] R. Sun, T. Liu, X. Chen, L. Yao, A biomass-coal co-firing based bi-level optimal approach for carbon emission reduction in China, J. Clean. Prod. (2020) 123318.

[118] J. Xu, Q. Huang, C. Lv, Q. Feng, F. Wang, Carbon emissions reductions oriented dynamic equilibrium strategy using biomass-coal co-firing, Energy Policy 123 (2018) 184–197.

[119] D. Yue, F. You, Stackelberg-game-based modeling and optimization for supply chain design and operations: a mixed integer BL programming framework, Comput. Chem. Eng. 102 (2017) 81–95.

[120] S. Medina-González, L.G. Papageorgiou, V. Dua, A reformulation strategy for mixed-integer linear bi-level programming problems, Comput. Chem. Eng. (2021) 107409.

[121] N. Zhao, F. You, Dairy waste-to-energy incentive policy design using Stackelberg-game-based modeling and optimization, Appl. Energy 254 (2019) 113701.

[122] A.L. Bukar, C.W. Tan, L.K. Yiew, R. Ayop, W.S. Tan, A rule-based energy management scheme for long-term optimal capacity planning of grid-independent microgrid optimized by multi-objective grasshopper optimization algorithm, Energy Convers. Manag. 221 (2020) 113161.

276 Advances of Artificial Intelligence in a Green Energy Environment

[123] M.K. Kiptoo, O.B. Adewuyi, M.E. Lotfy, T. Senjyu, P. Mandal, M. Abdel-Akher, Multi-objective optimal capacity planning for 100% renewable energy-based microgrid incorporating cost of demand-side flexibility management, Appl. Sci. 9 (18) (2019) 3855.

[124] Y. Lei, D. Wang, H. Jia, J. Chen, J. Li, Y. Song, J. Li, Multi-objective stochastic expansion planning based on multi-dimensional correlation scenario generation method for regional integrated energy system integrated renewable energy, Appl. Energy 276 (2020) 115395.

[125] A.M. Haidar, A. Fakhar, A. Helwig, Sustainable energy planning for cost minimization of autonomous hybrid microgrid using combined multi-objective optimization algorithm, Sustain. Cities Soc. 62 (2020) 102391.

[126] E. Bakhtavar, T. Prabatha, H. Karunathilake, R. Sadiq, K. Hewage, Assessment of renewable energy-based strategies for net-zero energy communities: a planning model using multi-objective goal programming, J. Clean. Prod. 272 (2020) 122886.

[127] Z. Luo, S. Yang, N. Xie, W. Xie, J. Liu, Y.S. Agbodjan, Z. Liu, Multi-objective capacity optimization of a distributed energy system considering economy, environment and energy, Energy Convers. Manag. 200 (2019) 112081.

[128] B. Yang, Y. Guo, X. Xiao, P. Tian, Bi-level capacity planning of wind-PV-battery hybrid generation system considering return on investment, Energies 13 (12) (2020) 3046.

[129] G. Ma, Z. Cai, P. Xie, P. Liu, S. Xiang, Y. Sun, C. Guo, G. Dai, A bi-level capacity optimization of an isolated microgrid with load demand management considering load and renewable generation uncertainties, IEEE Access 7 (2019) 83074–83087.

[130] S. Haghifam, K. Zare, M. Dadashi, Bi-level operational planning of microgrids with considering demand response technology and contingency analysis, IET Gener., Transm. Distrib. 13 (13) (2019) 2721–2730.

[131] H.U. Abdelhady, Y.E. Imam, Z. Shawwash, A. Ghanem, Parallelized bi-level optimization model with continuous search domain for selection of run-of-river hydropower projects, Renew. Energy (2020).

[132] J. Zhu, L. Chen, X. Wang, L. Yu, Bi-level optimal sizing and energy management of hybrid electric propulsion systems, Appl. Energy 260 (2020) 114134.

[133] M. Wu, L. Kou, X. Hou, Y. Ji, B. Xu, H. Gao, A bi-level robust planning model for active distribution networks considering uncertainties of renewable energies, Int. J. Electr. Power Energy Syst. 105 (2019) 814–822.

[134] S. Haghifam, M. Dadashi, K. Zare, H. Seyedi, Optimal operation of smart distribution networks in the presence of demand response aggregators and microgrid owners: a multi follower Bi-Level approach, Sustain. Cities Soc. 55 (2020) 102033.

[135] L. Liu, Y. Zhang, C. Da, Z. Huang, M. Wang, Optimal allocation of distributed generation and electric vehicle charging stations based on intelligent algorithm and bi-level programming, Int. Trans. Electr. Energy Syst. 30 (6) (2020) e12366.

[136] X. Luo, J. Liu, Y. Liu, X. Liu, Bi-level optimization of design, operation, and subsidies for standalone solar/diesel multi-generation energy systems, Sustain. Cities Soc. 48 (2019) 101592.

[137] N. Nasiri, A.S. Yazdankhah, M.A. Mirzaei, A. Loni, B. Mohammadi-Ivatloo, K. Zare, M. Marzband, A bi-level market-clearing for coordinated regional-local multi-carrier systems in presence of energy storage technologies, Sustain. Cities Soc. 63 (2020) 102439.

[138] H. Yang, Q. Yu, J. Liu, Y. Jia, G. Yang, E. Ackom, Z.Y. Dong, Optimal wind-solar capacity allocation with coordination of dynamic regulation of hydropower and energy intensive controllable load, IEEE Access 8 (2020) 110129–110139.

[139] Z. Yi, Y. Xu, H. Sun, Self-adaptive hybrid algorithm based bi-level approach for virtual power plant bidding in multiple retail markets, IET Gener. Transm. Distrib. 14 (18) (2020) 3762–3773.

[140] M.B. Mollah, J. Zhao, D. Niyato, K.Y. Lam, X. Zhang, A.M. Ghias, L.H. Koh, L. Yang, Blockchain for future smart grid: a comprehensive survey, IEEE Internet Things J. 8 (1) (2020) 18–43.

[141] S. Nandi, J. Sarkis, A.A. Hervani, M.M. Helms, Redesigning supply chains using blockchain-enabled circular economy and COVID-19 experiences, Sustain. Prod. Consum. 27 (2021) 10–22.

[142] B.Q. Tan, F. Wang, J. Liu, K. Kang, F. Costa, A blockchain-based framework for green logistics in supply chains, Sustainability 12 (11) (2020) 4656.

[143] B. Cao, X. Wang, W. Zhang, H. Song, Z. Lv, A many-objective optimization model of industrial internet of things based on private blockchain, IEEE Netw. 34 (5) (2020) 78–83.

Chapter 14

Hybrid optimization and artificial intelligence applied to energy systems: a review

Gilberto Pérez Lechuga[1], Karla N. Madrid Fernández[1] and Ugo Fiore[2]

[1]*Instituto de Ciencias Basicas e Ingenieria-AAIA, Universidad Autónoma del Estado de Hidalgo, Pachuca, Hidalgo, México;* [2]*Parthenope University, Naples, Italy*

14.1 Introduction

Global warming and the problems of supplying energy to countries with few natural resources for its production have forced technology to change its perspectives regarding its production, administration, and distribution. In practice, it sought to combine two or more alternative sources of energy to face the increasingly growing demand. To date, there are in the literature several types of models associated with energy systems depending on what is to be represented. For example, there are several kinds of models related to the planning, operation, and maintenance of hydroelectric, geothermal, and nuclear plants, for forecasting the demand and projecting the needs of energy markets and more with various different classes of objective functions and associated technological constraints. As in any optimization model, the analysis of energy systems aims to make the most of the scarce resources of the system under natural conditions or imposed in the form of technological constraints. The optimal configuration of a power producing system often requires the use of hybrid algorithms that optimize operating costs and environmental emissions. In these models, the real uncertainty associated with them is included through the use of random variables that represent the uncertainty of the energy demand and production costs, for example, or other variables that vary over time.

To deal with the analysis, operation, and optimization of these systems, modern resources related to artificial intelligence (AI) and hybrid optimization (HO) are used. AI is an interdisciplinary science with multiple approaches; it is a branch of computer science that deals with building machines capable of performing tasks that normally require human intelligence due to complex

280 Advances of Artificial Intelligence in a Green Energy Environment

problems associated with decision-making. They are based on machine learning (ML) and deep learning and have a wide application in practically all sectors of the technology industry. HO constitutes an interesting class of compiler heuristics. An algorithm is conceived as an ordered set of sequential, systematic, and repetitive operations that solve a problem through approximations in each repetition. In a hybrid algorithm, two or more simple algorithms are combined to solve a complex problem. In these, the available information plays an important role since, depending on its availability, a selection is made from the participating algorithms or a mixture of these to achieve convergence. Normally the selection is made through heuristics and/or through the use of AI to make its performance more efficient. An exact definition can be found in Ref. [1].

It is because of that a hybrid algorithm aims to combine two or more algorithms that solve the same problem, either by choosing one (based on the data) or by switching between them during the course of the optimization process [2]. The combination through the use of AI of the best properties of the chosen algorithm allows the overall algorithm to be better than the individual components. A large number of real engineering problems require taking advantage of the benefits of these in the implementation of recursive algorithms under the philosophy of "divide and conquer", a phrase that applies fundamentally to the reduction of data size. Generally, the technique makes use of a generic algorithm that handles large amounts of information (data), alternately changing to other different algorithms that are normally more efficient in the use of a low volume of information (data). HO combines the benefits of stochastic and deterministic optimization by implementing heuristics that do extensive searches, which are ultimately refined using deterministic procedures.

In contrast, a deterministic artificial intelligence model (DAIM) is based on self-learning models where most of the information is known (not necessarily based on deterministic algorithms), and therefore, what is fundamental is the nature of the problem as well as the ease of expressing it through equations that represent a change in time or any other continuous magnitude; for example, through difference equations or differential equations and/or solutions to variants of Euler's equation. In this context, two factors inherent to the subject take on special importance, self-awareness, and autonomous learning.

Self-awareness is defined as the property that a machine has of being aware of its own existence, while autonomous learning is the property that the machine has to learn or improve its performance from the information provided to it. In the deterministic case, the increase in these attributes is basically based on linear regression methods and more.

In DAIM, self-awareness is especially important when forcing self-learning based on the response received from the environment, that is, feedback received through an impulse signal that modifies the original action based on the error made or the deviation detected. Here, the learning process becomes iterative by continuously improving the probability of success (evaluated through a utility function) in each new attempt and minimizing the error made in relation to said function, for which a proximity metric is defined based on norms of the volatility obtained in relation to the approximation that the machine has and its objective, to later generate a reparameterization of the process in search of a better approximation [3].

For the case at hand and focusing on AI not necessarily deterministic, the complexity of the models lies in the lack of information or the total ignorance of it. Formally, this uncertainty is supplied through the use of probability models. In energy systems, there are a large number of sources of uncertainty. Perhaps the most important is the energy demand required during the planning horizon of the operation, the electromagnetic radiation received in a certain area, the state of the charge of a set of batteries, the wind speed reported by the anemometers farm of a determined area, energy prices in consumer markets, and the amount of heat that must be withdrawn from a given place to cool it or to reduce or maintain the desired temperature (see Ref. [4]).

For example, a classic application is the use of hybrid algorithms in classification problems when the classification algorithm specifies how to organize the data in a particular order and the most common formats are in numerical or lexicographic order, normally requiring a more flexible format. Here, the search for information must be optimized with great efficiency.

This document focuses on the integration of AI with mathematical programming to solve problems in Operations Research and Combinatorial Optimization applied to various areas of knowledge. It provides a very cursory analysis (discussion and comparison) of the development, techniques, authors, and results of the use of HO as an alternative tool to solve complex instances related to energy systems. Also included are cases in which AI has been used to help HO produce good results in applications, based on the precision and speed of convergence achieved by the technique. We consider some of the most representative cases of traditional optimization, stochastic programming, and its variant known as robust optimization. The results suggest that energy systems tend to grow toward large-scale models with multiple integration requiring for their study, diverse algorithms (based on metaheuristics) working together and coordinated by an AI model to attend to their complexity.

14.2 Stochastic programming

A stochastic program model is an optimization problem where one or more of its parameters are unknown or only partially known. The information provided by the uncertain parameters is covered through theoretical or empirical probability distributions.

14.2.1 The general model

Formally, the general structure of a stochastic programming model is given by

$$\text{Minimize } F_0\,(x) = \mathbb{E}[f_0\,(x,\,\omega)] \tag{14.1}$$

Subject to $x \in \mathcal{S} \subset \mathbb{R}^n$

where $\mathcal{S} = \{x/f_i\,(x,\,\omega) \leq 0,\, i = 1,\, 2,\, \cdots,\, m\,\}$, and $\mathbb{E}$ is the mathematical expectation operator with regard of some probability space $(\Omega,\, A,\, P)$ and $\omega \in \Omega$.

The structure of a stochastic model can vary considerably depending on the information available for its construction. Its classical form is described through the mathematical expectation of a utility function or objective function accompanied by a set of constraints that may or not contain random variables with known or unknown densities. Another classic format is the existence of objective functions and/or opportunity constraints that must occur with a certain predetermined probability. In the literature there is a large number of algorithms developed expressly to solve instances associated with Problem (14.1). Among the most representative are the following: Monte Carlo method, climbing procedures, greedy descent methods, taboo search, simulated annealing, stochastic tunneling, genetic procedures, ant colony, firefly method, Bat method, bee colony, Search for Harmony, grey wolves, etc. [5].

Within the simplest methods are those that use randomness for the search process to locate extreme value candidates more quickly but not necessarily efficiently. Frequently, these methods do not define the nature of the optimum found and the search can be prolonged indefinitely, thereby reducing the efficiency of the process. Perhaps the oldest method of this type is named random search [6] or one of its variants based on the modification of the step size fundamentally. Some widely used variants of stochastic methods for optimizing random functions with or without constraints are [7] stochastic approximation, stochastic gradient descent, finite difference, simultaneous perturbation, stochastic quasi gradients, and scenario optimization.

A more extensive list of these methods is as follows: ant colony optimization, artificial bee colony, artificial immune systems, artificial neural network (ANN), automatic computing, bacterial foraging, Bat algorithm, biological computing, chaos optimization, combinatorial optimization, computational intelligence, continuous optimization, differential evolution, direct search and random search, direct search, evolutionary computing, fuzzy optimization, genetic algorithms (GAs), granular computing, grey wolves, greedy algorithms, hybrid algorithms, local and global Search, memetic algorithms, metaheuristic methods, natural computing, particle swarm optimization (PSO), pattern search, quasi-Newton methods, simulated annealing, soft computing techniques, swarm intelligence, taboo search, and variable neighborhood search. A specific classification of methods used in energy under uncertainty can be found in Ref. [4].

14.2.2 Software for stochastic programming instances

It is difficult to write and/or recommend general purpose software to optimize stochastic programming instances. Usually, each model has its own custom aspects that make it unique and therefore difficult to fix with a given generic method. However, some computer programs incorporate some of the following methods (in alphabetical order):

Software for stochastic linear programming, (1) stochastic linear programming solvers on NEOS Server is used, and for multistage programming models, (2) SAMPL: an AMPL-based stochastic programming modeling language translator is used (Stochastic Programming, without Date). These are problems that must be solved in stages because the solution of one alters the decisions of another with the realization of stochastic data. The goal is to minimize the total expected costs of all decisions. The main sources of code (not necessarily in the public domain) depend on how the data are distributed and the number of stages (decision points) in the problem.

For discretely distributed multistage problems, a good program is MSLiP. For problems other than large discrete distribution problems, an equivalent deterministic model can be created and solved with a standard solver, for example, LINGO optimization modeling software for linear, nonlinear, and integer programming. Another interesting alternative is the STOPGEN program, available via FTP. STOPGEN is a program that forms equivalent deterministic models. The most recent program for continuously distributed data is BRAIN by K. Frauendorfer [8]. An excellent alternative for multistage stochastic programs with recourse and scenarios construction is given in Ref. [9].

14.3 Optimization in energy systems

Energy systems optimization models frequently relate the ways of generating, supplying, and managing energy demand in an interconnected network. In some cases, we have complex generation and cogeneration systems linked in a network that involves generating plants, maintenance, resources, and consumer markets. In others, for example, it is necessary to optimize the design and duration of energy storage systems (batteries) used in the construction of hybrid vehicles; also we can mention the redesign the layout of their accommodation within the automobile and minimize the total length of wiring used, as well as mass, volume, and material costs used in the vehicle without diminishing the performance of the total energy system.

In any case, the complexity of these and other problems require, for their optimization, the creation of a general algorithm that combines two or more partial algorithms capable of solving the same problem and, later, choosing one according to the available information or interacting between them during the development of the algorithm.

284 Advances of Artificial Intelligence in a Green Energy Environment

This is usually done to combine the desired characteristics of each, so that the overall algorithm is better than the individual components. Therefore, while the build is being developed, HO uses a heuristic to select the best algorithm from a set of algorithms to be implemented in the model to be optimized [10].

14.3.1 Energy system models and their optimization processes

The application of hybrid algorithms to the industry and especially to the energy area has been well received. However, the quandary of its application is that it is difficult to evaluate its efficiency due to the forced use of simulation based on the Monte Carlo method, and therefore, its results can only be evaluated through average values of the associated computational runs. Hybrid models of optimization in energy systems necessarily include stochastic programming, robust optimization, fuzzy programming, and interval methods [4]. In the literature there is a large number of proposals on models, techniques, advantages, and disadvantages obtained from the large amount of technology available in this regard.

14.3.2 A classification according to the applications

Some of the interest problems where HO and AI apply are the following:

1. Image recognition technology and its application in the security management of the electrical system
2. Smart electric power equipment
3. Intelligent optimization and its application in energy system planning, market trading, and dispatch
4. Big data—driven smart prediction and decision assistant
5. Integration of intelligence of renewable energy resources
6. Application of AI in the security and stability of the electrical system
7. Maintenance plan for energy equipment based on AI
8. Management and energy consumption based on AI

A first approach to the type of models used as well as their applications is shown in Table 14.1 [11].

In an attempt to make our own classification, we proceeded to carry out a sampling and classification of the most recent literature on emerging areas that represent an opportunity for the application of HO and AI.

14.3.2.1 Energy, manufacturing, and production

- An application based on the use of hybrid algorithms in large-scale problems associated with planning models in the aerospace and electroplating industries and energy can be found in Ref. [12].

Hybrid optimization and artificial intelligence Chapter | 14   **285**

TABLE 14.1 Energy system models and their optimization processes.

Typology of the model	Scope	Techniques used
Analysis and optimization of decision-making	Planning of production and operations of thermoelectric plants	LP, MILP, PIL, NLP, SP, AI, H, MA
	Operation planning in systems with cogeneration	LP, MILP, PIL, NLP, SP, AI, H, MA
	Hydrothermal coordination between thermoelectric plants and hydraulic power generation plants	LP, MILP, PIL, NLP, SP, AI, H, MA, HO
	Supply chain models	LP, MILP, PIL, SP, H, HO
	Market allocation models	LP, H, MA, AI
Energy efficiency	Demand forecast models	TS, AI
	Time series associated with winds for planning the operation in wind systems	TS, AI
	Design of compact systems for energy providers (batteries)	LP, MILP, PIL, NLP, SP, AI, H, MA
Simulation models	PSE, EE	Discrete and continuous system simulation
Device and battery design	Application to compact devices	LP, MILP, PIL, NLP, SP, AI, MA, H
Distributed energy systems	Smart microgrids, integrated energy systems	ML, AI, NLP

- Flexible manufacturing systems [13] develop an interesting proposal for the modeling and optimization of nonlinear networks in RFID using AI techniques. The author proposes a novel AI algorithm, called "hybrid AI optimization technique", to develop RFID network planning from a difficult learning problem perspective. In his proposal, the author presents two different optimization algorithms: RAE and RPL to solve the instance. From an energy perspective, the work reduces energy efforts in locating and designing storage and production systems in a manufacturing plant.
- An interesting example of optimization of the ethanol production rate using MOGA can be found in Ref. [14]. In this instance, the authors obtain the optimal parameters of the size of the reactor for the production of the input as well as the efficiency associated with the process.

- Applications for optimizing fuel savings in the transport of perishable products can be found in Ref. [15].
- In Ref. [16], the article presents a novel methodology to optimize a hybrid photovoltaic-diesel-autonomous battery energy system and subsequently to develop a sensitivity analysis of the model. To do this, the author used an algorithm based on PSO and the ε constraint method to simultaneously minimize total system cost, unsatisfied load, and CO_2 emissions.
- In the optimization applied to the exploitation of oil wells, the authors [17] propose an optimization model with a discontinuous and multidimensional objective function. The authors analyze techniques based on metaheuristics and inspired by nature in order to maximize some economic factors defined in the instance. They also present their results in the form of reserve simulation models.
- An interesting review about the learning-assisted power system optimization technique can be found in Ref. [18]. This document addresses the coordination between ML approaches and optimization models and assesses how such data-driven analysis can improve rule-based optimization. The authors make a review by categorizing the references into four groups: the improvement of the limit parameter, the selection of the optimization option, the surrogate model, and the hybrid model. The proposed taxonomy provides them with a new perspective in the elaboration of the latest advances and research developments on the subject.
- In the document prepared by Ref. [19], the authors make an exhaustive review about planning techniques, configurations, modeling, and optimization of hybrid renewable energy systems for off-grid applications. The article presents a review of several mathematical models proposed by different researchers who have contributed with different models where a large number and variety of design parameters are involved.
- In Refs. [20,21], the authors introduce the charge density method in order to optimize the capacity of charging stations in electric vehicles. They also propose a network planning model at a minimum cost; The proposal involves operating costs, fixed system investments, and maintenance costs, as well as variability in electricity prices.
- A proposal was prepared by Ref. [22] for the optimization of energy management strategies for an autonomous photovoltaic and wind energy system integrated with fuel cells. The goal is to optimize the combination of system components in order to obtain low operating costs. The model coordinates the energy flow of system components while meeting load demand and other constraints.
- In the interesting proposal of Ref. [23], the authors develop a model to optimize the operation of a microgrid considering distributed energy resources (stochastic generation and variable demand over time), as well as cooperation limitations of the microgrid. In its model, the objective function minimizes the operating costs of the microgrid considering the classic

generation capacities and the exchange of energy with the upstream network, as well as the operational limitations. The analyzed microgrid is made up of a thermal engine, a photovoltaic system, an electrochemical storage system, and critical and interruptible loads. The analysis is carried out from a deterministic and stochastic perspective with several instances (case study) considering different operating states.

- In an interesting document, Engels et al. [24] propose an approach to optimize the control of a battery that is used both for self-consumption and for the control of the primary frequency. The model addresses both problems from a random perspective. In the first, the primary control of the frequency is optimized and in the second, the use of the battery for self-consumption is included. The authors propose a linear recharge policy to regulate the state of charge of the battery while obtaining primary frequency control. The model includes a self-consumption policy based on rules that allow optimizing the battery power reserve during peak periods.
- A very complete work related to metaheuristic optimization for insulators in gas turbines can be found in Ref. [25].
- Regarding optimal planning models of multiperiods for thermal generation through cross decomposition, it is recommended to see Ref. [26].

14.3.2.2 Energy and forecast models

- A model applied to the forecast of the energy requirement from voltaic plants can be found in Ref. [27]. In their proposal, the authors develop a forecasting method for the next day's energy production optimizing the management of renewable energy sources in smart grid applications. The model is mainly focused on optimizing the operating costs of energy production in wind farms by analyzing time series obtained from wind speed measurements in generator. In their proposal, the authors use a physical hybrid artificial neural network by means of an ANN.
- In Ref. [28], the model of the authors incorporates the randomness and intermittence of the wind. They develop an instance to study and to develop a set that includes a data analysis module, a data preprocessing module, a parameter optimization module, and a wind speed forecast module for the analysis of time series applied to management and decision-making in the design of a wind farm.
- A similar result is found in Ref. [29]. Here, authors develop a forecast model of wind speed over a 10—30-min interval. They propose a hybrid model based on the decomposition of the associated time series in the form of partial functions. Backpropagation neural networks (BP) are used in each function to make a forecast. In addition, they propose an algorithm to optimize the initial BP thresholds and weights.
- Similar results can be found in Ref. [30] where the authors propose a combined model, consisting of a data preprocessing strategy, an

optimization model, a theory without negative restrictions, and several prediction models in order to optimize the forecast of wind speeds. The tool proposed is a quasi-Newton algorithm to optimize the weight coefficients of individual models using AI. It highlights the use of hybrid models that combine multiobjective optimization, data preprocessing technology, and the use of neural networks and the Elman neural network to forecast ultrashort-term wind speed.

- In the review developed by Ref. [31], the authors assess the importance of monitoring-based ML algorithms for forecasting demand for various types of energy (wind, solar, and thermal). The proposal includes a comparative study to determine the effectiveness of the methods based on their precision. The study's conclusion suggests that Bayesian regularization backpropagation neural networks and Levenberg Marquardt backpropagation neural networks are the most powerful tools for forecasting with a correlation of up to 0.971.

- In the research of Ref. [32], the document proposes the design of a sustainable hybrid renewable energy system and a reverse osmosis desalination system with fluctuating renewable energy supply and changing water demand. The model begins with the development of a forecast using neural networks to assess the needs for energy from renewable sources and the demand for water. This is considered as a random variable in addition to ambient temperature, solar radiation, and wind speed. The improvement alternative is carried out through multicriteria optimization using extended mathematical programming to minimize total annual costs and greenhouse gas emissions. The stochastic part also evaluates the probability of potential loss of electricity supply to illustrate the sustainability of the proposed scenarios.

- In Ref. [33], the document presents an analysis of the literature on forecasting techniques using AI in order to forecast cargo demand. The document reviews, identifies, evaluates, and analyzes the performance of AI-based cargo forecasting models and open areas of research. in the subject. The precision of the forecasting model is done using ANNs, which depends on the number of parameters used as well as the architecture of the forecasting model, the combination of inputs, the activation functions, and the network training algorithm and other exogenous variables that affect the inputs of the forecasting model.

- In the paper of Ref. [34], they propose intelligent modeling techniques using fuzzy logic (FL), ANN, and ANFIS models based on sky conditions, namely clear sky/sunny, hazy sky, partly cloudy/hazy sky, and completely cloudy/hazy sky conditions to forecast the global solar energy available. The proposed model was implemented for the forecasting of photovoltaic energy in the short term in composite climatic conditions. Their results suggest the ANFIS model provides supremacy for PV forecasting compared to other models.

An interesting contribution is found in Ref. [35]. Here, they unfold an algorithm based on a convolutional neural network for the short-term forecast of wind speed in models of wind power production; the model is applied in Taiwan. The authors develop a comparison with four AI-based algorithms. Their results suggest good efficiency of the method through traditional indicators such as mean absolute error and root mean square error.

14.3.2.3 Energy computation and mathematical programming

- In Ref. [36], the authors use the Quesada Grossman algorithm (QG) and taboo search simultaneously to optimize a mixed integer nonlinear programming model using a heterogeneous parallel structure. The proposal develops a communication system between algorithms exchanging important information such as obtaining upper bounds of the solution for the QG algorithm from taboo search. They also make an update of the linearization of the model through adjustments to the upper bounds of the model and reduce the search space by integrating additional integer variables to the process. The results obtained suggest a great speed of convergence and precision in this type of model.
- In Ref. [37], a review of the main problems and challenges posed by the multiagent system and smart microgrids is presented. Future applications are considered, with attention to the integration of renewable energy resources in emerging scenarios. Several combinatorial optimization problems are integrated into the proposal that represent challenges to be improved and discussed over the next few years.
- The article of Ref. [38] presents an analysis of the application of ML models to the optimization of energy systems together with a new taxonomy of models and applications. The authors propose a novel methodology about ML models, their identification and classification according to the ML modeling technique, the type of energy, and the area of application. This document further concludes that there is an outstanding increase in the accuracy, robustness, precision, and generalizability of ML models in power systems using hybrid ML models.
- In their proposal, the authors [39] provide a general description of the state of the art and the developments related to the increase in intelligence used in the study of future smart grids. The integration of renewable sources and storage systems in electrical networks is analyzed. Demand response and energy management methods are also reviewed, as well as important automation paradigms and domain standards.
- In their document [40] they classify the optimization models applied to distributed power generation systems in autonomous and independent systems (connected to a network) with special emphasis on AI, hybrid algorithms, and PSO. The models analyzed consider objective functions

related to maximizing the reliability of the system and/or minimizing operating and resource costs.
- In Ref. [41], the author elaborates an extensive analysis on the applications of soft computing to problems in the energy sector, focusing his work on the analysis, design, and control of heat exchangers, among others. The proposal highlights the use of classical tools such as ANNs, GAs, FL, and cluster analysis. The work constitutes an excellent consultation document in the area of thermal engineering. See also [42] and Ganesan et al. [25].
- A very interesting application about bilevel optimization applied to virtual machine locations in cloud computing is available in Ref. [43].

14.3.2.4 Energy and construction

- In Ref. [44], and in order to solve a soil−structure interaction energy problem, which involves a complex load transfer mechanism from the pile to the supporting geological environment, PSO and BP algorithms were used in combination to create robust hybrid training with both local and global search capabilities. The model was used to predict the effect of energy on the load deformation of axially loaded piles.
- Regarding energy sustainability in cities and urban areas, Tai et al. [45] analyze the application of AI in energy sustainability in smart cities. The document discusses smart metering and nonintrusive load monitoring as alternatives for profiling the electricity consumption of electrical appliances. Likewise, a hybrid GA is proposed to analyze learning multicore vector machines (GA-SVM-MKL) through a multiobjective optimization model mixing elements of big data analysis.
- An interesting proposal on the use of HO models is found in Ref. [46]. In their paper, the authors propose a hybrid AI model with robust performance, to predict bench blast-induced ground vibration. In an interesting mix of tools, the paper describes the combination of an ANN with the firefly algorithm as well as a classification and regression tree, a k-nearest neighbor, and a support vector machine. The model was tested with 83 explosions in a quarry in Vietnam to assess the danger of ground vibration. The results suggest that such a combination provides reliable results in the prediction of blast-induced ground vibration.
- A highly interesting approach is found in Ref. [47]. Here, an AI proposal for forecasting slope failures in open pit mines is presented. The proposal is based on a hybridization between a GA and the M5Rules algorithm to estimate the stability of the slopes among others. The paper reports the application of the model to 450 observations of slopes in an open pit mine in Vietnam using Geo-Studio software. The results suggest that the technique is robust and reliable in slope stability analysis.

14.3.2.5 Management energy

- In the management of energy consumption in an efficient way, the application of two AI algorithms called the firefly algorithm and the GA [48] stands out. In this proposal, the GA feeds on information such as lighting, temperature, and air quality in order to maximize user comfort with a minimum of energy consumption within an intelligent building. The research results highlight the need for a preliminary forecast of the heating energy demand required in a building in order to understand the heat balance inside the building. Typically, this can be achieved using dynamic simulation software.
- In the same way, Ciulla et al. [49] focus again on the problem of forecasting the demand for heating energy in a building through the use of dynamic simulation software based on the thermal balance of the building from the use of neural networks. The authors explain how the network training was conducted from an energy database in a TRNSYS environment based on the energy standards and requirements of buildings in seven European countries.
- A similar development can be found in Ref. [50]. Here, the authors propose a decision model applied to wind energy producing systems based on its optimization through swarm intelligence and data preprocessing. The objective is to carry out a study of the energy potential available in an area in order to reduce the operating cost of wind farms. The proposal consists of two modules: wind energy potential analysis and wind speed forecast. In the first, the authors optimize the Weibull distribution through swarm intelligence in order to assess usable wind energy. In the second, they eliminate the noise associated with the time series by reprocessing the information in order to refine the energy demand forecasting models.
- An interesting approach is found in Ref. [51]. Here, the authors show the potential of supervised and transferred learning techniques to optimize an energy system. To do this, they create a surrogate model with the support of a supervised learning technique (through the use of an ANN), thereby avoiding the computationally intensive real engineering model (AEM). His proposal consists of eight different neural network architectures in the surrogate model development process. Subsequently, they propose hybrid Pareto optimization that combines surrogate and AEM to accelerate the optimization process while maintaining precision considering the net present value and the level of integration of the network as objective functions.
- Regarding the operation of the energy system and the electricity market, using AI with a holistic vision [52], a proposal is developed that presents an architectural logic of the platform that provides technical and economic alternatives for the development of controlled energy platforms through AI and their application in the study of the energy market based on massive and distributed renewable energies. In this development, a constructive and

292 Advances of Artificial Intelligence in a Green Energy Environment

inductive approach is used in the construction of theories for the conceptual proposal of the AI energy platform through the use of aggregated data from a Horizon 2020 project of the European Union and an innovation project Finnish national.

- In the article of Ref. [53], the authors elaborate an analysis of several energy management strategies applicable to renewable hybrid systems (photovoltaic, wind, hydrogen) based on hydrogen backup. In their document, they include the most outstanding technical and economic optimization criteria reported in the literature in relation to the hybridization of renewable energy systems. Its evaluation highlights the need to know the most important criteria to define efficient and effective strategies for the management of such systems.

- This article presents an analysis of the optimization techniques used in the sizing and energy management of hybrid photovoltaic/wind/battery systems. The authors [54] optimize the model by approaching it as a multi-objective problem with economic, technical, and environmental limitations. The proposal includes the use of a parameter called network power absorption probability (GPAP). The results of the model are applied to residential load and agricultural load. These suggest that the proposed technique has higher precision and faster convergence and requires less computational time compared to the PSO technique.

- In their proposal, these authors [55] analyze cutting-edge research related to wind energy, solar energy, geothermal energy, hydro energy, ocean energy, bioenergy, hydrogen energy, and hybrid energy. They conduct a comprehensive study of the role that unique AI and hybrid algorithms play in the research and development of ER sources.

- In the proposal of Ref. [56], a method is developed to carry out the management of renewable energy and the control of generation costs for its use in houses. The main idea is based on proposing a flexible allocation strategy based on a two-stage FL controller with decision limits and the corresponding sensitivity analysis. The model is applied to a hybrid system in the South China Sea. The instance achieves maximum synergy between power supply and cooling demand demonstrating better reliability, economy, and speed of response for a real-time system.

- In this proposal, the authors Ref. [57] propose the management of a hybrid renewable energy system based on the explicit consideration of uncertainty. The resulting programming problem for the ESS operation was formulated as a two-stage stochastic programming model in this study. Later, this was transformed into a mixed integer linear programming problem based on multiple equivalent scenarios.

- In the proposal of Ref. [58], they address the problem of exploring the long-term operating rules for an integrated system through a stochastic optimization model considering the uncertainty in the inflow of the reservoir and the photovoltaic energy. A long-term multiobjective optimization

model maximizes total energy production and system reliability. Later the authors apply dynamic programming in order to obtain the optimal trajectory obtained using a linear fit method. Finally, they use simulation to optimize the parameters of the rules.
- An excellent work related to the problem of economic cargo dispatch considering a valve-point effect through the bat algorithm is found in Ref. [59].

14.4 Conclusions

In this chapter we have reviewed the optimization of stochastic models applied to the analysis of energy systems and/or related systems. A random sample of the most used methods and models was obtained in five areas common to engineering. In most of the proposals analyzed, it is common to find highly complicated models in their mathematical structure and with a high degree of difficulty for their computational solution (NP-hard). In all of them, the authors fundamentally address emerging methodologies combined to achieve optimal solutions through the use of hybrid algorithms and AI. It should be noted that the ML technique constitutes an important source of application in the design and operation of energy systems.

As a result of this experience, it is important to note the wide range of possibilities offered by this field of research for the development of new methods and stochastic optimization algorithms applied to the energy sector and related fields such as manufacturing, design, and construction of forecasting models.

In particular, AI and ML are two of the most promising alternatives to address the growing complexity inherent in the subject.

Finally, we wish to highlight the importance of stochastic optimization in the analysis of complex systems of all types, since the inclusion of randomness makes the solutions provided more real and, therefore, the decisions taken in this regard more useful. There is no doubt that there is a huge vein of possibilities for this precious field of mathematical programming.

Abbreviations

AHPM	Analytical hierarchy process method
AI	Artificial intelligence
ANFIS	Adaptive fuzzy neural inference system
ANN	Artificial neural network
BP	Backpropagation neural network
CEEMD	Complementary set decomposition
DEG	Distributed energy generation
EE	Energy economics
EES	Energy storage system

EMP	Extended mathematical programming
EMS	Energy management strategies
EV	Electric vehicles
FFA	Firefly algorithm
FL	Fuzzy logic
FMS	Flexible manufacturing systems
GA	Genetic algorithm
GA-SVM-MKL	Vector machine multicore learning approach
H	Heuristics
HO	Hybrid optimization
HOA	Hybrid Pareto optimization
IMF	Intrinsic model functions
LMBNN	Levenberg Marquardt backpropagation on neural networks
LP	Linear programming
MA	Monte Carlo analysis
MILP	Mixed integer linear programming
MINLP	Mixed integer nonlinear programming
ML	Machine learning
MOGA	Multi-objective genetic algorithm
NILM	Nonintrusive load monitoring
NLP	Nonlinear programming
PHANN	Physical hybrid artificial neural network
PIL	Pure integer linear programming
PSE	Process system engineering
PSO	Particle swarm optimization
QG	Quesada Grossmann
RAE	Redundant antenna elimination
RFID	Radio frequency identification systems
RNN	Recurrent neural networks
RPLNN	Ring probabilistic logical neural networks
SP	Stochastic programming
TS	Time series
TS	Taboo search

Acknowledgments

The authors are deeply grateful to Professors Pandian Vasant and Joshua. J. Thomas as well as their team of collaborators for the effort spent in the edition of this book. We also deeply appreciate the arbitration and the suggestions of the anonymous referees to improve this proposal. Many thanks also to Elsevier Editors.

References

[1] J. Cavazos, J.E.B. Moss, M.F.P. O'Boyle, in: A. Mycroft, A. Zeller (Eds.), Hybrid Optimizations: Which Optimization Algorithm to Use? In Compiler Construction, Springer Berlin Heidelberg, Berlin, Heidelberg, 2006, pp. 124−138, https://doi.org/10.1007/11688839_12.

[2] S. Bhattacharyya, P. Dutta, B. Chakraborty, Particle Swarm Optimization Algorithm and its Hybrid Variants for Feature Subset Selection, 2013, https://doi.org/10.4018/978-1-4666-2518-1.ch017.

Hybrid optimization and artificial intelligence **Chapter | 14 295**

[3] T. Sands, Development of deterministic artificial intelligence for unmanned underwater vehicles (UUV), J. Mar. Sci. Eng. 8 (2020) 578, https://doi.org/10.3390/jmse8080578.

[4] B. Liu, Y. Wang, Energy system optimization under uncertainties: a comprehensive review, in: J. Ren, Y. Wang, C. He (Eds.), Towards Sustainable Chemical Processes, Elsevier, 2020, pp. 149−170, https://doi.org/10.1016/B978-0-12-818376-2.00006-5. ISBN 9780128183762.

[5] D. Stefanoiu, P. Borne, D. Popescu, F.G. Filip, A. El Kamel, Optimization in Engineering Sciences: Metaheuristic, Stochastic Methods and Decision, Wiley, 2014.

[6] J. Kiefer, J. Wolfowitz, Stochastic estimation of the maximum of a regression function, Ann. Math. Stat. 23 (3) (1952) 462−466, https://doi.org/10.1214/aoms/1177729392.

[7] S. Andradóttir, A Review of Random Search Methods, 2015, https://doi.org/10.1007/978-1-4939-1384-8_10.

[8] D. Holmes, Software for Stochastic Programming, 2020. Recovered from: http://users.iems.northwestern.edu/∼jrbirge/html/dholmes/SPTSslpfaq.html.

[9] LINGO 19.0 - Optimization Modeling Software for Linear, Nonlinear, and Integer Programming, 2021. Recovered from: https://www.lindo.com/lindoforms/downlingo.html.

[10] I.E. Grossmann, I. Harjunkoski, Process system engineering: academic and industrial perspectives, Comput. Chem. Eng. 126 (2019) 474−484.

[11] I.E. Grossmann, R.M. Apap, A. Bruno, P. García-Herreros, Q. Zhang, Recent advances in mathematical programming techniques for the optimization of process systems under uncertainty, Comput. Chem. Eng. 91 (2016) 3−14, https://doi.org/10.1016/j.compchemeng.2016.03.002.

[12] P.N. Basan, C.A. Méndez, Hybrid MILP/simulation/heuristic algorithms to complex hoist scheduling problems, Comput. Aided Chem. Eng. 38 (2016) 1929−1934.

[13] A. Azizi, Introducing a novel hybrid artificial intelligence algorithm to optimize network of industrial applications in modern manufacturing, Complexity 2017 (2017) 18, https://doi.org/10.1155/2017/8728209. ID 8728209.

[14] E.M. De Medeiros, J.A. Posada, H. Noorman, R.M. Filhob, Modeling and multi-objective optimization of syngas fermentation in a bubble column reactor, in: A.A. Kiss, E. Zondervan, R. Lakerveld, L. Özkan (Eds.), Computer Aided Chemical Engineering, vol. 46, Elsevier, 2019, pp. 1531−1536, https://doi.org/10.1016/B978-0-12-818634-3.50256-3. ISSN 1570-7946, ISBN 9780128186343.

[15] G. Pérez Lechuga, Optimal logistics strategy to distribute medicines in clinics and hospitals, J. Math. Ind. 8 (2018) 2, https://doi.org/10.1186/s13362-018-0044-5.

[16] F. Fodhil, A. Hamidat, O. Nadjemi, Potential, optimization and sensitivity analysis of photovoltaic-diesel-battery hybrid energy system for rural electrification in Algeria, Energy 169 (2019) 613−624. ISSN 0360-5442.

[17] J. Islam, P.M. Vasant, B.M. Negash, M.B. Laruccia, Watada, M. Myint, Junzo, A holistic review on artificial intelligence techniques for well placement optimization problem, Adv. Eng. Softw. 141 (2020) 102767. ISSN 0965-9978.

[18] G. Ruan, H. Zhong, G. Zhang, Y. He, X. Wang, T. Pu, Review of learning-assisted power system optimization, CSEE J. Power Energy Syst. 7 (2) (2021) 221−231, https://doi.org/10.17775/CSEEJPES.2020.03070.

[19] R. Siddaiah, R.P. Saini, A review on planning, configurations, modeling and optimization techniques of hybrid renewable energy systems for off grid applications, Renew. Sustain. Energy Rev. 58 (2016) 376−396. ISSN 1364-0321.

[20] P. Vasant, J.A. Marmolejo, I. Litvinchev, R. Rodriguez Aguilar, Nature-inspired metaheuristics approaches for charging plug-in hybrid electric vehicle, Wireless Network 26 (7) (2020) 4753−4766.

296 Advances of Artificial Intelligence in a Green Energy Environment

[21] X. Lin, J. Sun, S. Ai, X. Xiong, Y. Wan, D. Yang, Distribution network planning integrating charging stations of electric vehicle with V2G, Int. J. Electr. Power Energy Syst. 63 (2014) 507−512. ISSN 0142-0615.

[22] A.L. Bukar, C.W. Tan, Review on stand-alone photovoltaic-wind energy system with fuel cell: system optimization and energy management strategy, J. Clean. Prod. 221 (2019) 73−88. ISSN 0959-6526.

[23] G.C. Lazaroiu, V. Dumbrava, G. Balaban, M. Longo, D. Zaninelli, Stochastic optimization of microgrids with renewable and storage energy systems, in: IEEE 16th International Conference on Environment and Electrical Engineering (EEEIC), 2016.

[24] J. Engels, B. Claessens, G. Deconinck, Combined stochastic optimization of frequency control and self-consumption with a battery, IEEE Trans. Smart Grid 10 (2) (March 2019) 1971−1981, https://doi.org/10.1109/TSG.2017.2785040.

[25] T. Ganesan, P. Vasant, I. Litvinchev, M. Aris, Extreme value metaheuristics and coupled mapped lattice approaches for gas turbine-absorption chiller optimization, in: P. Vasant, G. Weber, W. Punurai (Eds.), Research Advancements in Smart Technology, Optimization, and Renewable Energy, IGI Global USA, 2021, pp. 283−312 (Chapter 14).

[26] A. Marmolejo, I. Litvinchev, R. Aceves, J.M. Ramirez, Multiperiod optimal planning of thermal generation using cross decomposition, J. Comput. Syst. Sci. Int. 50 (5) (2011) 793−804.

[27] A. Gandelli, F. Grimaccia, S. Leva, M. Mussetta, E. Ogliari, Hybrid model analysis and validation for PV energy production forecasting, in: Proceedings of the International Joint Conference on Neural Networks, 2014, pp. 1957−1962, https://doi.org/10.1109/IJCNN. 2014.6889786.

[28] J. Wang, T. Niu, H. Lu, Z. Guo, W. Yang, P. Du, An analysis-forecast system for uncertainty modeling of wind speed: a case study of large-scale wind farms, Appl Energy 211 (2018) 492−512. ISSN 0306-2619.

[29] Z. Yang, J. Wang, A hybrid forecasting approach applied in wind speed forecasting based on a data processing strategy and an optimized artificial intelligence algorithm, Energy (2018).

[30] Z. Yang, J. Wang, A combination 1% forecasting approach applied in multistep wind speed forecasting based on a data processing strategy and an optimized artificial intelligence algorithm, Appl. Energy (2018).

[31] T. Ahmad, H. Chen, A review on machine learning forecasting growth trends and their real-time applications in different energy systems, Sustain. Cities Soc. 54 (2020) 102010. ISSN 2210-6707.

[32] Q. Li, J. Loy-Benitez, K. Nam, S. Hwangbo, J. Rashidi, C.H.K. Yoo, Sustainable and reliable design of reverse osmosis desalination with hybrid renewable energy systems through supply chain forecasting using recurrent neural networks, Energy 178 (2019) 277−292. ISSN 0360-5442.

[33] M. Qamar Raza, K. Abbas, A review on artificial intelligence-based load demand forecasting techniques for smart grid and buildings, Renew. Sustain. Energy Rev. 50 (2015) 1352−1372. ISSN 1364-0321.

[34] G. Perveen, M. Rizwan, N. Goel, An ANFIS-based model for solar energy forecasting and its smart grid application, Eng. Reports 1 (2019) e12070, https://doi.org/10.1002/eng2.12070.

[35] C.-J. Huang, P.-H. Kuo, A short-term wind speed forecasting model by using artificial neural networks with stochastic optimization for renewable energy systems, Energies 11 (10) (2018) 2777, https://doi.org/10.3390/en11102777.

[36] K. Zhou, Kai, W. Wan, Wey, Xi Chen, Z. Shao, L.T. Biegler, A parallel method with hybrid algorithms for mixed integer nonlinear programming, in: A. Kraslawski, I. Turunen (Eds.), Computer Aided Chemical Engineering, vol. 32, Elsevier, 2013, pp. 271−276, https://doi.org/10.1016/B978-0-444-63234-0.50046-4. ISSN 1570-7946, ISBN 9780444632340.

[37] V.N. Coelho, W.M. Cohen, I.M. Coelho, N. Liu, F.G. Guimarães, Multi-agent systems applied for energy systems integration: state-of-the-art applications and trends in micro-grids, Appl. Energy 187 (2017) 820−832. ISSN 0306-2619.

[38] M. Amir, S. Mohsen, S. Faizollahzadeh Ardabili, Rabczuk, S. Shamshirband, Timon, A.R. Varkonyi-Koczy, State of the art of machine learning models in energy systems, a systematic review, Energies 12 (7) (2019), https://doi.org/10.3390/en12071301.

[39] T. Strasser, et al., A review of architectures and concepts for intelligence in future electric energy systems, IEEE Trans. Ind. Electron. 62 (4) (2015) 2424−2438, https://doi.org/10.1109/TIE.2014.2361486.

[40] S. Twaha, M.A.M. Ramli, A review of optimization approaches for hybrid distributed energy generation systems: off-grid and grid-connected systems, Sustain. Cities Soc. 41 (2018) 320−331.

[41] A. Pacheco-Vega, Soft computing applications in thermal energy systems, in: K. Gopalakrishnan, S.K. Khaitan, S. Kalogirou (Eds.), Soft Computing in Green and Renewable Energy Systems. Studies in Fuzziness and Soft Computing, vol. 269, Springer, Berlin, Heidelberg, 2011.

[42] T. Ganesan, P. Vasant, I. Litvinchev, Multiobjective optimization of a biofuel supply chain using random matrix generators, in: J.J. Thomas, P. Karagoz, B.B. Ahamed, P. Vasant (Eds.), Deep Learning Techniques and Optimization Strategies in Big Data Analytics, IGI Global USA, 2020, pp. 206−232 (Chapter 13).

[43] T. Ganesan, P. Vasant, I. Litvinchev, Chaotic simulator for bilevel optimization of virtual machine placements in cloud computing, J. Oper. Res. Soc. China (2021), https://doi.org/10.1007/s40305-020-00326-5.

[44] A. Ismail, D.-S. Jeng, L.L. Zhang, An optimized product-unit neural network with a novel PSO−BP hybrid training algorithm: applications to load−deformation analysis of axially loaded piles, Eng. Appl. Artif. Intell. 26 (10) (2013) 2305−2314, https://doi.org/10.1016/j.engappai.2013.04.007. ISSN 0952-1976.

[45] K.T. Chui, M.D. Lytras, A. Visvizi, Energy sustainability in smart cities: artificial intelligence, smart monitoring, and optimization of energy consumption, Energies 11 (2018) 2869, https://doi.org/10.3390/en11112869.

[46] Y. Shang, H. Nguyen, X.N. Bui, et al., A novel artificial intelligence approach to predict blast-induced ground vibration in open-pit mines based on the firefly algorithm and artificial neural network, Nat. Resour. Res. 29 (2020) 723−737, https://doi.org/10.1007/s11053-019-09503-7.

[47] X.N. Bui, H. Nguyen, Y. Choi, et al., Prediction of slope failure in open-pit mines using a novel hybrid artificial intelligence model based on decision tree and evolution algorithm, Sci. Rep. 10 (2020) 9939, https://doi.org/10.1038/s41598-020-66904-y.

[48] Wahid, L.H. Ismail, R. Ghazali, M. Aamir, An efficient artificial intelligence hybrid approach for energy management in intelligent buildings, KSII Trans. Internet Inf. Syst. 13 (12) (2019) 5904−5927, https://doi.org/10.3837/tiis.2019.12.007.

[49] G. Ciulla, A. D'Amico, V. Lo Brano, M. Traverso, Application of optimized artificial intelligence algorithm to evaluate the heating energy demand of non-residential buildings at European level, Energy 176 (2019) 380−391, https://doi.org/10.1016/j.energy.2019.03.168. ISSN 0360-5442.

298 Advances of Artificial Intelligence in a Green Energy Environment

[50] X. Zhao, C. Wang, J. Su, J. Wang, Research and application based on the swarm intelligence algorithm and artificial intelligence for wind farm decision system, Renew. Energy 134 (2019) 681−697. ISSN 0960-1481.

[51] A.T.D. Perera, P.U. Wickramasinghe, V.M. Nik, J.-L. Scartezzini, Machine learning methods to assist energy system optimization, Appl. Energy 243 (2019) 191−205. ISSN 0306-2619.

[52] Y. Xu, P. Ahokangas, J.-N. Louis, E. Pongrácz, Electricity market empowered by artificial intelligence: a platform approach, Energies 12 (2019) 4128, https://doi.org/10.3390/en12214128.

[53] F.F.J. Vivas, A. De las Heras, M.F. Segura, M.J.M. Andújar, A review of energy management strategies for renewable hybrid energy systems with hydrogen backup, Renew. Sustain. Energy Rev. 82 (Part 1) (2018) 126−155. ISSN 1364-0321.

[54] O. Nadjemi, T. Nacer, A. Hamidat, H. Salhi, Optimal hybrid PV/wind energy system sizing: application of cuckoo search algorithm for Algerian dairy farms, Renew. Sustain. Energy Rev. 70 (2017) 1352−1365. ISSN 1364-0321.

[55] S.K. Jha, B. Jasmin, A. Jha, P. Nilesh, H. Zhang, Renewable energy: present research and future scope of Artificial Intelligence, Renew. Sustain. Energy Rev. 77 (2017) 297−317. ISSN 1364-0321.

[56] L. Xu, Z. Wang, Y. Liu, L. Xing, Energy allocation strategy based on fuzzy control considering optimal decision boundaries of standalone hybrid energy systems, J. Clean. Prod. 279 (123810) (2021). ISSN 0959-6526.

[57] J. Yu, J.H. Ryu, I. Lee, A stochastic optimization approach to the design and operation planning of a hybrid renewable energy system, Appl. Energy 247 (2019) 212−220. ISSN 0306-2619.

[58] Z. Yang, L. Pan, L. Cheng, H. Wang, B. Ming, W. Gong, Deriving operating rules for a large-scale hydro-photovoltaic power system using implicit stochastic optimization, J. Clean. Prod. 195 (2018) 562−572. ISSN 0959-6526.

[59] P. Vasant, F.P. Mahdi, J.A. Marmolejo-Saucedo, I. Litvinchev, R.R. Aguilar, J. Watada, Quantum-behaved bat algorithm for solving the economic load dispatch problem considering a valve-point effect, Int. J. Appl. Metaheuristic Comput. 11 (3) (2020) 41−57.

Further reading

[1] Access IEEE, Artificial Intelligence Technologies for Electric Power Systems, December 31, 2019. Recovered from: https://ieeeaccess.ieee.org/closed-special-sections/artificial-intelligence-technologies-for-electric-power-systems/.

[2] T. Fu, C. Wang, A hybrid wind speed forecasting method and wind energy resource analysis based on a swarm intelligence optimization algorithm and an artificial intelligence model, Sustainability 10 (2018) 3913, https://doi.org/10.3390/su10113913.

[3] T. Ganesan, P. Vasant, P. Sanghvi, J. Thomas, I. Litvinchev, Random matrix generators for optimizing a fuzzy biofuel supply chain system, J. Adv. Eng. Comput. 4 (1) (2020) 33−50.

[4] J. Wang, Z. Yang, Ultra-short-term wind speed forecasting using an optimized artificial intelligence algorithm, Renew. Energy 171 (2021) 1418−1435. ISSN 0960-1481.

[5] P. Lammich, Efficient verified implementation of introsort and pdqsort, in: N. Peltier, V. Sofronie-Stokkermans (Eds.), Automated Reasoning. IJCAR 2020. Lecture Notes in Computer Science vol. 12167, Springer, Cham, 2020, https://doi.org/10.1007/978-3-030-51054-1_18.

Hybrid optimization and artificial intelligence **Chapter | 14 299**

[6] Neos Guide, Stochastic Programming (a), (without date), Recovered from: https://neos-guide.org/content/stochastic-programming.

[7] G. Perveen, M. Rizwan, N. Goel, Comparison of intelligent modelling techniques for forecasting solar energy and its application in solar PV based energy system, IET Energy Syst. Integr. 1 (1) (2019) 34–51, https://doi.org/10.1049/iet-esi.2018.0011. Online ISSN 2516-8401.

[8] Stochastic Programming (b), (without date), Recovered from: https://www.lindo.com/doc/online_help/lingo15_0/stochastic_programming_script_functions.htm.

[9] A.S.R. Subramanian, T. Gundersen, T.A. Adams, II. Modeling and simulation of energy systems: a review, Processes 6 (2018) 238, https://doi.org/10.3390/pr6120238.

[10] T. Ting, X.-S. Yang, S. Cheng, K. Huang, Hybrid metaheuristic algorithms: past, present, and future, in: X.S. Yang (Ed.), Recent Advances in Swarm Intelligence and Evolutionary Computation. Studies in Computational Intelligence, vol. 585, Springer, Cham, 2015, https://doi.org/10.1007/978-3-319-13826-8_4.

Chapter 15

A brief literature review of quantitative models for sustainable supply chain management

Pablo Flores-Sigüenza[1], Jose Antonio Marmolejo-Saucedo[2] and Roman Rodríguez-Aguilar[3]

[1]*Facultad de Ingeniería, Universidad Anáhuac México, Naucalpan de Juárez, Estado de México, México;* [2]*Universidad Panamericana, Facultad de Ingeniería, Mexico City, Mexico;* [3]*Facultad de Ciencias Económicas y Empresariales, Universidad Panamericana, Ciudad de México, México*

15.1 Introduction

Supply chain management (SCM) has now become the basis for executing operations, being regarded as the core of the business function in the 21st century [1]. According to Ref. [2], well-structured supply chains (SC) become strategic tools that help companies generate competitive advantages in today's market. Additionally, the interest in reducing environmental impact has caused governments and consumers to demand and pressure companies to be aware of and, above all, to reduce the pollution generated by their products and industrial processes [3]. To address these needs, especially during the last few years, sustainable supply chain management (SSCM) attracts significant attention among managers, researchers, and practitioners [4].

The World Commission on Environment and Development (WCED) defines and maintains that sustainable development is capable of meeting the needs of the present generation without compromising the ability of the future generation to meet its own needs [5]. Tautenhain et al. [6] mention that the inclusion of sustainable development in our lives and activities will become the only way to solve most of the global problems that exist today, such as climate change, water scarcity, inequality, poverty, hunger, among others.

As a result, companies are constantly researching new methods and strategies to make their operations sustainable. Research includes the use of renewable energy, development of environmentally friendly raw materials, green procurement, selection of green suppliers, reduction of plastic

Advances of Artificial Intelligence in a Green Energy Environment
https://doi.org/10.1016/B978-0-323-89785-3.00005-0
Copyright © 2022 Elsevier Inc. All rights reserved.

packaging, development of closed-loop supply chains (CLSC), reduction of carbon and greenhouse gas (GHG) emissions, remanufacturing, reverse logistics, among others [7].

These strategic decisions are the basis for the development of a sustainable supply chain network (SSCN) design, which is characterized by finding the best location of facilities, capacities, and flow between them, maximizing economic and social performance while minimizing environmental impact [8].

SSCM has always encouraged researchers to generate scientific output, initially based on environmental indicators, in addition to classical economic considerations. After a few years, studies began to timidly include social indicators, but it was not until 1994 that the concept of the "triple bottom line" (TBL) was first introduced by Elkington [9], which seeks to balance the three dimensions of business: economic, environmental, and social.

Rajeev et al. [10] showed in their literature review that in the last two decades there has been a considerable growth in the number of studies conducted on SSCM, empirical studies, conceptual models, and formal quantitative models highlighting. A formal quantitative model, unlike the others, is based on a set of strategies that allow obtaining and processing information through statistics and formal numerical techniques that are framed within a cause—effect relationship [11]. Quantitative models have been more used in CLSC; therefore, there are several review articles focused on these aspects [12]. A recent and exhaustive review of Ref. [13] shows that there are few studies looking at quantitative methods and the forward SC; recent works such as Ref. [14] demonstrate the importance of filling this gap and carry out a detailed review of the latest quantitative models applied in the forward SC that allow obtaining sustainability.

SC sustainability can be improved from multiple approaches; therefore, this chapter helps to visualize and understand some models used in the past and present to improve this area. Consequently, the main objective of this study is to provide a synthesis of the key elements of the quantitative model offerings that use sustainability indicators in the design and management of forward SC in order to finally identify trends, gaps, and lessons in the selected literature.

Studies such as this one guide companies and researchers to understand the current lines of research, the models being used and the solution approaches applied to find optimal and innovative results in terms of sustainability. For a better understanding, the document is structured as follows: Section 2 explains key concepts on SC and sustainability and an overview of literature reviews. Section 3 describes the methodology used to collect and analyze the literature. Section 4 presents the research findings by the content analysis and identifies directions for future research. Section 5 provides a brief discussion. And finally, Section 6 synthesizes the main conclusions.

15.2 Theoretical foundation and literature reviews

15.2.1 Supply chain management

The main elements that make up an SC are suppliers, plants, distribution centers, retailers, and customers, whose main function is to acquire raw materials, process them, and distribute the finished products to customers. Then SC network (SCN) design optimizes the configuration of all these elements in order to minimize total costs while meeting service levels [15] and is an area of decision-making that considers parameters such as planning, costs, demand, and supply.

The decisions that guide an SCN design are born at a strategic and tactical management level [8], are usually made in the long term due to their importance, and have an immediate impact on the performance of an SC [16].

In our study, two types of SC configuration are mentioned, forward SC and CLSC; the first, also known as traditional SC, deals with the flow of products from raw materials to the manufacturer, to the retailer, and finally to the consumer [17], instead a CLSC according to Ref. [18] essentially combines the traditional supply chain with reverse logistics, considering the item after it served its original purpose, the manufacturer works to encourage the item's return once it's no longer functional or needed and the items can either be repaired and resold, or they can be broken down for reuse in future products.

SCM, therefore, encompasses a management of each and every one of the links that are generated between the elements of an SC [19]; these links include in their flows physical, financial, and informational variables that must be managed jointly among all participants [20].

15.2.2 Sustainable supply chain management

The Dow Chemical Company says sustainability is about making every decision with the future in mind. The probability of a company's success improves with the implementation of sustainability [21] and its application in all its activities, considering not only economic but also environmental and social variables.

The three basic dimensions of sustainable development are economic, social, and environmental, a concept known as TBL, industries wishing to achieve this result must have the capacity and commitment to design, plan, and operate their SC based on sustainability requirements, without compromising resources of other stakeholders [22]. The main obstacle is the level of complexity that this represents, as it involves several products, suppliers, materials, capabilities, and other variables.

SSCM has been created in order to integrate two general concerns that arise in the participants of a supply chain, obtaining the greatest amount of profit and reducing the environmental and social impact of operations [23]. Seuring [24] defines SSCM as "the management of flows of products, material, money, and information, as well as cooperation between SC elements,

304 Advances of Artificial Intelligence in a Green Energy Environment

considering the objectives of the three dimensions of sustainable development (environmental, social, and economic) arising from customer and stakeholder requirements."

Professionals, technicians, managers, and researchers use the word SSCM more frequently today [25], investing large amounts of resources in the development and implementation of sustainability, highlight the importance of an SSCM, and value its role by adjusting to different business needs, in addition to raising awareness about sustainable practices such as the selection of green suppliers, ethical sourcing, products with carbon footprints, social responsibility, etc. [26].

15.2.3 Literature reviews

In recent years, a significant number of literature review articles have been published on the various topics covered by SSCM. We can find systematic and narrative reviews. Systematic literature reviews involve a methodical process composed of structured steps to collect and analyze the material found; this type of literature review is considered exhaustive, objective, valid, verifiable, and reproducible [27]. Narrative reviews, on the other hand, depend mainly on the experience of the researchers, the surveys that are applied, and a less formal scheme [28].

To get an idea of the amount of literature reviews on January 10, 2021, a search was performed in the ScienceDirect database with the following equation "sustainable supply chain management" AND "literature review" yielding 915 articles, which is evidence that these types of articles are currently an abundant resource that allows researchers to know the state of the art regarding sustainability in SC and guide their current and future research.

According to Ref. [13] in their study of 198 literature reviews, it shows that this type of work is undergoing a change with respect to methodology; the majority of reviews have gone from being narrative to systematic. The majority of literature reviews include conceptual models and formal models. Regarding formal models, studies in CLSC such as in Ref. [18] stand out, and conversely, there are few reviews of formal models in forward SC, like in Ref. [25]. The literature of SSCM studies multiple viewpoints, from focal firms [29] to particular country profiles [30] or segments of the population [31].

From the beginning and for many years, the environmental perspective dominated sustainability research, but now it is changing; TBL is the new approach that most works try to include. Some highlights are as follows: SSCM [32], green supply chain [4], SSCN design [10], dynamics modeling for SSCM [33], sustainable supply chain finance [34], and sustainable network design under uncertainty [7].

The environmental perspective covers various topics such as GHG emissions, pollution, CO_2 emissions, natural resources, energy, and others. As for the social variables included in the studies, we have training, worker safety, ethical SC, social justice, decent work, and fair treatment.

15.3 Methodology

This chapter performs a systematic literature review on SSCM focusing on the use of formal quantitative models in forward SSC; for its correct execution, decision-making, quality, consistency, and procedures to follow, we rely on the work and methodologies of two authors [27,35]. Together they will allow us to collect the appropriate information and analyze its content in a structured way.

Therefore, our literature review will consist of four general steps: (1) material collection, (2) descriptive analysis, (3) category identification, and (4) material evaluation [35]. To obtain and filter the optimal material for step 1, four tasks of the Fink methodology will be followed: (1) selection of research questions, (2) definition of database sources, (3) selection of search terms, and (4) application of practical inclusion and exclusion criteria.

15.3.1 Material collection

The following four steps, taken from Fink's methodology, will allow us to collect and filter quality articles related to our study.

15.3.1.1 Selection of research questions

The research questions guiding our systematic literature review are:

- What quantitative models have been applied in the effective management of a forward SC considering sustainability indicators?
- What characteristics, indicators, solution approaches are used in these quantitative models?
- How these quantitative models have supported sustainability decisions?

15.3.1.2 Definition of database sources

The database sources to be used in the literature review are

- Scopus
- ScienceDirect
- SpringerLink
- Web of Science

15.3.1.3 Selection of search terms

The keywords and phrases extracted from the research questions, which will serve as search terms and search strings, are

- "Quantitative Models" AND "Sustainable Supply Chain"
- "Quantitative Models" AND "Green Supply Chain"
- "Quantitative Models" AND "Sustainable Supply Chain Network design"
- "Implementation" OR "Application" AND "Quantitative Models" AND "Sustainable Supply Chain"

15.3.1.4 *Application of practical inclusion and exclusion criteria*

The inclusion criteria that have been defined are

- Articles written in English language
- Articles published in peer-reviewed journals
- Articles published between January 2010 and December 2020
- Studies based on a set of strategies that obtain and process information using statistics and numerical techniques

On the other hand, the exclusion criteria to be considered are

- Duplicity
- Articles focusing on CLSC, remanufacturing or reverse logistics
- Empirical studies and conceptual models

15.3.2 Descriptive analysis

Once the steps for collecting the different articles have been established, the descriptive analysis is responsible for analyzing the time and journals in which they have been published by means of a temporal distribution over the study horizon and a 4W analysis (when, who, what, and where).

To perform the 4W analysis and answer these questions, the articles (when) are divided by year of publication, (who) the journals where the different articles have been published are identified, (what) the quantitative models applied in SSC are analyzed, and, finally, (where) the institutions to which the researchers belong and their host country are identified.

15.3.3 Category identification

In this section, three structural dimensions are defined and grouped by category: SCM, modeling, and sustainability. These will provide a comprehensive and deep understanding of how quantitative models have been used in forward SC, how they have supported sustainability decisions, and which are the main sustainability pillars used.

The dimensions of the first category, SCM, are taken from the "Supply Chain Operations Reference" (SCOR) model [36]. The modeling category is evaluated according to the purpose and type of the model. And the last category, sustainability, its classification is based on the three pillars established in the "" concept, which are environmental, economic, and social.

15.3.4 Material evaluation

In this step the collected items are coded according to the dimensions of each category described in Section 3.3. This allows reflecting the SC structure, the modeling dimension, and their interaction to manage sustainability results.

Subsequently, frequency of occurrence tables is produced, which are the basis for the content analysis, as they allow identifying dominant characteristics and gaps in existing research that can guide future studies. In general, the research process is documented step by step in a clear and transparent manner to increase its objectivity.

15.4 Results

15.4.1 Material collection

The steps of the selection process and the associated paper counts have been summarized in Fig. 15.1 allowing to visualize them in a global way.

After rigorously following all the established steps to ensure the quality and objectivity of our work, we have obtained 80 articles, which represent the sample of our systematic literature review.

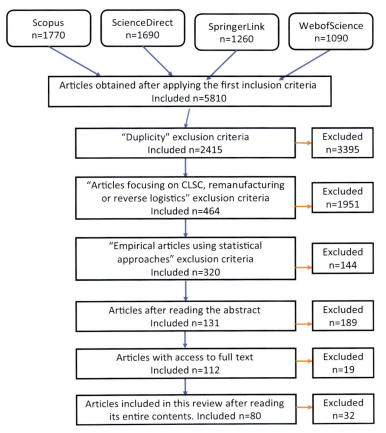

FIGURE 15.1 Diagram of material collection.

15.4.2 Descriptive analysis

In order to answer the 4W analysis proposed in Section 3.2 of the methodology, we start with the "when" field, for which we use a time distribution in Fig. 15.2, where we observe the number of articles published in the different years analyzed in our sample. There is an upward trend as the years progress, with a direct relationship between the variables, years, and the number of articles published. Most of the publications occurred in 2019 (15) and 2020 (13).

Regarding the "who" field, the articles in the sample have been published by 38 different journals, and only 11 of them have two or more publications.

Table 15.1 shows all the journals in the sample and the number of articles published in each of them in descending order and also includes a column indicating the types of quantitative models that have been applied and can be found in each journal, in response to the "what" analysis.

According to Table 15.1, the journals with the highest number of publications are the following: *Journal of Cleaner Production* (11 papers) tops the list, followed by *Computers & Industrial Engineering* (10 papers), *International Journal of Production Economics* (7), *Annals of Operations Research* (6), *Computers & Operations Research* (4), *European Journal of Operational Research* (4), *Clean Technologies and Environmental Policy* (3), *Energy* (2), *Journal of Manufacturing Technology Management* (2), *Transportation Research Part D* (2), and *Transportation Research Part E* (2). The others with 1 each one follow.

As evidenced in Table 15.1, SSCM-related research has been published in journals from diverse knowledge areas and topics, some with a specific focus on SC and operations management and others general and interdisciplinary, demonstrating once again its growing boom.

Regarding the types of quantitative models applied, it can be concluded that mathematical programming model is the most used (49 papers), followed by analytical (15), heuristics (5), hybrid (5), simulation (4), and various (2).

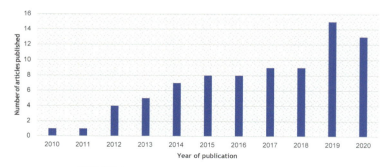

FIGURE 15.2 Time distribution of reference papers.

A brief literature review of quantitative models Chapter | 15 **309**

TABLE 15.1 Distribution of papers by journal.

Journal	Papers	Quantitative model type
Journal of Cleaner Production	11	Analytical (2), mathematical programming (9)
Computers & Industrial Engineering	10	Analytical (1), heuristics (2), hybrid (1), mathematical programming (6)
International Journal of Production Economics	7	Analytical (1), mathematical programming (5), simulation (1)
Annals of Operations Research	6	Hybrid (1), mathematical programming (4), simulation (1)
Computers & Operations Research	4	Analytical (1), hybrid (1), mathematical programming (2)
European Journal of Operational Research	4	Analytical (2), mathematical programming (2)
Clean Technologies and Environmental Policy	3	Hybrid (1), mathematical programming (1), Various (1)
Energy	2	Mathematical programming (2)
Journal of Manufacturing Technology Management	2	Analytical (1), mathematical programming (1)
Transportation Research Part D	2	Analytical (1), heuristics (1)
Transportation Research Part E	2	Mathematical programming (2)
American Institute of Chemical Engineers Journal	1	Mathematical programming
Biomass and Bioenergy	1	Mathematical programming
Biomass Conversion and Biorefinery	1	Mathematical programming
Canadian Journal of Forest Research	1	Simulation
Chaos, Solitons & Fractals	1	Mathematical programming
Computational Economics	1	Mathematical programming
Computer Aided Chemical Engineering	1	Simulation
Environmental Technology	1	Mathematical programming
Global Journal of Flexible Systems Management	1	Analytical
IFAC Proceedings Volumes	1	Heuristics
Industrial & Engineering Chemistry Research	1	Mathematical programming

Continued

310 Advances of Artificial Intelligence in a Green Energy Environment

TABLE 15.1 Distribution of papers by journal.—cont'd

Journal	Papers	Quantitative model type
Industrial Management & Data Systems	1	Analytical
International Journal of Advanced Manufacturing Technology	1	Mathematical programming
International Journal of Industrial Engineering Computations	1	Mathematical programming
International Journal of Production Research	1	Analytical
Journal of Intelligent & Fuzzy Systems	1	Mathematical programming
Journal of Intelligent Manufacturing	1	Heuristics
Journal of the Transportation Research Board	1	Mathematical programming
Management of Environmental Quality	1	Analytical
Mathematical Problems in Engineering	1	Mathematical programming
Operational Research	1	Analytical
Renewable and Sustainable Energy Reviews	1	Mathematical programming
Scientia Iranica	1	Hybrid
SpringerPlus	1	Analytical
Sustainability	1	Mathematical programming
Sustainable Energy Technologies and Assessments	1	Mathematical programming
Sustainable Production and Consumption	1	Various

Finally, the "where"" field is answered with the analysis of the geographical origin of the sample, i.e., the country of the institution associated with the researcher. Fig. 15.3 indicates that there are 24 countries, being those with the highest scientific production: Iran (14 papers), China (11), India (11), the United States of America (5), Turkey (5), Taiwan (4), Canada (3) France (3), the United Kingdom (3), and Chile (3).

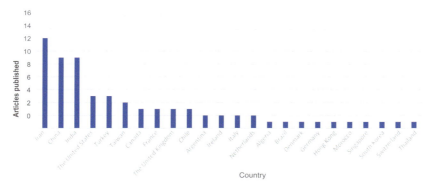

FIGURE 15.3 Distribution of the selected papers by country.

15.4.3 SCM dimension

The results of the first SCM dimension of the category identification section are presented in Table 15.2, detailing the absolute frequencies of four characteristics analyzed in each sample study: primary actor, organizational level, SCOR process, and application area.

Approximately 58 percent (46 papers) of the reviewed materials concentrate its study on manufacturers as the primary actor in planned SC, followed by industry/macroeconomics (28 papers), while distributors [37,38], retailers [39,40], suppliers [41], and warehousing [42] are seldom the focus of these studies. The primary actor of the analysis for our purpose is the subject who makes decisions and generates policies and procedures for the use of the quantitative models presented in the sample.

The level of organizational analysis shows a tendency for the interorganizational perspective with 36 papers between chain and network, followed by intraorganizational models with 23 papers considering a specific function or company, and finally a macroscopic perspective comprising 21 articles between industry and macroeconomy. The two most popular SCOR process are planning (54 papers) and making (11 papers).

As for the functional application area, SSCM modeling research targets general SCM (33 papers) or network design (29 papers), the latter being apparently a new trend in SSCM research that is gaining strength over time, because from 2015 to the present, there have been at least three publications per year, showing continuity, superiority, and a slight increase compared to other areas.

15.4.4 Modeling dimension

As part of the category identification, Table 15.3 provides an overview of the modeling dimension of the sample, where four categories are analyzed: (1) model data, (2) model type, (3) modeling technique, and (4) solution approach.

TABLE 15.2 Frequencies of the supply chain management (SCM) dimension.

Primary actor		Level		Process		Application area	
Distributor	2	Chain	9	Deliver	5	Logistics	7
Industry	28	Firm	12	Make	11	Network design	29
Manufacturer	46	Function	11	Plan	54	Planning	2
Retailer	2	Industry	16	Source	5	Production	7
Supplier	1	Macroeconomy	5	Various[a]	5	SCM	33
Warehousing	1	Network	27			Sourcing	2
Total	80	**Total**	80	**Total**	80	**Total**	80

[a]*Various refers to the fact that the study covers more than one particular process.*

TABLE 15.3 Frequencies of the modeling dimension.

Model data		Model type		Modeling technique		Solution approach	
Deterministic	48	Analytical	15	MCDM	6	AHP	16
Stochastic	32	Heuristics	5	Metaheuristic	2	DEA/IOA	2
		Hybrid	5	Multiobjective	32	Fuzzy programming	7
		Mathematical programming	49	Single objective	5	Genetic algorithm	3
		Simulation	4	System dynamics	4	Goal programming	2
		Various[a]	2	Systemic model	9	LP/MILP	20
				Various[a]	22	LCA	3
						Nonlinear programming	5
						Robust optimization	5
						Stochastic programming	8
						Various[a]	9
Total	80	**Total**	80	**Total**	80	**Total**	80

[a]Various refers to the fact that the model is composed of more than one category.

314 Advances of Artificial Intelligence in a Green Energy Environment

In the model data category, deterministic studies prevail with 60% (48 papers) compared to 40% (32 papers) stochastic studies, that is, in most articles the typical uncertainty of some variables is not considered. Regarding the modeling technique, it can be concluded that the multiobjective optimization technique is the most common (32 papers); although it is part of the multiple-criteria decision-making (MCDM) technique, we wanted to give it a particular space due to the number of papers found and its usefulness in generating a range of optimal solutions, which are considered equally good, such as in Ref. [43]. On the other hand, the MCDM technique, which includes the rest of the models that do not use multiobjective optimization in their approach or resolution, contains 6 papers, as shown in Ref. [40].

In addition, a considerable number of papers use more than one modeling technique (22 papers), for example, Ref. [44]. This is because sustainability problems integrate multiple variables and factors. Govindan et al. [45] propose a hybrid model, which includes a multiobjective metaheuristic technique to integrate sustainable order allocation.

In the solution approach category, we can highlight the following: the analytical and systemic models mainly employ analytic hierarchy process (AHP) (16 papers) or input—output analysis (IOA) (2 papers). The mathematical programming models deterministic usually use linear programming (LP) [46], mixed-integer linear programming (MILP) [47], or e-constraint method [48] (20 papers in total). On the other hand, the mathematical programming models with stochastic data, to be able to lead with uncertainty, use stochastic programming (8 papers) [49], fuzzy programming (7 papers) [50], and to a lesser extent robust optimization (5 papers).

Stochastic models in recent years have increased their scientific output and have focused on solving practical problems of the manufacturer such as aggregate production planning considering flexible lead times [51], handling market uncertainties and different risk attitudes [52], and production planning with stochastic demands and carbon variables [53].

In terms of industry, it can be seen that the models focused on the biofuels [11,47,54] and agriculture [55] sectors stand out, in which, through the design of their SC, they seek to implement sustainability measures while guaranteeing the level of production. The absence of research related to transportation is also noteworthy, due to its high contribution to GHG emissions.

To finish the modeling dimension, Table 15.4 illustrates in detail the models used in each of the 80 papers of the sample.

15.4.5 Sustainability dimension

The dimensions of the sustainability distributed in the studied sample can be seen in Fig. 15.4. Four investigations focus exclusively on environmental needs, analyzing environmental parameters and carbon emissions, like in Ref. [109]. Forty-four papers consider economic-environmental performance

A brief literature review of quantitative models Chapter | 15 **315**

TABLE 15.4 Modeling dimension by paper.

S. No.	Paper	Model purpose	Model type	Modeling techniques	Solution approach
1	[48]	Deterministic	Mathematical programming	Multiobjective	LP, augmented e-constraint method
2	[56]	Stochastic	Mathematical programming	Multiobjective	MILP, augmented e-constraint method
3	[57]	Deterministic	Mathematical programming	Multiobjective	Conventional e-constraint method
4	[58]	Deterministic	Mathematical programming	Multiobjective	MILP, conventional e-constraint method
5	[59]	Deterministic	Mathematical programming	Multiobjective	MILP, augmented e-constraint method
6	[17]	Deterministic	Mathematical programming	Multiobjective	MILP, conventional e-constraint method
7	[60]	Deterministic	Mathematical programming	Multiobjective	Conventional e-constraint method
8	[61]	Deterministic	Mathematical programming	Multiobjective	MILP, conventional e-constraint, LCA
9	[46]	Deterministic	Mathematical programming	Multiobjective	LP, conventional e-constraint method
10	[62]	Deterministic	Mathematical programming	Multiobjective	Conventional e-constraint method
11	[63]	Stochastic	Mathematical programming	Multiobjective	MILP, stochastic programming
12	[44]	Stochastic	Mathematical programming	Multiobjective	MILP, stochastic programming
13	[64]	Stochastic	Mathematical programming	Multiobjective	Multistage stochastic programming
14	[51]	Stochastic	Mathematical programming	Various	Stochastic nonlinear MIP
15	[49]	Stochastic	Mathematical programming	Chance-constrained	Stochastic programming, Benders decomposition
16	[65]	Stochastic	Mathematical programming	Various	Multistage stochastic dynamic programming
17	[54]	Stochastic	Mathematical programming	Various	Two-stage stochastic programming Lagrangian relaxation
18	[66]	Stochastic	Mathematical programming	Multiobjective	Fuzzy stochastic programming

TABLE 15.4 Modeling dimension by paper.—cont'd

S. No.	Paper	Model purpose	Model type	Modeling techniques	Solution approach
19	[67]	Deterministic	Mathematical programming	Multiobjective	Fuzzy goal programming
20	[68]	Stochastic	Mathematical programming	Multiobjective	Fuzzy e-constraint programming
21	[50]	Stochastic	Mathematical programming	Multiobjective	LP, fuzzy e-constraint, goal programming
22	[69]	Stochastic	Mathematical programming	Multiobjective	Fuzzy MILP, goal programming
23	[70]	Stochastic	Mathematical programming	Multiobjective	Fuzzy programming, Benders decomposition
24	[71]	Stochastic	Mathematical programming	Multiobjective	Fuzzy programming
25	[11]	Stochastic	Mathematical programming	Multiobjective	Robust possibilistic MILP
26	[52]	Stochastic	Mathematical programming	Multiobjective	Robust conventional e-constraint
27	[72]	Stochastic	Mathematical programming	Multiobjective	MILP, robust optimization
28	[73]	Stochastic	Mathematical programming	Bilevel programming	LP, MILP, robust optimization
29	[74]	Stochastic	Mathematical programming	Bilevel programming	MILP, robust optimization, Fuzzy programming
30	[75]	Stochastic	Mathematical programming	Multiobjective	AHP, robust optimization, nonlinear programming
31	[3]	Deterministic	Mathematical programming	Multiobjective	MILP, two-phase
32	[76]	Deterministic	Mathematical programming	Multiobjective	MILP, LCA

33	[43]	Deterministic	Mathematical programming	Multiobjective	Nonlinear MIP, genetic algorithm
34	[77]	Deterministic	Mathematical programming	Multiobjective	Nondominated sorting, genetic algorithm
35	[78]	Deterministic	Mathematical programming	Multiobjective	Weighted sum model, AHP
36	[47]	Deterministic	Mathematical programming	Multiobjective	MILP
37	[79]	Stochastic	Mathematical programming	Multiobjective	AHP, Fuzzy MILP
38	[55]	Deterministic	Mathematical programming	MCDM	MILP, AHP
39	[80]	Stochastic	Mathematical programming	MCDM	MILP, LCA
40	[81]	Deterministic	Mathematical programming	Various	MILP
41	[82]	Deterministic	Mathematical programming	Various	Speculation-postponement strategy
42	[83]	Stochastic	Mathematical programming	Various	LP, chance-constrained two-stage DEA model
43	[53]	Stochastic	Mathematical programming	Single objective	Two-stage stochastic programming
44	[39]	Deterministic	Mathematical programming	Single objective	Goal programming
45	[84]	Stochastic	Mathematical programming	Single objective	Chance-constrained programming
46	[85]	Deterministic	Mathematical programming	Single objective	MILP
47	[86]	Deterministic	Mathematical programming	Single objective	MILP
48	[87]	Deterministic	Mathematical programming	Single objective	MILP
49	[88]	Stochastic	Mathematical programming	Lagrangian relaxation	MILP
50	[42]	Deterministic	Heuristic	Multiobjective, metaheuristic	Genetic algorithm
51	[6]	Deterministic	Heuristic	Metaheuristic	MILP

Continued

TABLE 15.4 Modeling dimension by paper.—cont'd

S. No.	Paper	Model purpose	Model type	Modeling techniques	Solution approach
52	[89]	Deterministic	Heuristic	Metaheuristic	Nonlinear MIP, genetic algorithm
53	[90]	Deterministic	Heuristic	Multiobjective	Nonlinear MIP
54	[91]	Deterministic	Heuristic	Lagrangian relaxation	LP
55	[45]	Stochastic	Hybrid	Multiobjective, metaheuristic	AHP, ANP, DEA, MILP
56	[92]	Deterministic	Hybrid	Multiobjective, metaheuristic	Hybrid genetic Taguchi algorithm
57	[93]	Stochastic	Hybrid	Fuzzy MCMD	AHP, VIKOR
58	[94]	Deterministic	Hybrid	Systemic model	ANP
59	[95]	Deterministic	Hybrid	MCMD	AHP, Fuzzy TOPSIS
60	[96]	Deterministic	Simulation	System Dynamics	Nonlinear dynamics system
61	[97]	Deterministic	Simulation	System Dynamics	Nonlinear Dynamics System
62	[98]	Deterministic	Simulation	Various	LCA
63	[38]	Deterministic	Simulation	Various	Extended goal programming
64	[14]	Deterministic	Analytical, mathematical programming	Systemic model, multiobjective	AHP, augmented e-constraint
65	[99]	Stochastic	Analytical, mathematical programming	Systemic model, fuzzy TOPSIS	AHP, chance-constrained programming

66	[100]	Deterministic	Analytical		MCDM	AHP, Fuzzy, ILP
67	[40]	Deterministic	Analytical		MCMD	AHP, Fuzzy
68	[101]	Stochastic	Analytical		MCMD	AHP, ANP, Fuzzy
69	[102]	Stochastic	Analytical		MCDM, systemic model	Gray relational analysis, Fuzzy
70	[103]	Deterministic	Analytical		Systemic model	AHP, input−output−based life cycle assessment
71	[104]	Stochastic	Analytical		Systemic model	AHP
72	[37]	Deterministic	Analytical		Systemic model	AHP, genetic algorithm
73	[41]	Deterministic	Analytical		Systemic model	Multiagent, Fuzzy inference system
74	[105]	Deterministic	Analytical		Systemic model	Fuzzy game
75	[21]	Deterministic	Analytical		Systemic model	AHP, QFD
76	[106]	Deterministic	Analytical		Systemic model	IOA
77	[107]	Deterministic	Analytical		Systemic model	DEA
78	[108]	Deterministic	Analytical		Systemic model	AHP
79	[109]	Deterministic	Analytical		Systemic model	AHP, genetic
80	[110]	Deterministic	Analytical		Systemic model	AHP, Fuzzy

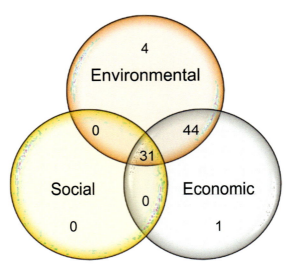

FIGURE 15.4 Distribution of reference papers with respect to the three sustainability dimensions.

focusing on the study of GHG emissions [63], carbon emissions [55], life cycle analysis (LCA) [76], amount of waste biomass [69], energy [82], nitrogen dioxide (NO_2) [58], water, and solid waste [98]. Finally, 31 articles develop SSCM models that consider the three dimensions of sustainability. As for the variables related to the social factor, the following stand out: population, training, noise, wages, job creation, occupational health, and safety [40,52,84].

15.4.6 Research gaps and future research perspectives

As part of the results of the systematic literature review developed throughout this study, we propose four ideas that explain the research gaps and future research prospects. These ideas may be useful in the academic or research area and will provide an overview for all stakeholders interested in the implementation and development of sustainability in their organizations.

First, a lack of industry focus is evident in the analyzed models of the sample, since around 30% of the sample (26 papers) does not apply their research in a particular sector, they only do it in a generic and empirical way. The industries most studied are biofuel and related products (15 papers), agriculture (6 papers), and food (4 papers). There is a need to consider applications in sectors like medical, automotive, chemical, and textile; they only have a total of 8 papers.

Second, while it is true that the concept of a product's life cycle is frequently used in SC and their operations, the environmental impacts they generate estimated through an LCA are not considered in the construction of any quantitative model of the sample; then including LCA in the design of SSC is a challenge that so far has not been fully developed.

Third, regarding the SSCM risk model, the findings confirm the predominance of economic variables, followed by environmental and, to a lesser extent, social variables. In addition, 60% of the quantitative models do not consider the implicit uncertainty of these variables and the risks they entail. Future research that analyzes the behavior of the model as a function of changes in its parameters should consider the explicit evaluation of the uncertainty in the variables, which would provide tools that are close to reality and would allow us to satisfy current needs.

Fourth, from the data acquired from the sample, the most studied dimensions of sustainability are the economic-environmental dimension; the social dimension has not yet been exploited but has already taken its first steps, since in recent years there has been a considerable increase in studies that consider the concept of the "triple bottom line", and this is the line to be followed by researchers. Considering economic, environmental, and social variables guarantees optimal development and implementation of sustainability in today's SC.

15.5 Discussion

Today's markets are increasingly dynamic, forcing companies to constantly reinvent themselves so as not to stagnate in the past [2], and they are obliged to know exactly the factors influencing their SC that they will have to work on to ensure their success. There are structural, operational design, technological, resource management, environmental, and economic factors [4]. Quantitative SC models then attempt to include one or more of these factors to support business management and decision-making.

The selection of a particular model depends exclusively on the needs of each company, which are influenced by the respective industrial context, for example, from what we have observed in the SCM dimension and the modeling dimension, we can say that managerial decisions are often supported by optimization methods, while in macroscopic contexts, models are more frequently used to analyze and explain the behavior and interaction of variables.

In several studies in our sample, it is observed that sustainability aspects to be considered are used as a moderating variable driving the purpose of modeling. Current research focuses on production processes and their respective environmental impacts; there is talk about improvements in planning and green chain initiatives, but there has not yet been a deepening of research into new production processes, machine design, and interface between actors, which would reduce these unsustainable sources.

Finally, it is necessary to expand research focused on critical industries such as transportation, chemicals, and textiles, which stand out for their environmental and social impact problems.

322 Advances of Artificial Intelligence in a Green Energy Environment

15.6 Conclusion

The main objective of our study has been met through the development of a systematic literature review, in which 80 papers were identified and critically, systematically, transparently, and reproducibly evaluated. Therefore, data concerning SSCM were collected and analyzed that provided a synthesis of the key elements of quantitative model offerings that use sustainability indicators in the design and management of forward SC.

The sample analysis evidences that research on quantitative modeling applied in SSC has increased its production continuously since 2012, driven by factors such as scarcity of natural resources, global warming, governmental agreements and policies, and above all the search for competitive advantages to attract environmentally friendly customers. Companies and organizations can find in the literature multiple types of quantitative models; studies like this one help them to focus on the right set of models for their needs, and the selection of the final model will depend on the actors involved and its adaptability in the current SC.

The findings show that deterministic models are the most popular in SSCM, evidencing the need for more stochastic approaches in modeling to relay a more realistic uncertain decision environment. Other future research directions that were analyzed through the gaps found include a greater focus and scope on modeling in the industry, the integration of LCA in an SSCN design, and to consider in greater measure social risk factors into modeling.

Finally, regarding the limitations of the study, we can say that one of them is the type of literature review used, since the systematic review is generally limited to collecting and analyzing research data, unlike the integrative review, which evaluates, criticizes, and synthesizes the literature to enhance the emergence of new theoretical frameworks and perspectives [111]. Other limitation was the subjectivity with which the content of the sample was analyzed, which depends to a great extent on the knowledge, judgment, experience, and number of researchers involved. On the other hand, despite having followed the methodologies of authors such as Refs. [27,35], the sample obtained is limited to the search keywords and databases used. Overcoming these details, we are sure that this work will serve as a decision support tool for all those who wish to incur, study, investigate, and implement sustainability in SC.

References

[1] S.M. Mirzapour Al-E-Hashem, H. Malekly, M.B. Aryanezhad, A multi-objective robust optimization model for multi-product multi-site aggregate production planning in a supply chain under uncertainty, Int. J. Prod. Econ. 134 (2011) 28–42, https://doi.org/10.1016/j.ijpe.2011.01.027.

[2] A. Baghalian, S. Rezapour, R.Z. Farahani, Robust supply chain network design with service level against disruptions and demand uncertainties: a real-life case, Eur. J. Oper. Res. 227 (2013) 199–215, https://doi.org/10.1016/j.ejor.2012.12.017.

A brief literature review of quantitative models **Chapter | 15 323**

[3] I. Moon, Y. Jeong, S. Saha, Fuzzy Bi-objective production-distribution planning problem under the carbon emission constraint, Sustainability 8 (2016) 798−815, https://doi.org/10.3390/su8080798.

[4] Z. Xu, A. Elomri, S. Pokharel, F. Mutlu, The design of green supply chains under carbon policies: a literature review of quantitative models, Sustainability 11 (2019) 3094, https://doi.org/10.3390/su11113094.

[5] WCED, Our Common Future, Oxford University Press, 1987.

[6] C.P. Tautenhain, A.P. Barbosa-Povoa, M.C. Nascimento, A multi-objective metaheuristic for designing and planning sustainable supply chains, Comput. Ind. Eng. 135 (2019) 1203−1223, https://doi.org/10.1016/j.cie.2018.12.062.

[7] R. Daghigh, M.S. Pishvaee, S.A. Torabi, Sustainable logistics network design under uncertainty, in: Springer Optimization and Its Applications, vol. 129, Springer International Publishing, 2017, pp. 115−151, https://doi.org/10.1007/978-3-319-69215-9_6.

[8] A. Chaabane, A. Ramudhin, M. Paquet, Design of sustainable supply chains under the emission trading scheme, Int. J. Prod. Econ. 135 (2012) 37−49, https://doi.org/10.1016/j.ijpe.2010.10.025.

[9] J. Elkington, Cannibals with Forks: the triple bottom line of 21st century, Altern. Manag. Obs. (1997) 1−16, https://doi.org/10.1002/tqem.3310080106.

[10] A. Rajeev, R.K. Pati, S.S. Padhi, K. Govindan, Evolution of sustainability in supply chain management: a literature review, J. Clean. Prod. 162 (2017) 299−314, https://doi.org/10.1016/j.jclepro.2017.05.026.

[11] H. Gilani, H. Sahebi, A multi-objective robust optimization model to design sustainable sugarcane-to-biofuel supply network: the case of study, Biomass Convers. Biorefinery (2020) 1−22, https://doi.org/10.1007/s13399-020-00639-8.

[12] H. Min, I. Kim, Green supply chain research: past, present, and future, Logist. Res. 4 (2012) 39−47, https://doi.org/10.1007/s12159-012-0071-3.

[13] C.L. Martins, M.V. Pato, Supply chain sustainability: a tertiary literature review, J. Clean. Prod. 225 (2019) 995−1016, https://doi.org/10.1016/j.jclepro.2019.03.250.

[14] H.G. Resat, B. Unsal, A novel multi-objective optimization approach for sustainable supply chain: a case study in packaging industry, Sustain. Prod. Consum. 20 (2019) 29−39, https://doi.org/10.1016/j.spc.2019.04.008.

[15] X. Bai, Y. Liu, Robust optimization of supply chain network design in fuzzy decision system, J. Intell. Manuf. 27 (2016) 1131−1149, https://doi.org/10.1007/s10845-014-0939-y.

[16] K. Devika, A. Jafarian, V. Nourbakhsh, Designing a sustainable closed-loop supply chain network based on triple bottom line approach: a comparison of metaheuristics hybridization techniques, Eur. J. Oper. Res. 235 (2014) 594−615, https://doi.org/10.1016/j.ejor.2013.12.032.

[17] Z. Zhang, A. Awasthi, Modelling customer and technical requirements for sustainable supply chain planning, Int. J. Prod. Res. 52 (2014) 5131−5154, https://doi.org/10.1080/00207543.2014.899717.

[18] K. Govindan, H. Soleimani, D. Kannan, Reverse logistics and closed-loop supply chain: a comprehensive review to explore the future, Eur. J. Oper. Res. 240 (2015b) 603−626, https://doi.org/10.1016/j.ejor.2014.07.012.

[19] D.M. Lambert, M.G. Enz, Issues in supply chain management: progress and potential, Ind. Market. Manag. 62 (2017) 1−16, https://doi.org/10.1016/j.indmarman.2016.12.002.

[20] C.J.C. Jabbour, A.B.L. de Sousa Jabbour, J. Sarkis, Unlocking effective multitier supply chain management for sustainability through quantitative modeling: lessons learned and

324 Advances of Artificial Intelligence in a Green Energy Environment

discoveries to be made, Int. J. Prod. Econ. 217 (2019) 11−30, https://doi.org/10.1016/j.ijpe.2018.08.029.

[21] Q. Zhang, N. Shah, J. Wassick, R. Helling, P. Van Egerschot, Sustainable supply chain optimization: an industrial case study, Comput. Ind. Eng. 74 (2014) 68−83, https://doi.org/10.1016/j.cie.2014.05.002.

[22] B. Mota, M.I. Gomes, A. Carvalho, A.P. Barbosa-Povoa, Sustainable supply chains: an integrated modeling approach under uncertainty, Omega 77 (2018) 32−57, https://doi.org/10.1016/j.omega.2017.05.006.

[23] M. Pagell, A. Shevchenko, Why research in sustainable supply chain management should have no future, J. Supply Chain Manag. 50 (2014) 44−55, https://doi.org/10.1111/jscm.12037.

[24] S. Seuring, M. Müller, From a literature review to a conceptual framework for sustainable supply chain management, J. Clean. Prod. 16 (2008) 1699−1710, https://doi.org/10.1016/j.jclepro.2008.04.020.

[25] M. Brandenburg, K. Govindan, J. Sarkis, S. Seuring, Quantitative models for sustainable supply chain management: developments and directions, Eur. J. Oper. Res. 233 (2014) 299−312, https://doi.org/10.1016/j.ejor.2013.09.032.

[26] P. Ghadimi, C. Wang, M.K. Lim, Sustainable supply chain modeling and analysis: past debate, present problems and future challenges, Resour. Conserv. Recycl. 140 (2019) 72−84, https://doi.org/10.1016/j.resconrec.2018.09.005.

[27] A. Fink, Conducting Research Literature Reviews: From the Internet to Paper, Ucla ed., SAGE Publications, Inc., Los Angeles, 2014.

[28] A. Cipriani, J. Geddes, Comparison of systematic and narrative reviews: the example of the atypical antipsychotics, Epidemiol. Psichiatr. Soc. 12 (2003) 146−153, https://doi.org/10.1017/S1121189X00002918.

[29] J. Klewitz, E.G. Hansen, Sustainability-oriented innovation of SMEs: a systematic review, J. Clean. Prod. 65 (2014) 57−75, https://doi.org/10.1016/j.jclepro.2013.07.017.

[30] F. Jia, L. Zuluaga-Cardona, A. Bailey, X. Rueda, Sustainable supply chain management in developing countries: an analysis of the literature, J. Clean. Prod. 189 (2018) 263−278, https://doi.org/10.1016/j.jclepro.2018.03.248.

[31] R.U. Khalid, S. Seuring, P. Beske, A. Land, S.A. Yawar, R. Wagner, Putting sustainable supply chain management into base of the pyramid research, Supply Chain Manag. 20 (2015) 681−696, https://doi.org/10.1108/SCM-06-2015-0214.

[32] R. Dubey, A. Gunasekaran, S.J. Childe, T. Papadopoulos, S.F. Wamba, World Class Sustainable Supply Chain Management: Critical Review and Further Research Directions, 2017, https://doi.org/10.1108/IJLM-07-2015-0112.

[33] T. Rebs, M. Brandenburg, S. Seuring, System Dynamics Modeling for Sustainable Supply Chain Management: A Literature Review and Systems Thinking Approach, 2019, https://doi.org/10.1016/j.jclepro.2018.10.100.

[34] F. Jia, T. Zhang, L. Chen, Sustainable supply chain Finance: towards a research agenda, J. Clean. Prod. 243 (2020) 118680, https://doi.org/10.1016/j.jclepro.2019.118680.

[35] S. Seuring, S. Gold, Conducting content-analysis based literature reviews in supply chain management, Supply Chain Manag. 17 (2012) 544−555, https://doi.org/10.1108/13598541211258609.

[36] Supply Chain Council, Supply Chain Operations Reference Model Revision 11.0, Technical Report, Supply Chain Council, 2012. URL: www.supply-chain.org.

A brief literature review of quantitative models **Chapter | 15 325**

[37] S. Validi, A. Bhattacharya, P.J. Byrne, A solution method for a two-layer sustainable supply chain distribution model, Comput. Oper. Res. 54 (2015) 204–217, https://doi.org/10.1016/j.cor.2014.06.015.

[38] D. Broz, G. Durand, D. Rossit, F. Tohmé, M. Frutos, Strategic planning in a forest supply chain: a multigoal and multiproduct approach, Can. J. For. Res. 47 (2017) 297–307, https://doi.org/10.1139/cjfr-2016-0299.

[39] S. Coskun, L. Ozgur, O. Polat, A. Gungor, A model proposal for green supply chain network design based on consumer segmentation, J. Clean. Prod. 110 (2016) 149–157, https://doi.org/10.1016/j.jclepro.2015.02.063.

[40] N. Kafa, Y. Hani, A. El Mhamedi, Evaluating and selecting partners in sustainable supply chain network: a comparative analysis of combined fuzzy multi-criteria approaches, Opsearch 55 (2018) 14–49, https://doi.org/10.1007/s12597-017-0326-5.

[41] P. Ghadimi, F. Ghassemi Toosi, C. Heavey, A multi-agent systems approach for sustainable supplier selection and order allocation in a partnership supply chain, Eur. J. Oper. Res. 269 (2018) 286–301, https://doi.org/10.1016/j.ejor.2017.07.014.

[42] F. Niakan, A. Baboli, V. Botta-Genoulaz, R. Tavakkoli-Moghaddam, J.P. Camapgne, A multi-objective mathematical model for green supply chain reorganization, in: IFAC Proceedings Volumes (IFAC-PapersOnline), IFAC Secretariat, 2013, pp. 81–86, https://doi.org/10.3182/20130522-3-BR-4036.00048.

[43] A.T. Espinoza Pérez, P.C. Narváez Rincón, M. Camargo, M.D. Alfaro Marchant, Multiobjective optimization for the design of phase III biorefinery sustainable supply chain, J. Clean. Prod. 223 (2019) 189–213, https://doi.org/10.1016/j.jclepro.2019.02.268.

[44] H. Ren, W. Zhou, M. Makowski, H. Yan, Y. Yu, T. Ma, Incorporation of life cycle emissions and carbon price uncertainty into the supply chain network management of PVC production, Ann. Oper. Res. (2019), https://doi.org/10.1007/s10479-019-03365-1.

[45] K. Govindan, A. Jafarian, V. Nourbakhsh, Bi-objective integrating sustainable order allocation and sustainable supply chain network strategic design with stochastic demand using a novel robust hybrid multi-objective metaheuristic, Comput. Oper. Res. 62 (2015) 112–130, https://doi.org/10.1016/j.cor.2014.12.014.

[46] M. Soysal, J.M. Bloemhof-Ruwaard, J.G. Van Der Vorst, Modelling food logistics networks with emission considerations: the case of an international beef supply chain, Int. J. Prod. Econ. 152 (2014) 57–70, https://doi.org/10.1016/j.ijpe.2013.12.012.

[47] Y. Huang, F. Xie, Multistage Optimization of Sustainable Supply Chain of Biofuels. Transportation Research Record: Journal of the Transportation Research Board 2502, 2015, pp. 89–98, https://doi.org/10.3141/2502-11. URL: http://journals.sagepub.com/.

[48] T. Vafaeenezhad, R. Tavakkoli-Moghaddam, N. Cheikhrouhou, Multi-objective mathematical modeling for sustainable supply chain management in the paper industry, Comput. Ind. Eng. 135 (2019) 1092–1102, https://doi.org/10.1016/j.cie.2019.05.027.

[49] K. Shaw, M. Irfan, R. Shankar, S.S. Yadav, Low carbon chance constrained supply chain network design problem: a Benders decomposition based approach, Comput. Ind. Eng. 98 (2016) 483–497, https://doi.org/10.1016/j.cie.2016.06.011.

[50] A. Mohammed, Q. Wang, The fuzzy multi-objective distribution planner for a green meat supply chain, Int. J. Prod. Econ. 184 (2017) 47–58, https://doi.org/10.1016/j.ijpe.2016.11.016.

[51] S.M. Mirzapour Al-E-Hashem, A. Baboli, Z. Sazvar, A stochastic aggregate production planning model in a green supply chain: considering flexible lead times, nonlinear purchase and shortage cost functions, Eur. J. Oper. Res. 230 (2013) 26–41, https://doi.org/10.1016/j.ejor.2013.03.033.

326 Advances of Artificial Intelligence in a Green Energy Environment

[52] L.E. Hombach, C. Büsing, G. Walther, Robust and sustainable supply chains under market uncertainties and different risk attitudes — a case study of the German biodiesel market, Eur. J. Oper. Res. 269 (2018) 302−312, https://doi.org/10.1016/j.ejor.2017.07.015.

[53] A. Rezaee, F. Dehghanian, B. Fahimnia, B. Beamon, Green supply chain network de- sign with stochastic demand and carbon price, Ann. Oper. Res. 250 (2017) 463−485, https://doi.org/10.1007/s10479-015-1936-z.

[54] C.W. Chen, Y. Fan, Bioethanol supply chain system planning under supply and demand uncertainties, Transp. Res. E Logist. Transp. Rev. 48 (2012) 150−164, https://doi.org/10.1016/j.tre.2011.08.004.

[55] Y. Tong, Model for evaluating the green supply chain performance under low-carbon agricultural economy environment with 2-tuple linguistic information, J. Intell. Fuzzy Syst. 32 (2017) 2717−2723, https://doi.org/10.3233/JIFS-16802.

[56] F. Mohebaliza, H. Zolfagharinia, S.H. Amin, Designing a green meat supply chain network: a multi-objective approach, Int. J. Prod. Econ. 219 (2020) 312−327, https://doi.org/10.1016/j.ijpe.2019.07.007.

[57] K. Tsai-Chi, C. Hsiao-Min, T. Ming-Lang, C. Ping-Shun, C. Po-Chen, Design and analysis of supply chain networks with low carbon emissions, Comput. Econ. 52 (2018) 1353−1374, https://doi.org/10.1007/s10614-017-9675-7.

[58] E. Huang, X. Zhang, L. Rodriguez, M. Khanna, S. de Jong, K.C. Ting, Y. Ying, T. Lin, Multi-objective optimization for sustainable renewable jet fuel production: a case study of corn stover based supply chain system in Midwestern U.S, Renew. Sust. Energ. Rev. 115 (2019) 109403, https://doi.org/10.1016/j.rser.2019.109403.

[59] R. Hosseinalizadeh, A. Arshadi Khamseh, M.M. Akhlaghi, A multi-objective and multi-period model to design a strategic development program for biodiesel fuels, Sustain. Energy Technol. Assess. 36 (2019) 100545, https://doi.org/10.1016/j.seta.2019.100545.

[60] A. Tognetti, P.T. Grosse-Ruyken, S.M. Wagner, Green supply chain network optimization and the trade-off between environmental and economic objectives, Int. J. Prod. Econ. 170 (2015) 385−392, https://doi.org/10.1016/j.ijpe.2015.05.012.

[61] F. You, L. Tao, D.J. Graziano, S.W. Snyder, Optimal design of sustainable cellulosic biofuel supply chains: multiobjective optimization coupled with life cycle assessment and input-output analysis, AIChE J. 58 (2012) 1157−1180, https://doi.org/10.1002/aic.12637.

[62] R. Ortiz-Gutierrez, S. Giarola, F. Bezzo, Optimal design of ethanol supply chains considering carbon trading effects and multiple technologies for side-product exploitation, Environ. Technol. 34 (2013) 2189−2199, https://doi.org/10.1080/09593330.2013.829111.

[63] Z. Ghelichi, M. Saidi-Mehrabad, M.S. Pishvaee, A stochastic programming approach to- ward optimal design and planning of an integrated green biodiesel supply chain network under uncertainty: a case study, Energy 156 (2018) 661−687, https://doi.org/10.1016/j.energy.2018.05.103.

[64] Z. Sazvar, S.M. Mirzapour Al-E-Hashem, A. Baboli, M.R. Akbari Jokar, A bi-objective stochastic programming model for a centralized green supply chain with deteriorating products, Int. J. Prod. Econ. 150 (2014) 140−154, https://doi.org/10.1016/j.ijpe.2013.12.023.

[65] T.M. Choi, Optimal apparel supplier selection with forecast updates under carbon emission taxation scheme, Comput. Oper. Res. 40 (2013) 2646−2655, https://doi.org/10.1016/j.cor.2013.04.017.

[66] T. Yu-Chung, T. Vo-Van, L. Jye-Chyi, Y. Vincent, Designing sustainable supply chain networks under uncertain environments: fuzzy multi-objective programming, J. Clean. Prod. 174 (2018) 1550−1565, https://doi.org/10.1016/j.jclepro.2017.10.272.

[67] K. Boonsothonsatit, S. Kara, S. Ibbotson, B. Kayis, Development of a Generic decision support system based on multi-objective optimization for green supply chain network design (GOOG), J. Manuf. Technol. Manag. 26 (2015) 1069−1084, https://doi.org/10.1108/JMTM-10-2012-0102.

[68] M.M. Saffar, G. Hamed Shakouri, J. Razmi, A new multi objective optimization model for designing a green supply chain network under uncertainty, Int. J. Ind. Eng. Comput. 6 (2015) 15−32, https://doi.org/10.5267/j.ijiec.2014.10.001.

[69] S.Y. Balaman, H. Selim, A fuzzy multiobjective linear programming model for design and management of anaerobic digestion based bioenergy supply chains, Energy 74 (2014) 928−940, https://doi.org/10.1016/j.energy.2014.07.073.

[70] M.S. Pishvaee, J. Razmi, S.A. Torabi, An accelerated Benders decomposition algorithm for sustainable supply chain network design under uncertainty: a case study of medical needle and syringe supply chain, Transp. Res. E Logist. Transp. Rev. 67 (2014) 14−38, https://doi.org/10.1016/j.tre.2014.04.001.

[71] C. Rout, A. Paul, R.S. Kumar, D. Chakraborty, A. Goswami, Cooperative sustainable supply chain for deteriorating item and imperfect production under different carbon emission regulations, J. Clean. Prod. 272 (2020), https://doi.org/10.1016/j.jclepro.2020.122170.

[72] H. Heidari-Fathian, S.H.R. Pasandideh, Green-blood supply chain network design: robust optimization, bounded objective function & Lagrangian relaxation, Comput. Ind. Eng. 122 (2018) 95−105, https://doi.org/10.1016/j.cie.2018.05.051.

[73] H. Golpîra, E. Najafi, M. Zandieh, S. Sadi-Nezhad, Robust bi-level optimization for green opportunistic supply chain network design problem against uncertainty and environmental risk, Comput. Ind. Eng. 107 (2017) 301−312, https://doi.org/10.1016/j.cie.2017.03.029.

[74] M. Jin, L. Song, Y. Wang, Y. Zeng, Longitudinal cooperative robust optimization model for sustainable supply chain management, Chaos, Solit. Fractals 116 (2018) 95−105, https://doi.org/10.1016/j.chaos.2018.09.008.

[75] M. Sherafati, M. Bashiri, R. Tavakkoli-Moghaddam, M.S. Pishvaee, Supply chain network design considering sustainable development paradigm: a case study in cable industry, J. Clean. Prod. 234 (2019) 366−380, https://doi.org/10.1016/j.jclepro.2019.06.095.

[76] F.D. Mele, A.M. Kostin, G. Guillén-Gosálbez, L. Jiménez, Multiobjective model for more sustainable fuel supply chains. A case study of the sugar cane industry in Argentina, Ind. Eng. Chem. Res. 50 (2011) 4939−4958, https://doi.org/10.1021/ie101400g.

[77] M.A. Brahami, M. Dahane, M. Souier, M. Sahnoun, Sustainable capacitated facility location/network design problem: a non-dominated Sorting Genetic Algorithm based multiobjective approach, Ann. Oper. Res. (2020) 1−32, https://doi.org/10.1007/s10479-020-03659-9.

[78] Z. Chen, S. Andresen, A multiobjective optimization model of production-sourcing for sustainable supply chain with consideration of social, environmental, and economic factors, Math. Probl Eng. 2 (2014) 1−11, https://doi.org/10.1155/2014/616107.

[79] L. Kumar, P.K. Jain, A.K. Sharma, A fuzzy goal programmed based sustainable Greenfield supply network design for tire retreading industry, Int. J. Adv. Manuf. Technol. 108 (2020) 2855−2880, https://doi.org/10.1007/s00170-020-05140-0.

[80] S. Giarola, F. Bezzo, N. Shah, A risk management approach to the economic and environmental strategic design of ethanol supply chains, Biomass Bioenergy 58 (2013) 31−51, https://doi.org/10.1016/j.biombioe.2013.08.005.

[81] C.V. Valderrama, E. Santibanez-González, B. Pimentel, A. Candia-Véjar, L. Canales-Bustos, Designing an environmental supply chain network in the mining industry to

328 Advances of Artificial Intelligence in a Green Energy Environment

reduce carbon emissions, J. Clean. Prod. 254 (2019), https://doi.org/10.1016/j.jclepro.2019.119688.

[82] S.D. Budiman, H. Rau, A mixed-integer model for the implementation of postponement strategies in the globalized green supply chain network, Comput. Ind. Eng. 137 (2019) 106054, https://doi.org/10.1016/j.cie.2019.106054.

[83] M. Izadikhah, R.F. Saen, Assessing sustainability of supply chains by chance-constrained two-stage DEA model in the presence of undesirable factors, Comput. Oper. Res. 100 (2018) 343−367, https://doi.org/10.1016/j.cor.2017.10.002.

[84] R. Das, K. Shaw, M. Irfan, Supply chain network design considering carbon footprint, water footprint, supplier's social risk, solid waste, and service level under the uncertain condition, Clean Technol. Environ. Policy 22 (2020) 337−370, https://doi.org/10.1007/s10098-019-01785-y.

[85] J. Jonkman, A. Kanellopoulos, J.M. Bloemhof, Designing an eco-efficient biomass-based supply chain using a multi-actor optimization model, J. Clean. Prod. 210 (2019) 1065−1075, https://doi.org/10.1016/j.jclepro.2018.10.351.

[86] C.V. Valderrama, E. Santibanez-González, B. Pimentel, A. Candia-Véjar, L. Canales-Bustos, Designing an environmental supply chain network in the mining industry to reduce carbon emissions, J. Clean. Prod. 254 (2020), https://doi.org/10.1016/j.jclepro.2019.119688.

[87] O. Kabadurmus, M.S. Erdogan, Sustainable, multimodal and reliable supply chain design, Ann. Oper. Res. 292 (2020) 47−70, https://doi.org/10.1007/s10479-020-03654-0.

[88] M. Zheng, W. Li, Y. Liu, X. Liu, A Lagrangian heuristic algorithm for sustainable supply chain network considering CO_2 emission, J. Clean. Prod. 270 (2020), https://doi.org/10.1016/j.jclepro.2020.122409.

[89] F. Barzinpour, P. Taki, A dual-channel network design model in a green supply chain considering pricing and transportation mode choice, J. Intell. Manuf. 29 (2018) 1465−1483, https://doi.org/10.1007/s10845-015-1190-x.

[90] V.K. Manupati, S.J. Jedidah, S. Gupta, A. Bhandari, M. Ramkumar, Optimization of a multi-echelon sustainable production-distribution supply chain system with lead time consideration under carbon emission policies, Comput. Ind. Eng. 135 (2019) 1312−1323, https://doi.org/10.1016/j.cie.2018.10.010.

[91] S. Elhedhli, R. Merrick, Green supply chain network design to reduce carbon emissions, Transp. Res. Transp. Environ. 17 (2012) 370−379, https://doi.org/10.1016/j.trd.2012.02.002.

[92] R. Jamshidi, S.M. Fatemi Ghomi, B. Karimi, Multi-objective green supply chain optimization with a new hybrid memetic algorithm using the Taguchi method, Sci. Iran. 19 (2012) 1876−1886, https://doi.org/10.1016/j.scient.2012.07.002.

[93] K. Sari, A novel multi-criteria decision framework for evaluating green sup- ply chain management practices, Comput. Ind. Eng. 105 (2017) 338−347, https://doi.org/10.1016/j.cie.2017.01.016.

[94] A. Chauhan, H. Kaur, S. Yadav, S.K. Jakhar, A hybrid model for investigating and selecting a sustainable supply chain for agri-produce in India, Ann. Oper. Res. 290 (2020) 621−642, https://doi.org/10.1007/s10479-019-03190-6.

[95] R.K. Sharma, P.K. Singh, P. Sarkar, H. Singh, A hybrid multi-criteria decision approach to analyze key factors affecting sustainability in supply chain networks of manufacturing organizations, Clean Technol. Environ. Policy 22 (2020) 1871−1889, https://doi.org/10.1007/s10098-020-01926-8.

A brief literature review of quantitative models **Chapter | 15 329**

[96] M. Song, X. Cui, S. Wang, Simulation of land green supply chain based on system dynamics and policy optimization, Int. J. Prod. Econ. 217 (2019) 317−327, https://doi.org/10.1016/j.ijpe.2018.08.021.

[97] G. Wang, A. Gunasekaran, Modeling and analysis of sustainable supply chain dynamics, Ann. Oper. Res. 250 (2017) 521−536, https://doi.org/10.1007/s10479-015-1860-2.

[98] E.S. Nwe, A. Adhitya, I. Halim, R. Srinivasan, Green supply chain design and operation by integrating LCA and dynamic simulation, Comput. Aided Chem. Eng. 28 (2010) 109−114, https://doi.org/10.1016/S1570-7946(10)28019-7.

[99] K. Shaw, R. Das, Uncertain supply chain network design considering carbon footprint and social factors using two stage approach, Clean Technol. Environ. Policy 19 (2017) 2491−2519, https://doi.org/10.1007/s10098-017-1446-6.

[100] H. Kaur, S.P. Singh, R. Glardon, An integer linear program for integrated supplier selection: a sustainable flexible framework, Glob. J. Flex. Syst. Manag. 17 (2016) 113−134, https://doi.org/10.1007/s40171-015-0105-1.

[101] C. Wu, C. Lin, D. Barnes, Y. Zhang, Partner selection in sustainable supply chains: a fuzzy ensemble learning model, J. Clean. Prod. 275 (2020), https://doi.org/10.1016/j.jclepro.2020.123165.

[102] W. Kuo-Jui, L. Ching-Jong, T.K. Minglang, C. Kuan-Shun, Multi-attribute approach to sustainable supply chain management under uncertainty, Ind. Manag. Data Syst. 116 (2016) 777−800, https://doi.org/10.1108/IMDS-08-2015-0327.

[103] N. Ghani, G. Egilmez, M. Kucukvar, S. Bhutta, From green buildings to green supply chains: an integrated input-output life cycle assessment and optimization framework for carbon footprint reduction policy making, Manag. Environ. Qual. Int. J. 28 (2017) 532−548, https://doi.org/10.1108/MEQ-12-2015-0211.

[104] T. Ming-Lang, L.K. Ming, W. Kuo-Jui, Improving the benefits and costs on sustainable supply chain finance under uncertainty, Int. J. Prod. Econ. 218 (2019) 308−321, https://doi.org/10.1016/j.ijpe.2019.06.017.

[105] K. Bhattacharya, S.K. De, A robust two layer green supply chain modelling under performance based fuzzy game theoretic approach, Comput. Ind. Eng. 152 (2020) 107005, https://doi.org/10.1016/j.cie.2020.107005.

[106] A. Acquaye, T. Ibn-Mohammed, A. Genovese, G.A. Afrifa, F.A. Yamoah, E. Oppon, A quantitative model for environmentally sustainable supply chain performance measurement, Eur. J. Oper. Res. 269 (2018) 188−205, https://doi.org/10.1016/j.ejor.2017.10.057.

[107] X. Ji, J. Wu, Q. Zhu, Eco-design of transportation in sustainable supply chain management: a DEA-like method, Transp. Res. Transp. Environ. 48 (2016) 451−459, https://doi.org/10.1016/j.trd.2015.08.007.

[108] V.K. Sharma, P. Chandana, A. Bhardwaj, Critical factors analysis and its ranking for implementation of GSCM in Indian dairy industry, J. Manuf. Technol. Manag. 26 (2015) 911−922, https://doi.org/10.1108/JMTM-03-2014-0023.

[109] B. He, Y. Liu, L. Zeng, S. Wang, D. Zhang, Q. Yu, Product carbon footprint across sustainable supply chain, J. Clean. Prod. 241 (2019) 118320, https://doi.org/10.1016/j.jclepro.2019.118320.

[110] O. Boutkhoum, M. Hanine, H. Boukhriss, T. Agouti, A. Tikniouine, Multi-criteria decision support framework for sustainable implementation of effective green supply chain management practices, SpringerPlus 5 (2016) 1−24, https://doi.org/10.1186/s40064-016-2233-2.

[111] H. Snyder, Literature review as a research methodology: an overview and guidelines, J. Bus. Res. 104 (2019) 333−339, https://doi.org/10.1016/j.jbusres.2019.07.039.

Chapter 16

Optimized designing spherical void structures in 3D domains

Tatiana Romanova[1,2], Georgiy Yaskov[1], Igor Litvinchev[3], Igor Yanchevskyi[4], Yurii Stoian[1] and Pandian Vasant[5]

[1]*Department of Mathematical Modelling and Optimal Design, Institute for Mechanical Engineering Problems of the National Academy of Sciences of Ukraine (IPMach NASA), Kharkiv, Ukraine;* [2]*Kharkiv National University of Radioelectronics, Kharkiv, Ukraine;* [3]*Faculty of Mechanical and Electrical Engineering (FIME), Nuevo Leon State University, Monterrey, Nuevo Leon, Mexico;* [4]*National Technical University of Ukraine "Igor Sikorsky Kyiv Polytechnic Institute", Department of Dynamics and Strength of Machines and Strength of Materials, Kiev, Ukraine;* [5]*Faculty of Science and Information Technology, Universiti Teknologi Petronas, Teronoh, Seri Iskandar, Malaysia*

16.1 Introduction

Additive manufacturing (AM) is considered as the first manufacturing "revolution" of the 21st century and plays a principal role in the Industry 4.0 [1]. It enables a design-driven manufacturing, where design determines production and not the other way around. AM, also known as a 3D printing, is a set of technologies for producing complex parts in a layer-by-layer material deposition. Unlike subtractive manufacturing starting with a solid block of material and then cutting away the excess to create a finished part, AM builds up a part layer by layer [2].

AM offers enormous geometrical freedom to create parts. Creating hollowing systems is a typical practice for reducing printing material and time in 3D printing of light-weighted parts [3,4]. Generating void structures can be considered as a packing problem with holes standing for packing items. Number of holes, their shapes and sizes, positions, and space orientations must be defined minimizing, e.g., the weight of the part without significant loss of its mechanical strength [5].

One of the approaches to design spherical void structures is as follows. In a 3D part strength engineers indicate one or more polyhedral zones suitable for hollowing. In a polyhedral hollowing zone, a number of nonintersected spherical holes can be allocated. These spherical holes must be sufficiently distant from one another to maintain a mechanical strength of the hollowing

332 Advances of Artificial Intelligence in a Green Energy Environment

part. The objective is to maximize the material savings (volume of all the holes) without significant loss of the overall strength.

This approach gives rise to a new packing problem. Optimized layout for a given number of variable-sized spheres (objects) in a disconnected polyhedral domain (container) is considered. The spheres must be fully arranged in the 3D region that contains the given polyhedral domain and the distance between the objects must be at least a certain given threshold. The objective is to find coordinates of the centers and radii of the spheres maximizing the total volume of the spheres.

Many applied problems can be formulated as packing unequal spherical objects [6,7]. One can find spherical layout problems in optimizing the topology of parts [3] and determining the proportions between different fractions of powder [8] in AM; in treating oncological tumors with gamma rays [9–11] and diabetic retinopathy using laser coagulation of the retina [12] in medicine; in space technology [13,14] and in material science [15–17].

Packing problems are NP-hard [18]. Many approximate and heuristic approaches have been proposed for sphere packing problems [19–32] in different domains, including cuboids, cylinders, spheres, and polyhedrons. However, to the best of our knowledge, packing variable-sized spheres in a polyhedral container subject to a (lower) bounded distance between the spheres and under balancing conditions is considered in this work for the first time.

In this chapter, a phi-functions modeling approach [33,34] is used to state containment and distance conditions for packing spheres in a convex polyhedron.

16.2 Problem formulation

Let $\Omega = \{v \in \mathbb{R}^3 : 0 \le x \le l, \ 0 \le y \le w, \ 0 \le z \le h\}$ be a 3D cuboid with known dimensions and $S_q = S_q(v_q) = \{v = (x, y, z) \in \mathbb{R}^3 : \| v - v_q \|^2 \le r_q^2\}$, $q \in I = \{1, 2, ..., N\}$ be spherical objects. Here the center $v_q = (x_q, y_q, z_q)$ and the radius r_q of the sphere S_q are considered as variables $0 < r_q^- < r_q < r_q^+$, where r_q^-, r_q^+ are known as lower and upper bounds for r_q.

Given is a disconnected domain

$$C = \bigcup_{l \in I_p} P_l \subset \Omega \subset \mathbb{R}^3,$$

with connected polyhedral convex components $P_l = \{(x, y, z) \in \mathbb{R}^3 : \phi_{ml}(x, y, z) \ge 0\}, m = 1, 2, ..., M_l, l \in I_p = \{1, 2, ..., n_p\}$, where $P_i \cap P_j, i, j \in I_p$, $i \ne j$, $\phi_{ml}(x, y, z) = a_{ml}x + b_{ml}y + c_{ml}z + d_{ml} = 0$ are normal equations of the faces of $P_l, l \in I_p$.

Optimized designing spherical void structures in 3D domains **Chapter | 16** **333**

A layout of the spherical objects S_q, $q \in I$ in the domain P_l must satisfy the following conditions:

- containment conditions

$$S_q(v_q) \subset P_l \Leftrightarrow \text{int} S_q(v_q) \cap P_l^* = \varnothing, q \in I, l \in I_p \qquad (16.1)$$

where $P_l^* = \mathbb{R}^3 \backslash \text{int} P_l$;
- distance conditions (the distance between the objects must be at least ρ)

$$\text{dist}\left(S_q(v_q), S_g(v_g)\right) \geq \rho, (q, g) \in \Xi_l, l \in I_p \qquad (16.2)$$

where

$$\text{dist}\left(S_q(v_q), S_g(v_g)\right) = \min_{a \in S_q(v_q), b \in S_g(v_g)} \rho(a, b),$$

$\rho(a, b)$ is the Euclidean distance between points $a, b \in \mathbb{R}^3$,

$$\Xi_l = \left\{(q, g): S_q(v_q) \subset P_l, S_g(v_g) \subset P_l, q > g\right\}, l \in I_p.$$

The optimized layout problem of spheres into disconnected polyhedral domain can be formulated as follows.

Find centers v_q and radii r_q of the spheres S_q, $q \in I$ placed into polyhedral P_l, $l \in I_p$, maximizing the total volume of spheres subject to the containment (16.1) and distance (16.2) conditions.

16.3 Mathematical model

To analytically describe the layout conditions (16.1) and (16.2), the phi-functions approach [33,34] is used. This way the optimized layout problem is reduced to the following nonlinear programming problem:

$$\max \kappa(\omega) \text{ s.t. } \omega \in W \subset R^{4N}, \qquad (16.3)$$

where $\omega = (v, r)$, $v = (v_1, v_2, ..., v_N)$, and $r = (r_1, r_2, ..., r_N)$ are unknown parameters (variables) of the spheres;

$$\kappa(\omega) = \frac{4}{3} \pi \sum_{q \in I} r_q^3 \qquad (16.4)$$

is the objective function;

$$W = \left\{ \omega \in \mathbb{R}^{4N}: \widehat{\Phi}_{qg}\left(v_q, v_g, r_q, r_g\right) \geq 0, (q, g) \in \Xi_l, l \in I_p, \right. \qquad (16.5)$$

$$\left. \Phi_{ql}\left(v_q, r_q\right) \geq 0, q \in I, r_q - r_q^- \geq 0, -r_q + r_q^+ \geq 0, q \in I, l \in I_p \right\}$$

Here inequalities $\widehat{\Phi}_{qg}(v_q, v_g, r_q, r_g) \geq 0$ are used to represent the distance conditions (16.2) for spheres S_q and S_g for $(q, g) \in \Xi_l$, $l \in I_p$, where

$$\widehat{\Phi}_{qg}(v_q, v_g, r_q, r_g) = \| v_q - v_g \|^2 - (r_q + r_g + \rho)^2$$

is an adjusted phi-function for S_q and S_g, i.e.,

$$\widehat{\Phi}_{qg}(v_q, v_g, r_q, r_g) \geq 0 \text{ implies } \mathrm{dist}(S_q(v_q), S_g(v_g)) \geq \rho.$$

Correspondingly, inequality $\Phi_{ql}(v_q, r_q) \geq 0$ is used to ensure the containment conditions (16.1) for $l \in I_p$. Here $\Phi_{ql}(v_q, r_q)$ is a phi-function for S_q and the set $P_l^* = \mathbb{R}^3 \backslash \mathrm{int} P_l$,

$$\Phi_{ql}(v_q, r_q) = \min_{m=1,2,\ldots,M}\{\phi_{ml}(v_q) - r_q\}, q \in I,$$

i.e.,

$$\Phi_{ql}(v_q, r_q) \geq 0 \text{ implies } \mathrm{int}\, S_q(v_q) \cap P_l^* = \varnothing.$$

Problem (16.3)–(16.5) is a nonconvex nonlinear programming problem with $O(N)$ variables and $O(N^2)$ constraints.

In the problem, several nonintersected polyhedrons are available for spheres layout and distance conditions must be fulfilled for the spheres inside each polyhedron. In this case the corresponding volume maximization problem decomposes into n_p independent subproblems corresponding to each polyhedron P_l, $l \in I_p$.

To reduce the computational complexity of the algorithm, the decomposition method and the strategy of active set of inequalities [26] were implemented. The multistart approach was performed to select a better local maximum. Corresponding starting point $\omega^0 = (v^0, r^0)$ to solve (16.3)–(16.5) was constructed as follows. The centers $v_1^0, v_2^0, \ldots, v_N^0$ were chosen randomly so that $v_1^0, v_2^0, \ldots, v_N^0 \in P$, while the staring values of radii were set to zero.

16.4 Mathematical model with balancing conditions

Let us introduce a set $T = \Omega \backslash \bigcup_{q \in I_N} S_q$. We define the gravity center of the set T in the form

$$X_\Omega = \frac{x_C m_C - \sum\limits_{q \in I_N} x_q m_q}{m_C - \sum\limits_{q \in I_N} m_q}, Y_\Omega = \frac{y_C m_C - \sum\limits_{q \in I_N} y_q m_q}{m_C - \sum\limits_{q \in I_N} m_q}, Z_\Omega = \frac{z_C m_C - \sum\limits_{q \in I_N} z_q m_q}{m_C - \sum\limits_{q \in I_N} m_q},$$

where (x_C, y_C, z_C) is the gravity center of Ω, $m_C = \rho l w h$ is the mass of Ω, and $m_q = \frac{4}{3}\rho r_q^3$ is the mass of S_q while ρ is the material density.

Optimized designing spherical void structures in 3D domains **Chapter | 16** **335**

Therefore

$$X_\Omega = \frac{x_C lwh - \frac{4}{3}\pi \sum_{q \in I_N} x_q r_q^3}{lwh - \frac{4}{3}\pi \sum_{q \in I_N} r_q^3}; Y_\Omega = \frac{y_C lwh - \frac{4}{3}\pi \sum_{q \in I_N} y_q r_q^3}{lwh - \frac{4}{3}\pi \sum_{q \in I_N} r_q^3}; Z_\Omega = \frac{z_C lwh - \frac{4}{3}\pi \sum_{q \in I_N} z_q r_q^3}{lwh - \frac{4}{3}\pi \sum_{q \in I_N} r_q^3}.$$

The balancing conditions are used in the form

$$|X_\Omega - x_C| \le \varepsilon, |Y_\Omega - y_C| \le \varepsilon, |Z_\Omega - z_C| \le \varepsilon, \varepsilon > 0. \tag{16.6}$$

For balanced problem, the containment conditions (16.1) will be relaxed providing containment of the centers of the τ_l spherical objects in the appropriate polyhedron P_l, for $l \in I_p$, $\sum_{l \in I_p} \tau_l = N$, with allowance for spheres protruding beyond the faces of P_l, $l \in I_p$, up to $\delta > 0$.

The optimized layout problem for spheres in the disconnected polyhedral domain is stated as follows.

Arrange the spherical objects $S_q(v_q)$, $q \in I_N$, in the domain Ω to maximize the total volume of spherical holes taking into account the distance constraints (16.2) and the balancing conditions (16.6).

Let $v_j \in P_l$, $j \in I_l = \{\tau_{l-1} + 1, \tau_{l-1} + 2, ..., \tau_{l-1} + \tau_l\}$, $l \in I_p$, $\tau_0 = 0$.

We extend the problem (16.3)−(16.5) to the following nonlinear programming problem with the balancing conditions:

$$\max_{\omega \in W \subset \mathbb{R}^{4n}} \kappa(\omega), \tag{16.7}$$

where

$$W = \left\{ \omega \in \mathbb{R}^{4N} : \widehat{\Phi}_{qg}(v_q, v_g, r_q, r_g) \ge 0, (q, g) \in \Xi_l, l \in I_p, \phi_j(v_j, 0) \ge 0, \right.$$

$$\phi_j(v_j, r_j) \ge \delta, j \in I_l,$$

$$\tag{16.8}$$

$$r_q - r_q^- \ge 0, q \in I_N, -r_q + r_q^+ \ge 0, q \in I_N, |X_\Omega - x_C| \le \varepsilon,$$

$$\left. |Y_\Omega - y_C| \le \varepsilon, |Z_\Omega - z_C| \le \varepsilon \right\};$$

Inequalities $\widehat{\Phi}_{qg}\left(v_q, v_g, r_q, r_g\right) \ge 0$, $(q, g) \in \Xi_l, l \in I_p$, assure the distance conditions (16.2), inequalities $\phi_j(v_j, 0) \ge 0$, $j \in I_l$, guarantee placing centers of S_j inside the polyhedron P_l, $l \in I_p$ (relaxed containment conditions (16.1)), inequalities $\phi_j(v_j, r_j) \ge \delta$, $j \in I_l$, $l \in I_p$, ensure protruding the spheres S_j beyond the faces of P_l at most δ,

$$\phi_j(v_j, r_j) = \max_{l \in I_p} \left\{ \min_{m=1,2,...,M_l} \{\phi_{ml}(v_j) - r_j\} \right\}, j \in I_l.$$

336 Advances of Artificial Intelligence in a Green Energy Environment

Problem (16.7) and (16.8) is a nonlinear programming problem, its feasible region W is a disconnected set with multiconnected components. The problem has $O(N)$ variables and $O(N^2)$ constraints.

Starting points $\omega_l^0 = \left(v^{(l)0}, r^{(l)0}\right)$ for the problem (16.7) and (16.8) are constructed as follows. The centers $v_{\tau_{l-1}+1}^0, v_{\tau_{l-1}+2}^0, ..., v_{\tau_l}^0$ are chosen randomly so that $v_{\tau_{l-1}+1}^0, v_{\tau_{l-1}+2}^0, ..., v_{\tau_l}^0 \in P_l, l \in I_p$. Starting values of radii are all set to zero, i.e., $r^{(l)0} = (0, 0, ..., 0)$. These starting values meet the balancing conditions since $X_\Omega - x_C = Y_\Omega - y_C = Z_\Omega - z_C = 0$ for the staring zero values of variable radii.

To reduce the computational complexity of the algorithm, the decomposition method and the strategy of active set of inequalities [26] are used.

16.5 Numerical experiments

Numerical experiments were performed to illustrate the proposed modeling and solution approach. All computations were executed on a computer with AMD FX (tm)-6100, 3.30 GHz processor in Windows 7, with C++ used to program codes. The open-source local solver IPOPT [35] is used to get a local maximum in both (16.3)−(16.5) and (16.7)−(16.8) NLP problems. In the multistart approach 50 starting points were generated for each problem instance. We consider two instances realizing model (16.3)−(16.5) and one instance based on the model (16.7) and (16.8).

Example 1. Cuboid Ω $(w \times l \times h)$ has the following dimensions: $w = 15, l = 25$, $h = 17$. Two nonintersected polyhedrons P_1 and P_2 are defined for hollowing. The number of spherical objects in P_1 is $n_1 = 22$, while for P_2 we set $n_2 = 24$. The lower and upper bounds for the radii are $r_q^- = 0.5$, $r_q^+ = 2$, while the minimal allowed distance between the spherical objects is $\rho = 0.25$.

Coordinates of poles of polyhedrons are $x_1 = 7.5$, $y_1 = 7.5$, $z_1 = 8.5$ for P_1 and $x_2 = 7.5$, $y_2 = 11.5$, $z_2 = 8.5$ for P_2. The polyhedron P_1 has six vertices with coordinates in the Eigen coordinate system defined as follows: $V_{11} = (-7.1, 1.4, 0.1)$, $V_{12} = (-2, 0.7, 8.2)$, $V_{13} = (0.45, 7.25, 0.2)$, $V_{14} = (7.15, 0.15, 0.25)$, $V_{15} = (0.7, -8.05, 0.9)$, $V_{16} = (1, 1, -7)$. The frontier of P_1 consists of eight faces laying on the following vertices: V_{11}, V_{12}, V_{13}; V_{12}, V_{14}, V_{13}; V_{11}, V_{15}, V_{12}; V_{12}, V_{15}, V_{14}; V_{11}, V_{16}, V_{13}; V_{13}, V_{16}, V_{14}; V_{11}, V_{16}, V_{15}; and V_{16}, V_{15}, V_{14}, respectively. The polyhedron P_2 has 15 vertices having the following coordinates: $V_{21} = (3, -2, -7)$, $V_{22} = (0, -4, -7)$, $V_{23} = (3, -4, 7)$, $V_{24} = (3, 0, -7)$, $V_{25} = (3, 4, 0)$, $V_{26} = (0, 4, -7)$, $V_{27} = (-2, 4, -7)$, $V_{28} = (-3, 3, -7)$, $V_{29} = (-3, 4, 0)$, $V_{2,10} = (0, 4, 7)$, $V_{2,11} = (-3, 3, 7)$, $V_{2,12} = (3, 0, 7)$, $V_{2,13} = (-3, 0, 7)$, $V_{2,14} = (0, -4, 7)$, and $V_{2,15} = (-3, -4, -7)$. The frontier of P_2 consists of 12 faces laying on the vertices V_{21}, V_{22}, V_{23}; V_{24}, V_{25}, V_{26}; V_{27}, V_{28}, V_{29}; $V_{29}, V_{2,10}, V_{2,11}$; $V_{2,13}, V_{2,14}, V_{2,15}$; $V_{25}, V_{2,10}, V_{2,12}$; $V_{22}, V_{2,15}, V_{2,14}, V_{23}$;

Optimized designing spherical void structures in 3D domains **Chapter | 16** **337**

$V_{21}, V_{24}, V_{25}, V_{2,12}, V_{23}$; $V_{25}, V_{26}, V_{27}, V_{29}, V_{2,10}$; $V_{28}, V_{29}, V_{2,11}, V_{2,13}, V_{2,15}$; $V_{21}, V_{24}, V_{26}, V_{27}, V_{28}, V_{2,15}, V_{22}$; and $V_{23}, V_{2,12}, V_{2,10}, V_{2,11}, V_{2,13}, V_{2,14}$, respectively.

Radii and centers of the spherical objects are given in Table 16.1. The corresponding objective value is $\kappa^*(\omega) = 530.07$. A graphical illustration of the optimized layout is shown in Fig. 16.1.

Example 2. The cuboid Ω bounds for the radii and the distance threshold ρ are the same as in Example 1. Three polyhedrons P_1, P_2, P_3 were defined for hollowing with $n_1 = n_2 = n_3 = 15$.

The polyhedron P_1 is the same as in Example 1. Coordinates of poles of other polyhedrons are $x_2 = 7.5$, $y_2 = 11.5$, $z_2 = 5$ for P_2 and $x_3 = 7.5$, $y_3 = 11.5$, $z_3 = 12$ for P_3. Both P_2 and P_3 have 15 vertices: $V_{21} = V_{31} = (3, -2, -3)$, $V_{22} = V_{32} = (0, -4, -3)$, $V_{23} = V_{33} = (3, -4, 3)$, $V_{24} = V_{34} = (3, 0, -3)$, $V_{25} = V_{36} = (3, 4, 0)$, $V_{26} = V_{36} = (0, 4, -3)$, $V_{27} = V_{37} = (-2, 4, -3)$, $V_{28} = (-3, 3, -3)$, $V_{29} = V_{39} = (-3, 4, 0)$, $V_{2,10} = V_{3,10} = (0, 4, 3)$, $V_{2,11} = V_{3,11} = (-3, 3, 3)$, $V_{2,12} = V_{3,12} = (3, 0, 3)$, $V_{2,13} = V_{3,13} = (-3, 0, 3)$, $V_{2,14} = V_{3,14} = (0, -4, 3)$, and $V_{2,15} = V_{3,15} = (-3, -4, -3)$. The frontier of P_2 (P_3) consists of 12 faces laying on the vertices $V_{21}, V_{22}, V_{23} (V_{31}, V_{32}, V_{33})$; $V_{24}, V_{25}, V_{26}(V_{34}, V_{35}, V_{36})$; V_{27}, V_{28}, $V_{29}(V_{37}, V_{38}, V_{39})$; $V_{29}, V_{2,10}, V_{2,11}$ $(V_{39}, V_{3,10}, V_{3,11})$; $V_{2,13}, V_{2,14}, V_{2,15}(V_{3,13}, V_{3,14}, V_{3,15})$; $V_{25}, V_{2,10}, V_{2,12}$ $(V_{35}, V_{3,10}, V_{3,12})$; $V_{22}, V_{2,15}, V_{2,14}, V_{23}$ $(V_{32}, V_{3,15}, V_{3,14}, V_{33})$; V_{21}, V_{24}, V_{25}, $V_{2,12}, V_{23}$ $(V_{31}, V_{34}, V_{35}, V_{3,12}, V_{33})$; V_{25}, V_{26}, $V_{27}, V_{29}, V_{2,10}$ $(V_{35}, V_{36}, V_{37}, V_{39}, V_{3,10})$; $V_{28}, V_{29}, V_{2,11}, V_{2,13}, V_{2,15}$ $(V_{38}, V_{39}, V_{3,11}, V_{3,13}, V_{3,15})$; $V_{21}, V_{24}, V_{26}, V_{27}, V_{28}, V_{2,15}, V_{22}$ $(V_{31}, V_{34}, V_{36}, V_{37}, V_{38}, V_{3,15}, V_{32})$, and $V_{23}, V_{2,12}, V_{2,10}, V_{2,11}, V_{2,13}, V_{2,14}$ $(V_{33}, V_{3,12}, V_{3,10}, V_{3,11}, V_{3,13}, V_{3,14})$, respectively.

Radii and centers of $S_{2q}, S_{3q} \subset P_1$, $q \in I_2$, $n_2 = n_3 = 15$ are summarized in Table 16.2. Radii and center coordinates of S_{1q}, $q \in I_1$, are given in Table 16.1. The corresponding objective value is $\kappa^*(\omega) = 491.48$. An illustration of the optimized layout is shown in Fig. 16.2.

Example 3. Cuboid Ω ($w \times l \times h$) has the following dimensions: $w = 30, l = 10$, $h = 18$; $C = P_1 \cup P_2 \cup P_3$, $S_j(v_j) \subset P_l$, $j \in I_l$, $l \in I_p$; the numbers of spheres to be packed into the polyhedrons are $\tau_1 = \tau_3 = 15$, $\tau_2 = 10$; radii bounds are $r_q^- = 0.8, r_q^+ = 3$; the minimal allowed distance between the spherical objects is $\rho = 0.5$; the protruding parameter is $\delta = 0.7$; the balance threshold is $\varepsilon = 0.1$.

Coordinates of poles of polyhedrons P_1, P_2, and P_3 are $(4, 5, 8)$, $(15, 4.5, 8)$, and $(26, 6, 8)$, respectively.

P_1 and P_3 have 15 vertices, coordinates of which in the Eigen coordinate system are $V_{11} = V_{31} = (3, -2, -7)$, $V_{12} = V_{32} = (0, -4, -7)$, $V_{13} = V_{33} = (3, -4, 9)$, $V_{14} = V_{34} = (3, 0, -7)$, $V_{15} = V_{35} = (3, 4, 0)$, $V_{16} = V_{36} = (0, 4, -7)$, $V_{17} = V_{37} = (-2, 4, -7)$, $V_{18} = V_{38} = (-3, 3, -7)$, $V_{19} = V_{39}$

TABLE 16.1 Radii and centers of the spherical objects.

No.	Radius	Center coordinates			Radius	Center coordinates		
q	r_{1q}	x_{1q}	y_{1q}	z_{1q}	r_{2q}	x_{2q}	y_{2q}	z_{2q}
1	2	0.3579	3.8766	0.0063	1.4679	−1.5321	2.5321	−1.5752
2	1.3791	0.8348	0.8436	−4.5131	0.9894	−2.0106	3.0106	1.0462
3	1.1466	1.2579	1.707	2.4598	0.9893	2.0107	0.9638	4.1711
4	1.7074	0.82	−1.4234	−2.0651	2	1	−2	5
5	1.9661	0.1818	−3.8332	0.9645	1.0426	0.899	−2.7918	−2.4294
6	1.1331	−2.1087	−1.6215	2.0036	1.9677	1.0323	2.0323	1.31
7	0.6075	−3.4557	−1.8706	0.5594	1.845	1.155	0.3349	−2.3791
8	1.5047	−1.2613	1.3379	−2.2366	0.8675	1.0685	3.1325	−3.35
9	0.907	3.154	−3.0137	0.4664	0.7747	1.7655	−1.996	−6.2253
10	0.7566	−2.6369	3.7538	0.2427	0.6172	−2.3828	3.1406	2.8604
11	0.5713	0.8031	−3.9324	−1.7509	1.2828	1.6651	0.2527	−5.7172
12	1.114	−1.8026	−1.579	−0.4742	1.3384	−1.6616	0.0175	0.1571
13	2	−1.3246	0.6424	4.3921	0.7013	−0.7624	−2.9842	2.847
14	1.0078	−0.618	−2.478	3.7784	0.8182	1.6242	0.7618	6.1818
15	1.3555	−0.1827	0.3528	0.5454	1.9213	1.0717	−2.0787	0.83

Optimized designing spherical void structures in 3D domains Chapter | 16 339

16	2	3.6822	0.092	0.2607	1.0422	−1.9578	−0.0123	−2.5566
17	1.0144	2.8862	0.5546	−2.8712	0.8411	−2.1589	1.2772	2.1872
18	2	−3.7344	0.9732	0.5649	1.2842	−1.46	−2.3132	−1.5099
19	1.2883	1.2985	−0.9708	2.6498	2	−1	1.8249	5
20	0.7352	3.2196	3.039	0.1497	0.9271	2.0729	−1.896	−4.1994
21	0.6573	0.708	−6.6509	0.7642	1.8768	−1.1232	2.1232	−5.1232
22	0.9471	−0.6435	3.4444	3.0116	0.8946	1.848	3.1054	−1.4952
					2	−1	−2	−5
					1.2744	−1.5837	−0.9156	2.8623

340 Advances of Artificial Intelligence in a Green Energy Environment

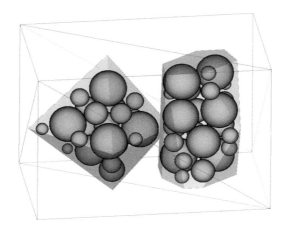

FIGURE 16.1 Packing of spherical objects in Example 1.

TABLE 16.2 Radii and center coordinates of spherical objects.

Number	Radius	Coordinates		
q	r_{1q}	x_{1q}	y_{1q}	z_{1q}
1	1.8867	−1.1133	2.1133	−1.1133
2	0.9932	0.3077	3.0068	1.8473
3	0.8773	−0.5415	−2.0884	2.1227
4	0.5321	−0.1309	−3.4679	1.2967
5	1.0576	−1.9424	2.5359	1.9424
6	0.7575	2.2425	−2.4066	−0.4611
7	0.6478	−2.3522	−0.0508	−2.3522
8	0.8134	2.1312	−1.8358	−2.1866
9	1.2678	1.7322	−2.7322	1.7322
10	0.5031	−2.4484	−3.4969	−2.4969
11	2	−0.6882	−2	−1
12	1.6331	1.3669	0.3761	1.3669
13	1.267	−1.733	−0.0484	1.733
14	1.2535	1.7465	0.4061	−1.7465
15	1.1086	1.8914	2.8914	−0.1654

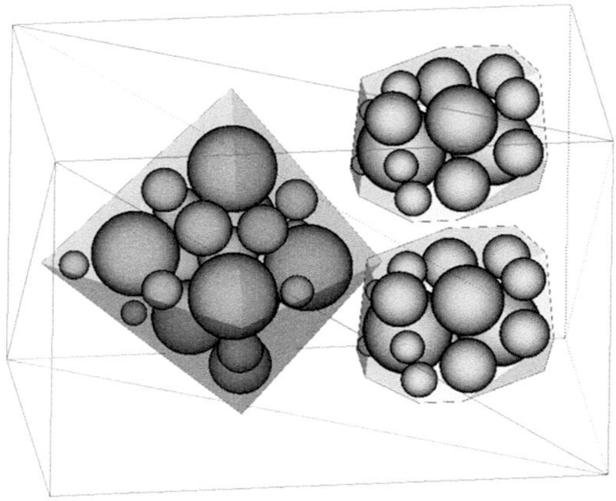

Optimized designing spherical void structures in 3D domains **Chapter | 16   341**

FIGURE 16.2   Packing of spherical objects in Example 2.

$= (-3, 4, 0)$, $V_{1,10} = V_{3,10} = (0, 4, 9)$, $V_{1,11} = V_{3,11} = (-3, 3, 9)$, $V_{1,12} = V_{3,12} = (3, 0, 9)$, $V_{1,13} = V_{3,13} = (-3, 0, 9)$, $V_{1,14} = V_{3,14} = (0, -4, 9)$, and $V_{1,15} = V_{3,15} = (-3, -4, -7)$. The frontiers of P_2 and P_3 consist of 12 faces, which lie on the vertices $V_{11}, V_{12}, V_{13}(V_{31}, V_{32}, V_{33})$; $V_{14}, V_{15}, V_{16}(V_{34}, V_{35}, V_{36})$; $V_{17}, V_{18}, V_{19}(V_{37}, V_{38}, V_{39})$; $V_{19}, V_{1,10}, V_{1,11}(V_{39}, V_{3,10}, V_{3,11})$; $V_{1,13}, V_{1,14}, V_{1,15}(V_{3,13}, V_{3,14}, V_{3,15})$; $V_{15}, V_{1,10}, V_{1,12}$ $(V_{35}, V_{3,10}, V_{3,12})$; $V_{12}, V_{1,15}, V_{1,14}, V_{13}$ $(V_{32}, V_{3,15}, V_{3,14}, V_{33})$; $V_{11}, V_{14}, V_{15}, V_{1,12}, V_{13}$ $(V_{31}, V_{34}, V_{35}, V_{3,12}, V_{33})$; $V_{15}, V_{16}, V_{17}, V_{19}, V_{1,10}$ $(V_{35}, V_{36}, V_{37}, V_{39}, V_{3,10})$; $V_{18}, V_{19}, V_{1,11}, V_{1,13}, V_{1,15}$ $(V_{38}, V_{39}, V_{3,11}, V_{3,13}, V_{3,15})$; $V_{11}, V_{14}, V_{16}, V_{17}, V_{18}, V_{1,15}, V_{12}$ $(V_{31}, V_{34}, V_{36}, V_{37}, V_{38}, V_{3,15}, V_{32})$,   and $V_{13}, V_{1,12}, V_{1,10}, V_{1,11}, V_{1,13}, V_{1,14}$ $(V_{33}, V_{3,12}, V_{3,10}, V_{3,11}, V_{3,13}, V_{3,14})$, respectively.

P_2 has six vertices, coordinates of which in the Eigen coordinate system are $V_{11} = (-7.1, 1.4, 0.1)$, $V_{22} = (-2, 0.7, 9.2)$, $V_{23} = (0.45, 4.25, 0.2)$, $V_{24} = (7.15, 0.15, 0.25)$, $V_{25} = (0.7, -4.05, 0.9)$, $V_{26} = (1, 1, -8)$. The frontier of P_1 consists of eight faces, which lie on the vertices V_{11}, V_{12}, V_{13}; V_{12}, V_{14}, V_{13}; V_{11}, V_{15}, V_{12}; V_{12}, V_{15}, V_{14}; V_{11}, V_{16}, V_{13}; V_{13}, V_{16}, V_{14}; V_{11}, V_{16}, V_{15}, and V_{16}, V_{15}, V_{14}, respectively.

The objective value calculated is $\kappa^*(\omega) = 1232.88$. Radii and coordinates of the centers of the spherical objects are given in Table 16.3. A graphical illustration of the corresponding packing of spherical objects is shown in Fig. 16.3.

342 Advances of Artificial Intelligence in a Green Energy Environment

TABLE 16.3 Radii and center coordinates of spherical objects.

No.	Radius	Center coordinates		
q	r_q	x_q	y_q	z_q
1	1.4155	4.2281	7.8441	1.7155
2	2.9919	3.2919	3.2919	3.2919
3	2.3226	5.3774	7.3774	6.8654
4	1.0155	6.6845	6.2563	3.4352
5	1.4869	1.7869	8.2131	9.0975
6	1.08	1.38	6.9786	1.38
7	1.5796	1.8796	8.0995	4.2916
8	2.6146	3.5555	2.9146	9.3811
9	1.2517	1.5517	5.8727	6.8714
10	1.9335	2.2335	7.2408	15.7665
11	1.1202	6.5798	2.5477	6.4396
12	0.9913	1.2913	8.2728	12.6397
13	2.2427	4.8601	7.4573	11.9036
14	2.6944	5.0056	2.9944	15.0056
15	1.3296	1.6296	5.0271	12.7839
16	1.7814	13.9205	5.4768	4.7252
17	1.307	17.6672	4.821	11.4042
18	1.5306	17.7084	5.0666	4.6044
19	0.971	15.7822	2.8737	5.3049
20	1.3415	15.9613	5.353	1.7344
21	2.2412	13.9041	5.0724	12.8755
22	1.8931	15.7961	7.0423	8.1101
23	1.939	19.4919	4.7231	8.1341
24	2.1947	15.4847	2.5416	8.9435
25	2.2486	11.4689	5.4174	8.534
26	1.2669	23.5669	3.2757	5.3048
27	3	25.3	7.7	3.3
28	1.657	28.043	9.043	7.4554

Continued

TABLE 16.3 Radii and center coordinates of spherical objects.—cont'd

No.	Radius	Center coordinates		
q	r_q	x_q	y_q	z_q
29	1.4398	26.0341	9.0073	16.2602
30	0.9226	23.2226	8.863	16.7774
31	1.2858	23.5858	5.6518	11.3005
32	1.7669	24.0669	8.6538	13.1385
33	2.8353	26.8647	4.2356	8.3682
34	3	26.7	4.3679	14.7
35	1.3619	28.3381	4.2758	1.6619
36	1.9343	24.2343	8.7657	8.942
37	1.6375	23.9375	2.9375	1.9375
38	0.9838	23.2838	6.4582	16.7162
39	1.8026	27.7032	8.3437	11.3379
40	1.2979	26.7155	2.83	3.9558

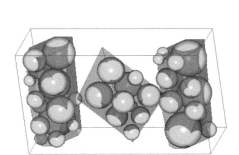 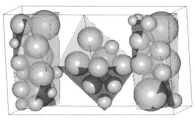

FIGURE 16.3 Packing of spherical objects in Example 3.

16.6 Conclusions

A new approach for optimized packing variable-sized spherical objects in a polyhedral disconnected domain is proposed. The model takes into account the bounds for the distance between the spheres and can be used in designing spherical void structures in AM. The nonlinear programming problem arising in the proposed approach has a large number of constraints and variables and thus direct solution is time-consuming. To cope with this problem, aggregation and/or decomposition techniques [36,37] can be implemented to make use of

344 Advances of Artificial Intelligence in a Green Energy Environment

the special structure of the constraints, as well as parallel computing techniques [38]. Optimized designing spherical void structures in 3D domains considering mechanical characteristics (similar to 2D case [39]) are on the way.

Acknowledgments

The study was partially supported by the National Research Fund of Ukraine (Grant No. 02.2020/167).

References

[1] I. Gibson, D. Rosen, B. Stucker, Additive Manufacturing Technologies, 3D Printing, Rapid Prototyping, and Direct Digital Manufacturing, Springer Science + Business Media, " New-York, 2015.

[2] R. Lachmayer, R.B. Lippert, Additive Manufacturing. Quantifiziert − Visionäre Anwendungen und Stand der Technik, Springer Vieweg Verlag, " Berlin, 2017.

[3] J. Liu, Y. Ma, A survey of manufacturing oriented topology optimization methods, Adv. Eng. Softw. 100 (2016) 161−175, https://doi.org/10.1016/j.advengsoft.2016.07.017.

[4] M. Lee, Q. Fang, Y. Cho, J. Ryu, L. Liu, D.-S. Kim, Support-free hollowing for 3D printing via Voronoi diagram of ellipses, Comput. Aided Des. 101 (2018) 23−36.

[5] J.P. Araújo Luiz, E. Özcan, A.D. Atkin Jason, M. Baumers, Analysis of irregular three-dimensional packing problems in additive manufacturing: a new taxonomy and dataset, Int. J. Prod. Res. (2018) 5920−5934, https://doi.org/10.1080/00207543.2018.1534016.

[6] M. Hifi, R. M'Hallah, A literature review on circle and sphere packing problems: models and methodologies, Adv. Oper. Res. 2009 (2009), https://doi.org/10.1155/2009/150624.

[7] L. Burtseva, B. Valdez Salas, R. Romero, F. Werner, Recent advances on modelling of structures of multi-component mixtures using a sphere packing approach, Int. J. Nano-technol. 13 (2016) 44−59, https://doi.org/10.1504/IJNT.2016.074522.

[8] Z. Duriagina, I. Lemishka, I. Litvinchev, J.A. Marmolejo, A. Pankratov, T. Romanova, G. Yaskov, Optimized filling of a given cuboid with spherical powders for additive manufacturing, J. Oper. Res. Soc. China (2020), https://doi.org/10.1007/s40305-020-00314-9.

[9] J. Wang, Packing of unequal spheres and automated radiosurgical treatment planning, J. Combin. Optim. 3 (1999) 453−463.

[10] O. Blyuss, L. Koriashkina, E. Kiseleva, R. Molchanov, Optimal placement of irradiation sources in the planning of radiotherapy: mathematical models and methods of solving, Comput. Math. Methods Med. (2015), https://doi.org/10.1155/2015/142987. Article ID 142987.

[11] J.R. Adler, A. Schweikard, Y. Achkire, O. Blanck, R.M. Bodduluri, L. Ma, H. Zhang, Treatment planning for self-shielded radiosurgery, Cureus 9 (9) (2017) e1663, https://doi.org/10.7759/cureus.1663.

[12] N. Ilyasova, A. Shirokanev, D. Kirsh, R. Paringer, A. Kupriyanov, E. Zamycky, Development of coagulate map formation algorithms to carry out treatment by laser coagulation, Procedia Eng. 201 (2017) 271−279, https://doi.org/10.1016/j.proeng.2017.09.623.

[13] Y. Stoyan, A. Pankratov, T. Romanova, G. Fasano, J.D. Pinter, Y.E. Stoian, A. Chugay, Optimized packings in space engineering applications: part I, in: G. Fasano, J. Pinter (Eds.), Modeling and Optimization in Space Engineering, vol. 144, Springer, Cham, 2019, pp. 395−437, https://doi.org/10.1007/978-3-030-10501-3_15.

Optimized designing spherical void structures in 3D domains **Chapter | 16 345**

[14] Y. Stoyan, I. Grebennik, T. Romanova, A. Kovalenko, Optimized packings in space engineering applications: part II, in: G. Fasano, J. Pinter (Eds.), Modeling and Optimization in Space Engineering, vol. 144, Springer, Cham, 2019, pp. 439–457, https://doi.org/10.1007/978-3-030-10501-3_15.

[15] G.E. Mueller, Numerically packing spheres in cylinders, Powder Technol. 159 (2005) 105–110.

[16] A.J. Otaru, A.R. Kennedy, The permeability of virtual macroporous structures generated by sphere packing models: comparison with analytical models, Scripta Mater. 124 (2016) 30–33.

[17] J. de Bono, G. McDowell, On the packing and crushing of granular materials, Int. J. Solid Struct. 187 (2020) 133–140.

[18] B. Chazelle, H. Edelsbrunner, L.J. Guibas, The complexity of cutting complexes, Discrete Comput. Geom. 4 (1989) 139–181.

[19] A. Sutou, Y. Day, Global optimization approach to unequal sphere packing problems in 3D, J. Optim. Theor. Appl. 114 (2002) 671–694.

[20] S. Yamada, J. Kanno, M. Miyauchi, Multi-sized sphere packing in containers: optimization formula for obtaining the highest density with two different sized spheres, Inf. Media Technol. 6 (2011) 493–500.

[21] T. Kubach, A. Bortfeldt, T. Tilli, H. Gehring, Greedy algorithms for packing unequal spheres into a cuboidal strip or a cuboid, Asia Pac. J. Oper. Res. 28 (2011) 739–753.

[22] J. Liu, Y. Yao, Y. Zheng, H. Geng, G. Zhou, An effective hybrid algorithm for the circles and spheres packing problems, Combin. Optim. Appl. Lect. Notes Comput. Sci. 5573 (2009) 135–144.

[23] A. Kazakov, A. Lempert, T. Thanh Ta, On the algorithm for equal balls packing into a multi-connected set, in: Proc. of the VIth International Workshop "Critical Infrastructures: Contingency Management, Intelligent, Agent-Based, Cloud Computing and Cyber Security" (IWCI 2019), 2019, pp. 216–222, https://doi.org/10.2991/iwci-19.2019.38.

[24] R. Torres-Escobar, J.A. Marmolejo-Saucedo, I. Litvinchev, Binary monkey algorithm for approximate packing non-congruent circles in a rectangular container, Wirel. Netw. 26 (7) (2020) 4743–4752.

[25] I. Litvinchev, E.L. Ozuna, Approximate packing circles in a rectangular container: valid inequalities and nesting, J. Appl. Res. Technol. 12 (4) (2014) 716–723.

[26] Y. Stoyan, G. Yaskov, T. Romanova, I. Litvinchev, S. Yakovlev, J.M.V. Cantú, Optimized packing multidimensional hyperspheres: a unified approach, Math. Biosci. Eng. 17 (6) (2020) 6601–6630, https://doi.org/10.3934/mbe.2020344.

[27] E.G. Birgin, F.N.C. Sobral, Minimizing the object dimensions in circle and sphere packing problems, Comput. Oper. Res. 35 (2008) 2357–2375, https://doi.org/10.1016/j.cor.2006.11.002.

[28] J.M. Martínez, L. Martínez, Packing optimization for automated generation of complex system's initial configurations for molecular dynamics and docking, J. Comput. Chem. 24 (2003) 819–825, https://doi.org/10.1002/jcc.10216.

[29] M. Hifi, L. Yousef, A local search-based method for sphere packing problems, Eur. J. Oper. Res. 274 (2019) 482–500, https://doi.org/10.1016/j.ejor.2018.10.016.

[30] Y.G. Stoyan, G. Scheithauer, G.N. Yaskov, Packing unequal spheres into various containers, Cybern. Syst. Anal. 52 (2016) 419–426, https://doi.org/10.1007/s10559-016-9842-1.

[31] Z.Z. Zeng, W.Q. Huang, R.C. Xu, Z.H. Fu, An algorithm to packing unequal spheres in a larger sphere, Adv. Mater. Res. 546–547 (2012) 1464–1469, https://doi.org/10.4028/www.scientific.net/AMR.546-547.1464.

346 Advances of Artificial Intelligence in a Green Energy Environment

[32] Y. Stoyan, G. Yaskov, Optimised packing unequal spheres into a multiconnected domain: mixed-integer non-linear programming approach, Int. J. Comput. Math. Comput. Syst. Theory (2020), https://doi.org/10.1080/23799927.2020.1861105.

[33] Y. Stoyan, T. Romanova, Mathematical models of placement optimisation: two- and three-dimensional problems and applications, in: G. Fasano, J. Pintér (Eds.), Modeling and Optimization in Space Engineering, vol. 73, Springer, New York, 2012, pp. 363—388, https://doi.org/10.1007/978-1-4614-4469-5_15.

[34] I.V. Grebennik, A.A. Kovalenko, T.E. Romanova, et al., Combinatorial configurations in balance layout optimization problems, Cybern. Syst. Anal. 54 (2018) 221—231, https://doi.org/10.1007/s10559-018-0023-2.

[35] A. Wächter, L.T. Biegler, On the implementation of a primal-dual interior point filter line search algorithm for large-scale nonlinear programming, Math. Program. 106 (2006) 25—57.

[36] I. Litvinchev, M. Mata, S. Rangel, J. Saucedo, Lagrangian heuristic for a class of the generalized assignment problems, Comput. Math. Appl. 60 (4) (2010) 1115—1123.

[37] I. Litvinchev, S. Rangel, Localization of the optimal solution and a posteriori bounds for aggregation, Comput. Oper. Res. 26 (1999) 967—988.

[38] T.E. Romanova, P.I. Stetsyuk, A.M. Chugay, S.B. Shekhovtsov, Parallel computing technologies for solving optimization problems of geometric design, Cybern. Syst. Anal. 55 (2019) 894—904, https://doi.org/10.1007/s10559-019-00199-4.

[39] T. Romanova, Y. Stoyan, A. Pankratov, I. Litvinchev, K. Avramov, M. Chernobryvko, I. Yanchevskyi, I. Mozgova, J. Bennell, Optimal layout of ellipses and its application for additive manufacturing, Int. J. Prod. Res. (2019), https://doi.org/10.1080/00207543.2019.1697836.

Chapter 17

Swarm-based intelligent strategies for charging plug-in hybrid electric vehicles

Pandian Vasant[1], Anirban Banik[2], J. Joshua Thomas[3], Jose Antonio Marmolejo-Saucedo[4], Timothy Ganesan[5], Elias Munapo[6] and Mukhdeep Singh Manshahia[7]

[1]*MERLIN Research Centre, TDTU, Ho Chi Minh City, Vietnam;* [2]*National Institute of Technology Agartala, India;* [3]*UOW Malaysia KDU Penang University College, George Town, Pulau Pinang, Malaysia;* [4]*Universidad Panamericana, Facultad de Ingeniería, Mexico City, Mexico;* [5]*Member of American Mathematical Society, AB, Canada;* [6]*North West University, Mahikeng, South Africa;* [7]*Department of Mathematics, Punjabi University, Patiala, Punjab, India*

17.1 Introductions

Green technology research in the transportation sector is gaining momentum among researchers from various disciplines. In this context, PHEVs have a brilliant future because of their charging storage and recharging systems from traditional grids. Some researchers have demonstrated that transport electrification can lead to considerable emission depletion of greenhouse gases [1]. Internal combustion fuel cars, hybrid electric vehicles, and all-electric vehicles have all seen significant commercial expansion in the last 10 years [2]. Better implementation of PHEVs will play an important role in the incorporation of renewable energy in conventional grid networks, as all similar strategies can be employed by hybrid plug-in vehicles with a simple smart grid link [3]. Effective mechanisms are required to resolve advanced problems such as cost reductions, electricity storage, efficient charging stations through diverse goals and system restrictions [4].

As per the European Electric Power Research Institute (EPRI), by 2050, about 62% of vehicles in the United States will be plug-in hybrids [5]. A complicated control device would be needed to manage multiple battery loads from a variety of PHEVs correctly for maximizing consumer content and for minimizing the strain of the grid [6]. Due to the differing demands of the PHEVs parked in the deck at varying times, the aggregate demand pattern will have a considerable impact on electricity generation [7].

Advances of Artificial Intelligence in a Green Energy Environment
https://doi.org/10.1016/B978-0-323-89785-3.00015-3
Copyright © 2022 Elsevier Inc. All rights reserved.

348 Advances of Artificial Intelligence in a Green Energy Environment

A PHEV's effectiveness is contingent upon its effective use of electric power, which is strictly determined by the battery's state of charge (SoC). The SoC of the battery is a key factor in PHEVs, since it determines the amount of power stored in the battery. It is comparable to the fuel gauge on a typical IC engine [8]. The assessment of SoC is becoming an exceedingly important concern in all environments that include a battery. Presently, battery operation is shifting toward what may be referred to as battery management rather than merely safety. The battery SoC is an important element in this increased battery control [9].

Swarm intelligence was inspired by the simulation of a real swarm, like ants and birds that dominate swarming for shelter construction or food gathering. Based on this, researchers developed many of the algorithms simulating living colonies such as Ant Colony Optimization (ACO) algorithm, Artificial Bee Colony (ABC) algorithm, particle swarm optimization (PSO) algorithm, and gravitational search algorithm (GSA) [10–12]. Both population-based metaheuristic algorithms such as PSO and GSA must maintain a good balance for solving difficult optimization problems [12].

One option for a power supplier to balance electricity production with the energy demand is to use a charging station, both of which fluctuate at random over time. PHEVs are made to pay at charging points when production reaches demand and discharged when usage exceeds production[13]. An in-depth investigation is required in this analysis to enhance the average SoC in order to make intelligent energy distribution convenient for PHEVs in a charging point.

17.1.1 Study objectives

The objective of this research is to improve SoC in relation to charging time, using the current SoC. This objective is divided into two subobjectives:

a. Using swarm intelligence techniques, maximize the SoC of a plug-in hybrid electric vehicle (PHEV).
b. Swarm intelligence strategies' performance is evaluated in terms of fitness value and computing time.

17.2 Problem formulation

Suppose a charge point with a maximum power called P and a maximum of N PHEVs have to be served within a single day. The proposed system would allow PHEVs to exit the charging station prior to the actual time of departure, to boost system performance. The objective is to extract energy intelligently for each PHEV reaching the charging point. The SoC must be maximized to optimize power allocation. In order to optimize the average SoC, the fitness function described in this section assigns resources to PHEVs on the next step.

The following constraints are assumed: recharge time, present SoC, and energy price. The fitness function is denoted by the following:

$$\text{Max } J(k) = \sum_i w_i (k) SoC_i(k+1) \tag{17.1}$$

$$w_i (k) = f \left(C_{r,i}(k), T_{r,i}(k), D_i(k) \right) \tag{17.2}$$

$$C_{r,i}(k) = (1 - SoC_i(k)) \cdot C_i \tag{17.3}$$

where $C_{r,i}(k)$ illustrates the battery space remaining for i PHEV number to be filled at the time step k; C_i shows the rated battery capacity of the i number of PHEV; the remaining time for charging a particular PHEV at k time step is expressed as $T_{r,i}(k)$; the difference in price between the real-time energy price and the price that a particular customer at the specified i number of PHEV chargers is willing to pay at a given time stage is represented by $D_i(k)$; $w_i(k)$ denotes the charging terms of the i number of PHEVs at k time steps; SoC_i $(k+1)$ illustrates the SoC of i number of PHEVs at $k+1$ steps.

17.3 Swarm-based intelligence approaches

The swarm intelligence develops by imitating a living swarm in nature, such as an ant, bird, or fish that shows unparalleled superiority in the swarm, rather than in individuals looking for food or nest buildings [14]. In order to ensure that readers understand well the features of the corresponding algorithms with regard to correct and intelligent optimization, four swarm intelligence algorithms selected for the charging optimization problem can be discussed briefly.

17.3.1 Particle swarm optimization

PSO is a well-known evolutionary computational technique proposed by Kennedy and Eberhart [15]. Bird flocking offers a detailed source of inspiration for this process. In the algorithm setting, there are many particles that float through the search space searching for the best solution. Meanwhile, they are all influenced by the best solution throughout their journey. In general, each particle can update its position by focusing on its current velocity, current position, and position difference according to *pbest* and *gbest* separately. PSO is briefly run over a randomly distributed set of particles (solutions), and then it runs over looking for the best solution(s) by modifying generations. Whereas, the "best" value obtained by any particle in the relevant population so far is known as global best (*gbest*) (Table 17.1). Mathematically, PSO can be represented as

$$V_i^{t+1} = wv_i^t + c_1 \cdot rand \cdot (pbest_i - x_i^t) + c_2 \cdot rand \cdot (gbest - x_i^t) \tag{17.4}$$

$$x_i^{t+1} = x_i^t + V_i^{t+1} \tag{17.5}$$

350 Advances of Artificial Intelligence in a Green Energy Environment

TABLE 17.1 Parameter settings for particle swarm optimization (PSO).

Parameters	Values
Swarm size	100
Maximum number of steps	100
PSO parameter, c1	1.4
PSO parameter, c2	1.4
Inertia weight (w)	0.9
Maximum number of iteration	100
Number of runs	50

In the above equations, v_i^t shows the particle velocity at the t iteration and w is the inertial weight function and can be defined by using Eq. (17.6):

$$w = w_{max} - \frac{w_{max} - w_{min}}{t_{max}} \times t \qquad (17.6)$$

The most suitable values for w_{min} and w_{max} are 0.4 and 0.9, respectively [16]. It is 1−2 for C_1 and C_2 [7], with 2 being the most recommended value for different scenarios [17]. On the other hand, *rand* denotes the random variable between 0 and 1 [7], x_i^t represents the present position of particle at t iteration, *pbest* represents the personal best of the individual particles at t iterations. The *gbest*, on the other hand, is the best global solution identified so far. The PSO algorithm operates by maintaining many particles or possible solutions in the search space at the same time. Fig. 17.1 depicts the key stages of PSO.

17.3.1.1 Advantages and disadvantages of PSO

The below are the specific benefits of PSO: (1) The algorithm is straightforward, with few customizable parameters and is quick to apply. (2) The random initialization of the population has a high global searching capacity, comparable to the genetic algorithm. (3) It employs the assessment function to determine the individual's searching speed. (4) It is highly scalable. However, the drawbacks of the PSO are as follows: (1) The algorithm is unable to fully use the system's input knowledge. (2) It has a limited potential to solve combinatorial optimization problems. (3) Obtaining a local optimum solution is simple for this algorithm.

17.3.2 Accelerated particle swarm optimization

Xin-She Yang created Accelerated Particle Swarm Optimization (APSO) to speed up the convergence of the algorithm by only using the global best [18].

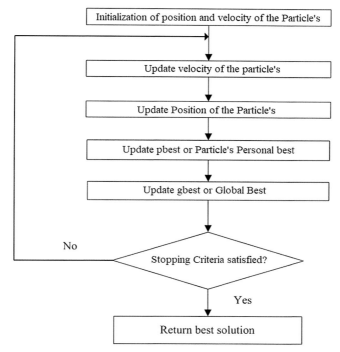

FIGURE 17.1 Particle swarm optimization flowchart.

Individual of the swarm are referred to as particles in APSO algorithmic system, whereas the population itself is referred to as the swarm. APSO starts off as a randomly distributed population, with each particle moving randomly based on certain parameters. Each particle of the population travels in the search solution space, and at this stage, it remembers its own best positions so far, as well as the positions of its neighbors, and its own velocity. A swarm's particles transmit best position and velocity to one another and dynamically set their position and velocity reproduce from the earlier experience. Finally, during the searching process, all particles are inclined to fly for better positions before the swarm moves close to an optimum of the fitness function. The convergence speed of the simpler variant of the PSO algorithm could be accelerated using global best only. As a result, in the APSO [19], the velocity vector is generated using Eq. (17.7) in which *randn* is in the range of (0, 1). To enhance the convergence even further, the modification of the particle position can be done in a single step, as follows:

$$V_i^{t+1} = V_i^t + \alpha \cdot randn(t) + \beta \cdot (g^* - x_i^t) \qquad (17.7)$$

$$x_i^{t+1} = (1-\beta)x_i^t + \beta \cdot g^* + \alpha r \qquad (17.8)$$

352 Advances of Artificial Intelligence in a Green Energy Environment

TABLE 17.2 Parameter settings of accelerated particle swarm optimization.

Parameters	Values
Size of the swarm	100
Maximum no. of steps	100
Alpha, α	0.2
Beta, β	0.5
Maximum iteration	100
Number of runs	50

In the present computation, α can be defined by using Eq. (17.9) [20]:

$$\alpha = 0.7^t \tag{17.9}$$

The general values for the APSO are $\alpha \approx 0.1 \sim 0.4$ and $\beta \approx 0.1 \sim 0.7$. However, $\alpha \approx 0.2$ and $\beta \approx 0.5$ were recommended [21]. In general, evolutionary search algorithm performs better with a larger population. A large population, on the other hand, would cost more in terms of fitness function evaluation without making substantial changes. The population size in the present simulation is considered to be 100. The settings of the parameters for APSO are illustrated in Table 17.2. Moreover, Fig. 17.2 summaries the general steps involved in APSO.

17.3.2.1 Advantages and disadvantages of APSO

The following are the benefits of APSO: (1) It is very effective in terms of solving the particular fitness function. (2) A simplified formula is used to generate the velocity vector. (3) Position updates are completed in only one step. (4) The willingness of local exploitation to search optimum solutions. Furthermore, APSO has the following drawbacks: (1) It is characterized by early convergence in the initial stages. (2) The alpha and beta variables affect algorithm efficiency, and no fixed value will guarantee higher fitness values.

17.3.3 Gravitational search algorithm

Rashedi et al. have developed GSA, an optimization method based on the population [12]. Each mass (or agent) in this algorithm has four parameters: position, inertial mass, and passive and active gravitational mass. The position in relation to the mass refers to a solution to a specific problem. The masses (both gravitational and inertial) are determined in this context using a fitness function. The gravitational forces that occur within them attract all of the

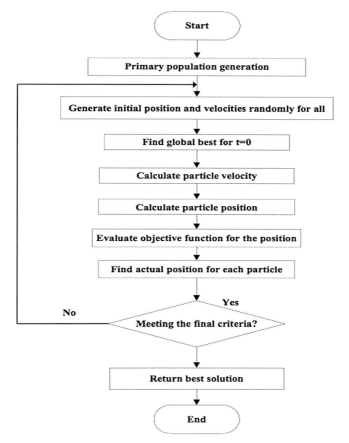

FIGURE 17.2 Flow diagram of accelerated particle swarm optimization.

masses to each other over generations. A heavier mass means a greater attraction force in this situation. The heavier masses seem to have an effect on other masses that is proportional to their distances from the global optimum value in this situation. The gravitational force is denoted by Eq. (17.10):

$$F_{ij}^d(t) = G(t) \frac{M_{pi}(t) \times M_{aj}(t)}{R_{ij}(t) + \varepsilon} \left(x_j^d(t) - x_i^d(t) \right) \quad (17.10)$$

where M_{aj} and M_{pi} denote the active and passive gravitational mass associated with agent j and agent i, respectively; $G(t)$ illustrates gravitational constant; ε denotes small constant; and $R_{ij}(t)$ represents the Euclidean distance between two agents i and j. The $G(t)$ can be evaluated using Eq. (17.11):

$$G(t) = G_0 \times \exp(-\alpha \times \text{iter} / \text{maxiter}) \quad (17.11)$$

354 Advances of Artificial Intelligence in a Green Energy Environment

where α and G_0 denote the coefficient of descending and primary value, respectively. Current and maximum iteration numbers are illustrated using *iter* and max *iter*, respectively. The total force acting on agent i is determined as follows in a problem domain with d-dimension implementing Eq. (17.12).

$$F_i^d(t) = \sum_{j=1, j \neq i}^{N} rand_j F_{ij}^d(t) \tag{17.12}$$

Here, $rand_j$ denotes the random number in the range [0, 1]. From law of motion, acceleration agents are directly proportional to the resultant force and inversely proportional to its mass. So, the acceleration of all agents can be evaluated using Eq. (17.13):

$$ac_i^d(t) = \frac{F_i^d(t)}{M_{ii}(t)} \tag{17.13}$$

where t denotes specific time and M_{ii} represents the mass of the object i. The velocity and position of agents can be evaluated using Eqs. (17.14) and (17.15):

$$vel_i^d(t+1) = rand_i \times vel_i^d(t) + ac_i^d(t) \tag{17.14}$$

$$x_i^d(t+1) = x_i^d(t) + vel_i^d(t+1) \tag{17.15}$$

In the above Eq. (17.14), $rand_i$ represents the random number in the range [0, 1]. Fig. 17.3 depicts the steps involved in implementation of GSA. The agents are first found using random values in the GSA methodology, as each of them is recognized as a candidate solution. Following that, all agents' velocities are modified in accordance with Eq. (17.13), while the total forces, gravitational constant, and accelerations are selected in accordance with Eqs. (17.10)−(17.12). The gravitational force is an information-transferring method since each entity will observe the output of the others. Table 17.3 illustrates the parameter settings for GSA.

17.3.3.1 Advantages and disadvantages of GSA

The GSA algorithm has the following advantages: (1) it offered a high-quality solution in terms of best fitness. (2) It has a high degree of convergence. (3) It is capable of local exploitation. The drawbacks of this algorithm, on the other hand, are as follows: (1) it has a longer computational time than APSO and PSO. (2) It has more parameter tuning options as mass and speed particles are considered.

17.3.4 Hybrid PSOGSA algorithm

The hybrid PSOGSA was created by Seyedali Mirjalili [22] to combine PSO's optimization ability with that of GSA. PSOGSA is heterogeneous due to the

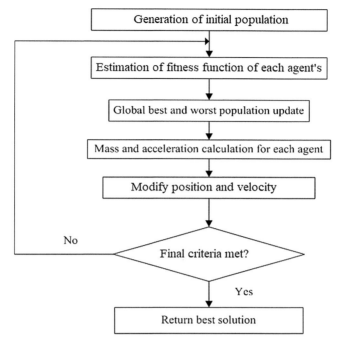

FIGURE 17.3 Flow diagram of gravitational search algorithm.

TABLE 17.3 Parameter settings of gravitational search algorithm [12].

Parameters	Values
Initial parameter, G_0	100
Number of mass agents, n	100
Coefficient of acceleration, α	20
Constant variable, ε	0.01
Power "R"	1
Maximum number of iteration	100
Number of runs	50

use of two different algorithms (PSO and GSA) in order to achieve final solution. PSOGSA saves the best results achieved so far in memory so that it can be accessed at any time [22]. PSOGSA's basic concept is to merge PSO's potential for social cognition (*gbest*) with GSA's local search capabilities. To

356 Advances of Artificial Intelligence in a Green Energy Environment

merge these two algorithms, velocity modification is done by using the following Eq. (17.16);

$$v_i(t+1) = w \times v_i(t) + \alpha' \times rand \times ac_i(t) + \beta' \times rand \times (gbest - x_i(t))$$
(17.16)

where $v_i(t)$ represents the agent's velocity at t number of iteration. However, w denotes weight factor, $rand$ illustrates the randomized number ranging from [0−1], $ac_i(t)$ denotes agent's acceleration at iteration t. Lastly, $gbest$ illustrates the global best solution achieved so far. In the above equation, α' and β' denote the weighting factors [23]. The global and local search ability can be balanced by adjusting α' and β'. For each iteration, particle position $x_i(t+1)$ is modified by using Eq. (17.17):

$$x_i(t+1) = x_i(t) + v_i(t+1)$$
(17.17)

Fig. 17.4 depicts the flowchart of the hybrid PSOGSA method. PSOGSA (adjusted based on current parameter settings in Table 17.4) was used for a similar fitness function, and its performance compared to that of the GSA by evaluating average best fitness. Size of the swarm and maximum iterations are selected to match those of the PSO and GSA approaches in order to achieve an analytical comparative approach. The parameters c_1, c_2, and α were set to regular values of 0.5, 1.5, and 23, respectively [22].

17.3.4.1 Advantages and disadvantages of PSOGSA

The following benefits are associated with hybrid PSOGSA: (1) the objective function for optimizing PHEV SoC is constrained by various restrictions that restrict the search domain to a specific feasible area. PSOGSA accommodates the restrictions independently, removing the requirement for extra parameters. (2) As opposed to other solo approaches, it offers a high-quality approach in terms of optimal fitness. (3) PSOGSA combines the ability to exploit in PSO with the ability to exploit in GSA to combine the strengths of both algorithms [24]. (4) PSOGSA's convergence speed is higher than that of PSO, APSO, and GSA. Finally, the drawbacks of the PSOGSA are as follows: (1) It needs more processing time than single optimization strategies. (2) It has more parameter tuning options due to the combination of PSO and GSA techniques.

17.4 Results and discussions

17.4.1 Particle optimization swarm findings

Fig. 17.5 illustrates the convergence action of the PSO methodology for each problem scenario. The PSO was updated to run for 100 number of iterations despite the fact that the objective function converges and becomes constant before 5 number of iterations for all five cases. As a result, an early convergence can result in the fitness function trapping in local optima. This can be

Swarm-based intelligent strategies Chapter | 17 357

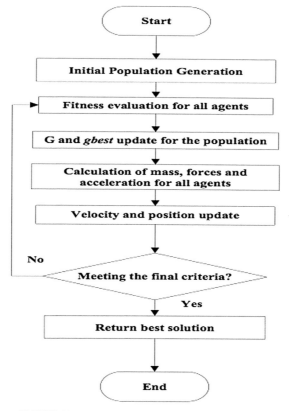

FIGURE 17.4 Flowchart of hybrid PSOGSA algorithm.

TABLE 17.4 Settings of PSOGSA parameters.

Parameters	Values
Swarm size	100
Maximum iteration	100
PSO variables, C1	0.5
PSO variables, C2	1.5
Constant of gravitational, G0	1
Constant variable of GSA, α	23
Number of runs	50

358 Advances of Artificial Intelligence in a Green Energy Environment

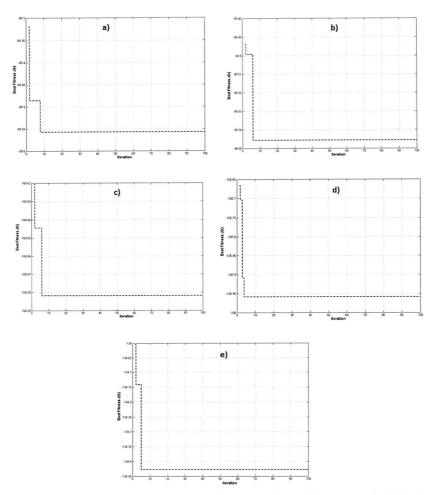

FIGURE 17.5 Plot between iteration versus fitness value for particle swarm optimization algorithm for five scenarios: (A) 50 PHEVs, (B) 100 PHEVs, (C) 300 PHEVs, (D) 500 PHEVs, and (E) 1000 PHEVs.

averted by maximizing the swarm's size, which also increases the computational time. Here, a trade-off between proper convergence and computing time should be considered. Fig. 17.6 illustrates the PSO simulation findings for the best value of the fitness function denoted by J for 50, 100, 300, 500, and 1000 PHEVs, respectively. At this stage, each scenario is run 50 times to determine the performance, as well as the supremacy and efficiency of the PSO algorithm.

Owing to the fact that PSO is a population-based optimization strategy with a nonlinear fitness function, the fitness values vary with each iteration

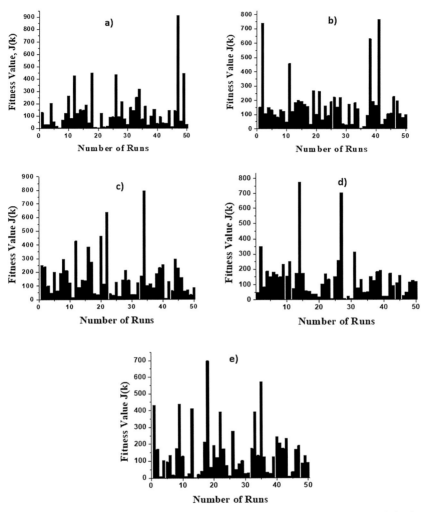

FIGURE 17.6 Plot between fitness value and number of runs for particle swarm optimization algorithm subjected to five scenarios: (A) 50 PHEVs; (B) 100 PHEVs; (C) 300 PHEVs; (D) 500 PHEVs; and (E) 1000 PHEVs.

[25–27]. However, maximum best value of fitness remains between 650 and 950, while the minimum best value of fitness remains between 0.70 and 8. The maximum, minimum, and average values of fitness forecasted by PSO algorithm when subjected to five different scenarios were demonstrated in Table 17.5. From Table 17.5, the average best fitness is within a nearly identical importance range for five considered cases. The average computing time required for the PSO algorithm is illustrated in Table 17.6. The total

TABLE 17.5 Evaluation for fitness for particle swarm optimization algorithm.

Fitness function	50 PHEVs	100 PHEVs	300 PHEVs	500 PHEVs	1000 PHEVs
Maximum best fitness	910.75	767.87	793.09	774.56	697.11
Minimum best fitness	4.84	5.38	5.22	7.18	0.73
Average best fitness	142.84	171.10	169.31	144.80	156.80

TABLE 17.6 Average time required for computation for particle swarm optimization.

PHEV numbers	Time required for computation (s)
50	1.62
100	1.67
300	1.76
500	1.95
1000	2.33

computing time for 50 PHEVs is 1.62 s, although it increases to 2.33 s for 1000 PHEVs.

17.4.2 Accelerated PSO findings

Fig. 17.7 illustrates APSO technique's nature of convergence for each scenario. Since the algorithm was updated to operate for a maximum of 100 iterations, the fitness value converges and remains constant after 10th iterations. As a result of this early convergence, the fitness function can become trapped in local minima. This can be prevented by increasing the swarm size and thereby increasing the computational time. Fig. 17.8 illustrates simulation findings of APSO for determining the optimal fitness value for a fitness function for 50, 100, 300, 500, and 1000 PHEVs, respectively. To get a grasp of the supremacy and efficiency of the corresponding algorithm, each scenario is repeated 50 times.

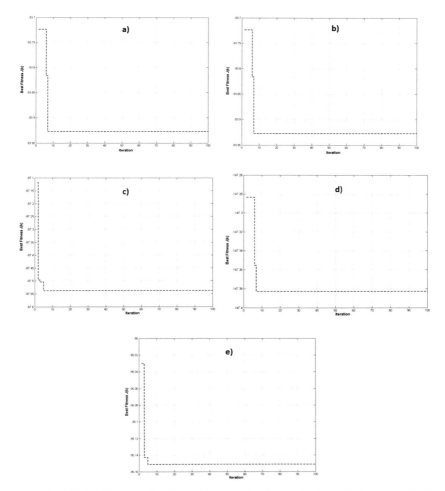

FIGURE 17.7 Plot between iteration and fitness value for accelerated particle swarm optimization considering five scenarios: (A) 50 PHEVs, (B) 100 PHEVs, (C) 300 PHEVs, (D) 500 PHEVs, and (E) 1000 PHEVs.

The maximum, minimum, and average fitness values for APSO under five different scenarios such as 50, 100, 300, 500, and 1000 PHEVs are demonstrated using Table 17.7. However, the maximum best fitness values remain between 450 and 700, whereas the minimum best value of fitness lies between 0.5 and 10. The average computing time required for the APSO approach is shown in Table 17.8. The total computational time for 50 PHEVs is 1.69 s, although it increases to 2.09 s for 1000 PHEVs. As compared to the standard form of PSO, APSO needs additional parameter modification. However, APSO can solve a greater amount of PHEVs in less time than PSO.

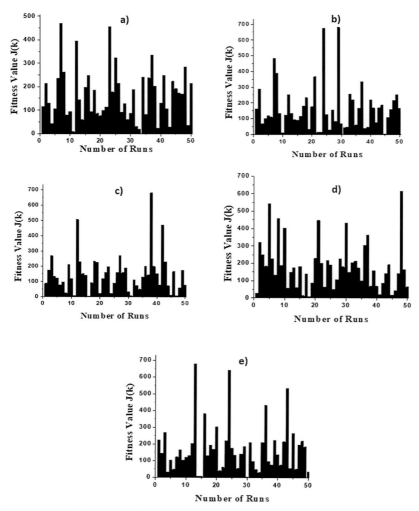

FIGURE 17.8 Plot between fitness value and total runs for accelerated particle swarm optimization assuming five scenarios: (A) 50 PHEVs, (B) 100 PHEVs, (C) 300 PHEVs, (D) 500 PHEVs, and (E) 1000 PHEVs.

17.4.3 Gravitational search algorithm findings

Graphical illustration of the convergence trend regarding GSA was provided in Fig. 17.9. The convergence of best fitness function was achieved after 35 number of iterations for 50 and 100 PHEVs. However, 500 and 1000 PHEVs, a premature convergence is observed. Fig. 17.10 highlights the findings of GSA

Swarm-based intelligent strategies Chapter | 17 **363**

TABLE 17.7 Maximum, minimum, and average fitness value evaluation for accelerated particle swarm optimization.

Fitness function	50 PHEVs	100 PHEVs	300 PHEVs	500 PHEVs	1000 PHEVs
Maximum best fitness	469.75	679.71	679.55	615.83	678.92
Minimum best fitness	7.65	3.46	3.54	5.96	0.99
Average best fitness	162.70	168.23	147.42	184.15	171.16

TABLE 17.8 Average accelerated particle swarm optimization computational time.

No. of PHEVs	Computing time (s)
50	1.69
100	1.71
300	1.76
500	1.83
1000	2.09

simulation for determining the optimal fitness value for the objective function for 50, 100, 300, 500, and 1000 PHEVs, respectively. Each scenario is run for 50 times to assess the typical performance and to get a grasp of the algorithm's dominance and efficiency.

The maximum and minimum best fitness values for 50 PHEVs were determined to be 781.13 and 0.22, respectively, whereas the overall value of optimal fitness is 158.83. For 100 PHEVs, the maximum and minimum ideal fitness values were 872.65 and 1.01, respectively, and the average value of best fitness was 182.31. The maximum and minimal best fitness values for 300 PHEVs were 743.13 and 2.33, respectively, and the average value was 172.43. The maximum and minimum values of the fitness for 500 PHEVs were 836.27 and 0.98, respectively. However, the average fitness is 152.36. The maximum, minimum, and average ideal fitness values for 1000 PHEVs were 968.77, 7.27, and 161.52, respectively. Table 17.9 summarizes the findings and it can be inferred that average optimal fitness shows identical trend for five assumed

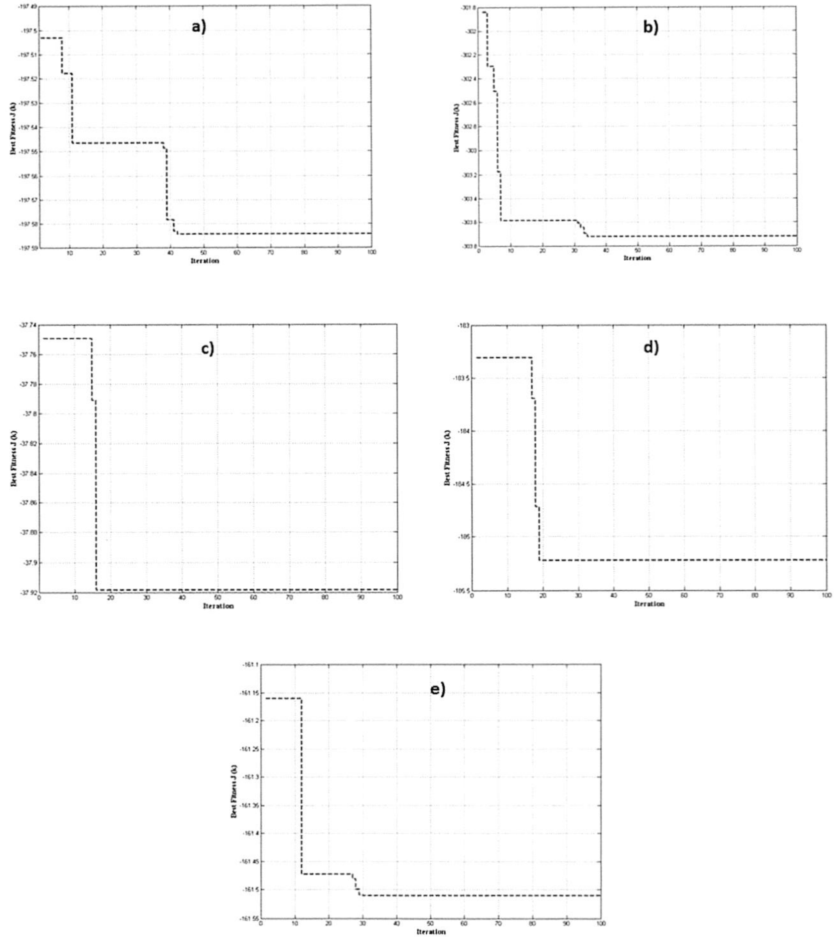

364 Advances of Artificial Intelligence in a Green Energy Environment

FIGURE 17.9 Graphical plots between iteration and fitness value for gravitational search algorithm under five different scenarios: (A) 50 PHEVs; (B) 100 PHEVs; (C) 300 PHEVs; (D) 500 PHEVs; (E) 1000 PHEVs.

scenarios. The average computing time required for GSA approach was shown in Table 17.10. The total computing time for 50 PHEVs is 2.72 s, although it increases to 2.092 s for 1000 PHEVs.

17.4.4 Hybrid PSO and GSA (PSOGSA) findings

Fig. 17.11 depicts the convergence behavioral patterns of the PSOGSA strategy (iteration vs. fitness value) for each scenario. After that, the formula was adapted to operate for 100 iterations, with the fitness value approaching

Swarm-based intelligent strategies Chapter | 17 365

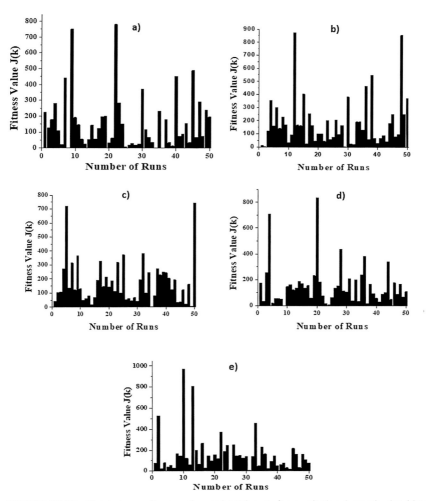

FIGURE 17.10 Plot between fitness value and total runs for gravitational search algorithm subjected to five scenarios such as (A) 50 PHEVs; (B) 100 PHEVs; (C) 300 PHEVs; (D) 500 PHEVs; (E) 1000 PHEVs.

after five number of iterations before being determined again. As a consequence, there is an early convergence that could trap the objective functions in local minima. By increasing swarm size and therefore increasing the computational time, this can be avoided. Fig. 17.12 depicts the PSOGSA simulation results for determining the optimal value of the fitness function for 50, 100, 300, 500, and 1000 PHEVs. Each scenario ran for 50 times to get a sense of the results as well as the supremacy and efficiency of the PSOGSA algorithm. The simulation behavior of the PSOGSA was evaluated based on

366 Advances of Artificial Intelligence in a Green Energy Environment

TABLE 17.9 Assessment of fitness of gravitational search algorithm.

Fitness function	50 PHEVs	100 PHEVs	300 PHEVs	500 PHEVs	1000 PHEVs
Maximum best fitness	781.13	872.65	743.13	836.27	968.77
Average best fitness	158.83	182.31	172.43	152.36	161.52
Minimum best fitness	0.22	1.01	2.33	0.98	7.27

TABLE 17.10 Average computation time for gravitational search algorithm.

No. of PHEVs	Time required for computation (s)
50	2.72
100	4.44
300	11.28
500	18.17
1000	36.28

the aforementioned numerical data. Since it is a population-based optimization strategy with a nonlinear fitness function, the fitness values change with each iterations [28–30]. The average ideal fitness, on the other hand, lies in the 400 to 950 band, while the minimum optimum fitness lies in the 0.1–8 range. Table 17.11 outlines findings and it is possible to infer that the overall best fitness pattern for five distinct situations follows a common trend. The average computing time required for the PSOGSA approach is shown in Table 17.12. The total time for computation for 50 PHEVs is 4.23 s, although it increases to 72.41 s for 1000 PHEVs.

17.4.5 Comparative analysis

A comparative study of all algorithms must be stressed in addition to the findings for each algorithm. To ensure the proper comparison, all four techniques were performed on the same computer with the same iterations (100)

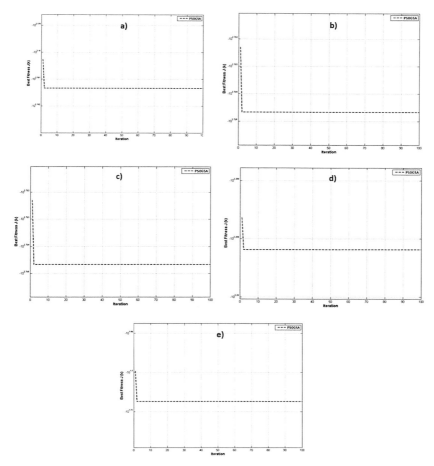

FIGURE 17.11 Plots between iteration and fitness value for PSOGSA considering five scenarios: (A) 50 PHEVs; (B) 100 PHEVs; (C) 300 PHEVs; (D) 500 PHEVs; and (E) 1000 PHEVs.

and a total of 50 individual runs [25,31]. The following are comparisons of applied swarm intelligence—based techniques.

17.4.5.1 Convergence analysis

Of the four techniques, the convergence of the PSO and APSO techniques follows a similar pattern, while the GSA technique requires a higher number of iterations to be converged. At this stage, the hybrid approach PSOGSA often needs the least iteration to reach convergence. As a result, there is an early convergence that may allow the fitness function to become trapped in local minima. This can be prevented by maximizing the swarm size and therefore by

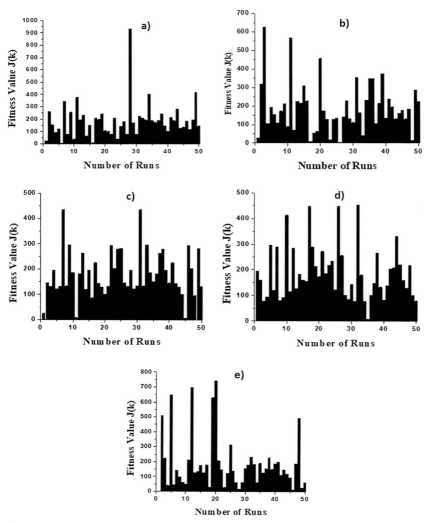

FIGURE 17.12 Fitness value and total runs for PSOGSA assuming five scenarios: (A) 50 PHEVs, (B) 100 PHEVs; (C) 300 PHEVs; (D) 500 PHEVs; and (E) 1000 PHEVs.

increasing the computational time. Table 17.13 indicates the iteration number required for each algorithm to converge in five considered situations.

17.4.5.2 Fitness value

Fig. 17.13 depicts a comparison of fitness values for all methods. PSOGSA yields the highest fitness values in all five scenarios (50, 100, 300, 500, and 1000 PHEVs). When opposed to the PSO method, single techniques such as GSA and APSO provide considerably better results.

Swarm-based intelligent strategies Chapter | 17 **369**

TABLE 17.11 Evaluation of fitness for PSOGSA.

Fitness function	50 PHEVs	100 PHEVs	300 PHEVs	500 PHEVs	1000 PHEVs
Maximum best fitness	931.03	625.82	434.16	454.04	740.40
Average best fitness	184.36	188.67	181.03	186.70	185.16
Minimum best fitness	3.39	3.71	7.43	7.23	0.17

TABLE 17.12 PSOGSA's average time for computation.

PHEV numbers	Time for computing (sec)
50	4.23
100	7.90
300	22.33
500	36.82
1000	72.41

TABLE 17.13 Analysis of convergence.

PHEV numbers	Number of convergent iterations			
	PSO	APSO	GSA	PSOGSA
50	<10	<10	35	<5
100	<10	<10	35	<5
300	<10	<10	15	<5
500	<10	<10	40	<5
1000	<10	<10	5	<5

17.4.5.3 Time of computation

Fig. 17.14 depicts a summary of the average time of computation of the four optimization strategies when five cases are considered. PSOGSA, the hybrid method, takes longer to complete 100 iterations, while both PSO and APSO techniques do well in terms of computing time.

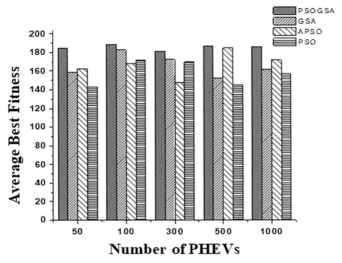

FIGURE 17.13 The average best fitness and no. of plug-in hybrid power vehicles (PHEVs).

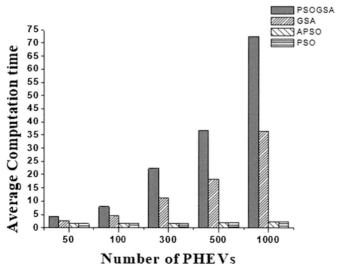

FIGURE 17.14 The average time of computation versus no. of plug-in hybrid power vehicles (PHEVs).

17.4.5.4 Robustness

Table 17.14 displays the standard deviations for four methods where all five PHEV scenarios are considered. PSOGSA has the highest robustness among the optimization methods, with a lower standard deviation than PSO, APSO, and GSA. The average optimal value of fitness of the PSOGSA system

Swarm-based intelligent strategies Chapter | 17 **371**

TABLE 17.14 Standard deviation for the implemented methods.

Optimization techniques	Average optimal fitness					Standard deviation
	50 PHEVs	100 PHEVs	300 PHEVs	500 PHEVs	1000 PHEVs	
PSO	142.84	171.10	169.31	144.80	156.80	11.83
APSO	162.70	168.23	147.42	184.15	171.16	11.95
GSA	158.83	182.31	172.43	152.36	161.52	10.62
PSOGSA	184.36	188.67	181.03	186.70	185.16	2.55

remains almost constant in all five scenarios, demonstrating the technique's robustness.

17.5 Conclusions

The aim of the current study was to optimize SoC in terms of time of charging and current SoC, and in this perspective, accomplished implementations addressed the associated objectives of optimizing the SoC of a PHEV utilizing swarm-based intelligence approach and a generic performance assessment of the techniques in terms of optimal solution and computational time is conducted. To accomplish this, the analysis used swarm-based strategies such as PSO, GSA, APSO, and a hybrid PSOGSA. Since the conclusions from this analysis are critical for an innovative actual optimization problem such as recharging a plug-in hybrid electric car, appropriate focus was placed on various types of case studies and the implemented swarm intelligence study works in this manner. The study's findings established the relative merits and demerits of various swarm intelligence strategies for the charging optimization problem and also allowed readers develop a better understanding of what it's like to use single or hybrid configured swarm intelligence applications for the same optimization problem structure. Moreover, this research demonstrated that swarm intelligence and the methods it employs are capable of solving the optimization challenge associated with charging PHEVs, which are a critical component of green technology.

17.5.1 Future research direction

The results of this research pave the way for future research. It is possible to see any potential research directions in this sense. Several of the most noteworthy can be discussed as follows: although approaches focused on swarm intelligence have shown their ability to navigate vast search domain, they are relatively inept at solution fine-tuning. This vulnerability is typically prevented by using a local search system on the population's individuals [32,33]. Future research may be conducted by combining PSO or GSA (an algorithm based on swarm intelligence) with the local search process. Another area of potential study is demand side management. Demand side management systems can be integrated into the current intelligent energy management system (iEMS) to eliminate voltage sags and blackouts and to optimize financial benefits [34].

References

[1] M. Caramanis, J.M. Foster, Management of electric vehicle charging to mitigate renewable generation intermittency and distribution network congestion, in: Decision and Control, 2009 Held Jointly with the 2009 28th Chinese Control Conference. CDC/CCC 2009. Proceedings of the 48th IEEE Conference on, 2009, pp. 4717–4722.

Swarm-based intelligent strategies **Chapter | 17   373**

[2] S.F. Tie, C.W. Tan, A review of energy sources and energy management system in electric vehicles, Renew. Sustain. Energy Rev. 20 (2013) 82−102.

[3] H. Lund, W. Kempton, Integration of renewable energy into the transport and electricity sectors through V2G, Energy Policy 36 (2008) 3578−3587.

[4] A.R. Hota, M. Juvvanapudi, P. Bajpai, Issues and solution approaches in PHEV integration to the smart grid, Renew. Sustain. Energy Rev. 30 (2014) 217−229.

[5] J. Soares, T. Sousa, H. Morais, Z. Vale, B. Canizes, A. Silva, Application specific modified particle swarm optimization for energy resource scheduling considering vehicle-to-grid, Appl. Soft Comput. 13 (2013).

[6] W. Su, M.-Y. Chow, Computational intelligence-based energy management for a large-scale PHEV/PEV enabled municipal parking deck, Appl. Energy 96 (2012) 171−182.

[7] W. Su, M.-Y. Chow, Performance evaluation of a PHEV parking station using particle swarm optimization, in: Power and Energy Society General Meeting, 2011, IEEE, 2011, pp. 1−6.

[8] J. Chiasson, B. Vairamohan, Estimating the state of charge of a battery, in: Control Systems Technology, IEEE Transactions on, vol. 13, 2005, pp. 465−470.

[9] S. Piller, M. Perrin, A. Jossen, Methods for state-of-charge determination and their applications, J. Power Sources 96 (2001) 113−120.

[10] M. Dorigo, Ant Colony Optimization and Swarm Intelligence: 5th International Workshop, ANTS 2006, Brussels, Belgium, September 4−7, 2006, Proceedings, vol. 4150, Springer, 2006.

[11] D. Karaboga, B. Basturk, A powerful and efficient algorithm for numerical function optimization: artificial bee colony (ABC) algorithm, J. Global Optim. 39 (2007) 459−471.

[12] E. Rashedi, H. Nezamabadi-Pour, S. Saryazdi, GSA: a gravitational search algorithm, Inf. Sci. 179 (2009) 2232−2248.

[13] J. Yang, L. He, S. Fu, An improved PSO-based charging strategy of electric vehicles in electrical distribution grid, Appl. Energy 128 (2014) 82−92.

[14] C. Jia-zhao, Z. Yu-xiang, L. Yin-sheng, A unified frame of swarm intelligence optimization algorithm, in: Knowledge Discovery and Data Mining, Springer, 2012, pp. 745−751.

[15] R.C. Eberhart, S. Yuhui, Particle swarm optimization: developments, applications and resources, in: Evolutionary Computation, 2001. Proceedings of the 2001 Congress on, vol. 1, 2001, pp. 81−86.

[16] X. Wu, B. Cao, J. Wen, Y. Bian, Particle swarm optimization for plug-in hybrid electric vehicle control strategy parameter, in: Vehicle Power and Propulsion Conference, 2008. VPPC'08, IEEE, 2008, pp. 1−5.

[17] J. Soares, H. Morais, Z. Vale, Particle Swarm Optimization based approaches to vehicle-to-grid scheduling, in: Power and Energy Society General Meeting, 2012, IEEE, 2012, pp. 1−8.

[18] X.-S. Yang, S. Deb, S. Fong, Accelerated particle swarm optimization and support vector machine for business optimization and applications, in: Networked Digital Technologies, Springer, 2011, pp. 53−66.

[19] S. Talatahari, E. Khalili, S. Alavizadeh, Accelerated particle swarm for optimum design of frame structures, in: Mathematical Problems in Engineering, vol. 2013, 2013.

[20] A.H. Gandomi, G.J. Yun, X.-S. Yang, S. Talatahari, Chaos-enhanced accelerated particle swarm optimization, Commun. Nonlinear Sci. Numer. Simulat. 18 (2013) 327−340.

[21] A.E. Fergany, Accelerated particle swarm optimization-based approach to the optimal design of substation grounding grid, Przegląd Elektrotechniczny 89 (2013) 30−34.

374 Advances of Artificial Intelligence in a Green Energy Environment

[22] S. Mirjalili, S.Z.M. Hashim, A new hybrid PSOGSA algorithm for function optimization, in: Computer and Information Application (ICCIA), 2010 International Conference on, 2010, pp. 374–377.

[23] C. Purcaru, R.-E. Precup, D. Iercan, L.-O. Fedorovici, R.-C. David, Hybrid PSO-GSA robot path planning algorithm in static environments with danger zones, in: System Theory, Control and Computing (ICSTCC), 2013 17th International Conference, 2013, pp. 434–439.

[24] T. Ganesan, I. Elamvazuthi, K.Z. Ku Shaari, P. Vasant, Swarm intelligence and gravitational search algorithm for multi-objective optimization of synthesis gas production, Appl. Energy 103 (2013) 368–374.

[25] M. Črepinšek, S.-H. Liu, M. Mernik, Replication and comparison of computational experiments in applied evolutionary computing: common pitfalls and guidelines to avoid them, Appl. Soft Comput. 19 (2014) 161–170.

[26] M. Yilmaz, P.T. Krein, Review of charging power levels and infrastructure for plug-in electric and hybrid vehicles, in: Electric Vehicle Conference (IEVC), 2012 IEEE International, 2012, pp. 1–8.

[27] E. Akhavan-Rezai, M. Shaaban, E. El-Saadany, A. Zidan, Uncoordinated charging impacts of electric vehicles on electric distribution grids: normal and fast charging comparison, in: Power and Energy Society General Meeting, 2012, IEEE, 2012, pp. 1–7.

[28] F. Jiménez, G. Sánchez, P. Vasant, A multi-objective evolutionary approach for fuzzy optimization in production planning, J. Intell. Fuzzy Syst. 25 (2013) 441–455.

[29] T. Ganesan, P. Vasant, I. Elamvazuthi, Hopfield neural networks approach for design optimization of hybrid power systems with multiple renewable energy sources in a fuzzy environment, J. Intell. Fuzzy Syst. 26 (2014) 2143–2154.

[30] L. Le Dinh, D. Vo Ngoc, P. Vasant, Artificial bee colony algorithm for solving optimal power flow problem, Sci. World J. 2013 (2013).

[31] J. Derrac, S. García, D. Molina, F. Herrera, A practical tutorial on the use of nonparametric statistical tests as a methodology for comparing evolutionary and swarm intelligence algorithms, Swarm Evol. Comput. 1 (2011) 3–18.

[32] A.C. Martínez-Estudillo, C. Hervás-Martínez, F.J. Martínez-Estudillo, N. García-Pedrajas, Hybridization of evolutionary algorithms and local search by means of a clustering method, in: Systems, Man, and Cybernetics, Part B: Cybernetics, IEEE Transactions on, vol. 36, 2005, pp. 534–545.

[33] K. Harada, K. Ikeda, S. Kobayashi, Hybridization of genetic algorithm and local search in multiobjective function optimization: recommendation of GA then LS, in: Proceedings of the 8th Annual Conference on Genetic and Evolutionary Computation, 2006, pp. 667–674.

[34] C. Gerkensmeyer, M.C. Kintner-Meyer, J.G. DeSteese, Technical Challenges of Plug-In Hybrid Electric Vehicles and Impacts to the US Power System: Distribution System Analysis, Pacific Northwest National Laboratory, 2010.

Index

'*Note*: Page numbers followed by "f" indicate figures and "t" indicate tables.'

A

Accelerated particle swarm optimization (APSO), 350–352, 363t
  advantages of, 352
  computational time, 363t
  disadvantages of, 352
  fitness value *vs.* total runs for, 360, 362f
  flow diagram of, 352, 353f
  general values for, 352
  iteration *vs.* fitness value for, 360, 361f
  parameter settings of, 352, 352t
Adaptive neuro-fuzzy inference system (ANFIS), 205
Additive manufacturing (AM), 331
Aerobic–anaerobic–aerobic treatment, 24–25
Aerobic–anaerobic digestion, 14
Aggregate production planning (APP)
  components, 194, 195f
  decision-making process, 198
  employee cognition, 198
  employee satisfaction, 198
  fuzzy programming, 197
  holistic approach, 196
  human factors (HFs), 194
  production conditions, 193
  reliability, 196
  robust optimization, 197
  trade-offs, 194
  uncertainty, 197
  workforce plan, 194
  workforce satisfaction, 196, 198–199
AHPD. *See* Anaerobic high pressure digestion (AHPD)
Air quality index (AQI), 205–208
Akaike information criterion, 153
Allergy-free goods, 42–43
AM. *See* Additive manufacturing (AM)
Anaerobic bioconversion, organic matter, 4f
  advantages, 28
  aerobic–anaerobic digestion, 14
  agro-industrial complex, 1–3

biogas, 1–2
biohydrogen production, 14–15
energy model, 27f
  agro-industrial complex, 25, 26f
  energy efficiency, 25
  types, 26
hydrolysis and acidogenesis process, 14–15
immobilization, methane-forming microorganisms, 18–19
internal combustion engines (ICEs), 15
manure treatment, 2
methane digestion process
  biochemical methods, 4, 11
  biogas production, 9
  biological and thermochemical gasification, 9
  biostimulants, 6–7
  chemical reagents, 6–7
  decomposition process, 8
  digester fermentation chamber, 5–6
  ecosystems, 7–8
  energy market, 8
  liquid organic waste, 10
  manure runoff, 3, 10
  mechanical effect, 4–5, 11
  methane-forming microorganisms, 7
  microbiological methods, 4, 11
  mixing methods, 6
  organic biomass, 10–11
  organic raw materials, 8–9
  renewable energy sources, 8
  temperature, 5
  thermal and electromagnetic treatment, 4, 11
  volatile solids, 10
methanogenesis, 1
organic fertilizers, 1–2
pretreatment, biological and physicochemical methods
  biogas production rate, 16
  efficiency of, 15, 17f

375

376 Index

Anaerobic bioconversion, organic matter
  (Continued)
  hydraulic retention time (HRT), 16
  solid phase, 17
  on processed substrate
    anaerobic high pressure digestion
      (AHPD), 19−20
    biogas recirculation, 20
    carbon dioxide concentration, 19
    conductive materials, 22
    direct interspecies electron transfer
      (DIET), 21−22
    electrophysical impact, 20
    heat exchanger, 22
    microwave radiation, 22
  sewage treatment, 2
  temperature
    mesophilic sediment stabilization
      methods, 11
    specific heat consumption, 12−13, 14f
    thermal energy, 12
    thermophilic treatment regime, 11
  thermal and electric energy, 25, 27
  thermal transformers
    aerobic−anaerobic−aerobic treatment,
      24−25
    aerobic pretreatment apparatus (AAP),
      24
    bioreactor−pump−heat exchanger-
      condenser−bioreactor, 23
    direct heat recovery, 22−23
    residual gas evolution, 23
  thermochemical gasification, 17−18
  two-stage fermentation process, 14
Anaerobic conversion, green biomass
  agro-industrial waste, 67
  biocatalyst, 69−71
  biomass pretreatment, 73
  butyric and acetic acids, 81
  chemical nature components, 67
  Clostridium acetobutylicum, 69, 71
  flame ionization detector, 72
  gas chromatography, 71−72
  lignocellulose, 68
  methane, 67
  methanogenesis, 68
  organic substances, 72
  oxidative depolymerization, 68−69, 79
  substrate, 71
  volatile fatty acids (VFAs), 68, 70f
    Clostridium acetobutylicum, 74−76, 77f
    glycerin, 76, 78f

  physicochemical treatment, 73, 75f
  pretreated sawdust, acidogenic
    fermentation, 74
  straw biological acidogenesis products,
    73−74, 75f
  substrates conversion, 79
Anaerobic high pressure digestion (AHPD),
  19−20
Anaerobic processing, organic waste
  activated sludge cells, 136−137
  anaerobic waste treatment, 131
  anthropogenic pollution, 129
  bioconversion waste, 138
  biomass, 137
  Cambi Process, 135−136
  fermentation−hydrolysis−fermentation,
    135−136
  fossil fuels, 129
  gas formation process
    anaerobic reactors, 133
    biochemical process, 133
    gastrointestinal tract, 134
    methane fermentation, 134
    syntrophic bacteria, 133−134
  hydrolysis−fermentation, 135−136
  hydrolysis process, 131−133
  methanogenesis, 130−131
  organic polysubstrates, 137
  preliminary aerobic treatment, 137
  preprocessing
    sediments, pretreatment, 135
    solid phase, 135
  solid retention time (SRT), 138
  vortex layer apparatus (VLA)
    acoustic radiation, 144
    biogas plants, 145−147
    capital expenditures, 143
    economic effect, 142
    electrochemical process, 144−145
    elements, 139, 139f
    Faraday's law, 144−145
    ferromagnetic particles, 139, 140f
    laboratory version, 139, 140f
    magnetic fields, 142
    magnetostriction, 144
    material processing, 141
    operation, 139
    physicochemical properties, 144
    process activation units, 142−143
    wastewater treatment, 138
Analysis of variance (ANOVA), 251−252
Analytic hierarchy process (AHP), 314

Index **377**

ANFIS. *See* Adaptive neuro-fuzzy inference system (ANFIS)
Ant Colony Optimization (ACO) algorithm, 348
APP. *See* Aggregate production planning (APP)
APSO. *See* Accelerated particle swarm optimization (APSO)
AQI. *See* Air quality index (AQI)
Artificial Bee Colony (ABC) algorithm, 348
Artificial intelligence (AI)
    adaptive neuro-fuzzy inference system (ANFIS), 205
    air quality index (AQI), 205−208
    basemap, 208, 209f
    content-based image retrieval system, 206
    convolutional neural network (CNN), 204, 216−217, 217f, 219f
    deep autoencoder techniques, 207
    digital elevation models (DEMs), 203−204, 212, 212f
    drainage, 214, 214f
    forecasting model, 205−206
    fossil fuels, 173
    fuzzy expert system (FES), 204−205
    fuzzy logic controller (FLC), 173−174
    gated recurrent unit (GRU) models, 207−208
    geographical information system (GIS), 203
    geospatial technology (GST), 204
    Landslide Susceptibility Zones (LSZ), 203−204, 214−216, 215f
    land use and land cover, 210−211, 211f
    lineaments, 213, 213f
    long short-term memory (LSTM), 206
    load side converter control, 181
    overfitting and underfitting, 206
    MATLAB/Simulink program, 173−174
    maximum power point tracking (MPPT), 173
        adaptive neuro-fuzzy inference system (ANFIS) architecture, 180, 181f
        algorithms, 177−178, 178f
        backpropagation algorithm, 180
        DC−DC boost converter, 179
        fuzzy logic controller (FLC), 178, 178f, 179t, 180f
        trained networks, 179−180, 181t
    performance analysis
        accuracy measure, 218, 219f
        landslide feature analysis, 218, 218f
        root mean square error, 218, 219f
    performance evaluation, 206
    proportional integral (PI) controller, 173−174
    recurrent neural networks (RNNs), 204, 218, 220f
    renewable resources (RERs), 173
    scene classification, 204−205
    slope, 212, 213f
        gradient, 212
    thematic maps preparation
        database creation, 209f−210f, 210
        lithology, 210, 211f
    train−test mixtures, 205
    wavelet transformation, 206
    wind energy harvesting system (WEHS)
        permanent magnet synchronous generator (PMSG), 174, 175f, 176−177, 177t
        turbine model, 174−175, 176t
    wind speed, 182, 182f, 185f
        active and reactive power, 187, 187f
        adaptive neuro-fuzzy inference system (ANFIS) approach, 187, 189t
        battery power, 183, 184f
        boost converter power, 183, 184f
        characteristics, 187, 188t
        generator speed, 182−183, 183f, 186f
        mechanical power, 182−183, 184f, 186f
        peak phase voltage, 187, 187f
        power coefficient, 182−183, 182f, 185f
        speed controller, 190, 190t
        tip speed ratio, 182−183, 183f, 186f
        trial and error, 185
    workflow, 208

B

Bayesian information criterion, 153
Biocatalyst
    acid-producing activity, 70−73
    characteristics, 69
    Clostridium acetobutylicum, 74−76, 77f
    green biomass, 73−74, 75f
        glycerin, 76, 78f
        temperature and time, 79
    Mettler Toledo potentiometer, 70
    nontarget metabolic products, 71
Bioenergy supply chains, 257−263, 262t
Biohydrogen production, 14−15
Bioreactor−pump−heat exchanger-condenser−bioreactor, 23
Bonferroni correction, 169

378 Index

C

Cambi Process, 135−136
Centre for Research on the Epidemiology of
    Disasters (CRED), 43
Closed-loop supply chains (CLSC), 301−302
    quantitative models, 302
Computational intelligence, 245−246
Content-based image retrieval system, 206
Convolutional neural network (CNN), 204,
    216−217, 217f, 219f
Current density, 63

D

DAIM. *See* Deterministic artificial
    intelligence model (DAIM)
Deep autoencoder techniques, 207
Deep learning, 279−280
Demand-supply model, 90, 90f
Deterministic artificial intelligence model
    (DAIM), 280
Digital elevation models (DEMs), 203−204,
    212, 212f
Direct heat recovery, 22−23
Direct interspecies electron transfer (DIET),
    21−22
Disasters impact assessment
    allergy-free goods, 42−43
    allergy medication, 42
    balanced approach, 47
    biological hazards, 39−41
    climate change, 44
    cooperation system, 47
    damage of, 37−38, 38f
    decision-making support systems, 36
    direct losses/restoration costs, 38
    economic losses, 38
    educational institutions, 45
    forecasting technologies, 46
    immunotherapy, 43
    infrastructure construction technologies, 46
    international organizations, 46
    man-made disasters, 36, 42
    mortality rates, 35
    natural disasters, 36−37, 39
    population growth, 48, 48f
    prevention/mitigation measures, 51
        algorithm selection, 54f
        loading and displaying input data, 54f
        losses and risks, 54f
        natural disasters, 52t
        normalized values, 53t
        risk mitigation, 52t

protective facilities, 46
risk management, 41, 43−44
risk reduction, 47
Sendai Framework, 44−45
socioeconomic approach, 36
    decision-making process, 49
    efficiency of, 50
    natural disasters, 50
    risk dynamics, 48−49
    risk mitigation, 49, 49f, 52t
sustainable development, 36
warning systems, 45
3D printing, 331

E

EAACI. *See* European Academy of Allergy
    and Clinical Immunology (EAACI)
Economic load/emission dispatch (ELED),
    251−257
    applications in, 253−254, 255t,
        258t
    IEEE30 and IEEE57 systems, 252
    Markov decision process (MDP), 256−257
    nondominated sorting genetic algorithm
        (NSGA-II), 253
    swarm-based metaheuristics, 253
Electricity supply industry (ESI)
    consumption, 84−85, 86f
    economic growth, 83−84
        allocative efficiency, 87
        demand-supply model, 90, 90f
        electric grid system, 92
        electricity demand, 94
        electricity markets, 87
        load curve, 92, 92f
        load profile, 90−92, 91f
        production efficiency, 87
        renewable energy technologies,
            93−94
        retail competition, 88, 89f
        single buyer, 88, 88f
        sustainable development, 93
        wholesale competition, 88, 89f
    electricity rates/prices, 94−97
    electric power, 85−86
        characteristics, 86
    energy consumption, 83−84
    International Energy Agency (IEA),
        84−85
    life cycle, 94, 95t−96t
    life cycle assessment (LCA)
        decarbonization, 101

Index   **379**

electricity and heat production systems, 98

electric mix, 100

environmental assessment tools, 99–100

environmental loads, 100

greenhouse gas (GHG) emissions, 98, 99f

ISO 14040 and ISO 14044 standards, 100

phases, 101–102

sustainability, 98–99

life cycle cost (LCC)

circular economy, 105–106

costs and benefits, 103–104

decision-making/evaluation process, 102

economic efficiency, 105, 105f

elements, 104

ISO 14040 standard, 102, 102f

phases, 104, 104f

sensitivity analysis, 105

stages, 107, 107f

marketing stages, 97

model, 94–97, 97t

primary energy, 84–85, 85f

quality of life, 84–85

Energy model, 27f

agro-industrial complex, 25, 26f

energy efficiency, 25

types, 26

Energy systems optimization models, 283–293

applications, 284, 285t

energy and construction, 290

energy computation, 289–290

flexible manufacturing systems, 285

forecast models and, 287–289

management energy, 291–293

mathematical programming, 289–290

Monte Carlo method, 284

Enzymatic bacteria, 130

ESI. *See* Electricity supply industry (ESI)

Ethyl alcohol

Clostridium acetobutylicum, 74–76, 77f

glycerin, 76, 78f

physicochemical treatment, 73, 75f

pretreated sawdust, acidogenic fermentation, 74

straw biological acidogenesis products, 73–74, 75f

substrates conversion, 79

Euclidean distance, 333

European Academy of Allergy and Clinical Immunology (EAACI), 42

European Electric Power Research Institute (EPRI), 347

Explanatory variables, 151

F

Fuel economy, 63

Fuzzy expert system (FES), 204–205

Fuzzy logic controller (FLC), 178, 178f, 179t, 180f

Fuzzy programming, 197

G

GAR. *See* Global Assessment Report (GAR)

Gas chromatography, 71–72

Gas formation process

anaerobic reactors, 133

biochemical process, 133

gastrointestinal tract, 134

methane fermentation, 134

syntrophic bacteria, 133–134

Gated recurrent unit (GRU) models, 207–208

Geographical information system (GIS), 203, 259

Geospatial technology (GST), 204

Global Assessment Report (GAR), 43

Global warming, 279

Gravitational search algorithm (GSA), 348, 352–354

advantages of, 354

computation time for, 363–364, 366t

disadvantages of, 354

Euclidian distance, 352–354

fitness assessment, 363–364, 366t

fitness value *vs.* total runs for, 362–363, 365f

flow diagram of, 354, 355f

gravitational force, 352–354

iteration *vs.* fitness value for, 362–363, 364f

parameter settings of, 355t

random number, 354

Greenhouse gases (GHG), 256–257

Grey relational analysis (GRA), 247–248

GRU models. *See* Gated recurrent unit (GRU) models

GSA. *See* Gravitational search algorithm (GSA)

GST. *See* Geospatial technology (GST)

380 Index

H

Hannan—Quinn information criterion, 153
HO. *See* Hybrid optimization (HO)
HOMER. *See* Hybrid optimization of
 multiple energy resources (HOMER)
HRT. *See* Hydraulic retention time (HRT)
Hybrid optimization (HO), 279—281
 energy systems optimization, 283—293
 stochastic programming, 281—283
  general model, 282
  software for, 283
Hybrid optimization of multiple energy
  resources (HOMER), 113, 114f, 119
 battery and inverter, 125
 calculation time, 125, 125f
 data components, 124
 load profile, 119—121, 121f
 solar radiation data input, 122, 122f
 wind speed data input, 122, 123f
Hybrid PSOGSA algorithm, 354—356
 advantages of, 356
 computation time for, 369t
 disadvantages of, 356
 fitness evaluation, 364—366, 369t
 fitness value and total runs for, 364—366,
  368f
 flowchart of, 356, 357f
 iteration *vs.* fitness value for, 364—366,
  367f
 parameter settings, 357t
Hybrid renewable energy systems
 components, 111—112
 design and optimization, 111—112
 energy-related data, 111—112
 genetic algorithms, 112—113
 multiobjective optimization tools, 120t. *See
  also* Multiobjective/multilevel
  optimization
  hybrid optimization of multiple energy
   resources (HOMER), 113, 114f, 119
  HYBRID-2 software, 116—119, 121
  iHOGA, 114—115, 119—121, 121f
  residential energy modeling problem,
   119
 solar radiation data input, 122, 122f
 state of charge (SOC), 112—113
 subalgorithm, 112—113
HYBRID-2 software
 algorithm flowchart, 116—117, 118f
 calculation time, 125, 127f
 data components, 124, 124f
 economic simulation, 116—117

load dispatch strategy, 118—119
load dumping, 116—117
load profile, 121
peak shaving, 118
performance simulation, 116—117
renewable battery, 118
renewable genset, 118
solar radiation data input, 122
wind speed data input, 122, 123f
Hydraulic retention time (HRT), 16
Hydrolysis process, 130
 anaerobic lipid hydrolysis, 132—133
 Clostridium thermocellum, 131—132
 monomeric aromatic lignin derivatives,
  133
 Thermobacteroides proteolyticus, 132

I

iHOGA
 battery life cycle, 124
 calculation time, 125, 126f
 data components, 124
 genetic algorithms, 114—115
 load dispatch strategies
  algorithm of, 116, 117f
  cycle charging, 115
  demand, 115
  genetic algorithm, 116
  NASA RET screen, 116
  options for, 115
 load profile, 119—121, 121f
 primary algorithm, 115
 renewable energy system, 119, 121f
 secondary algorithm, 115
 solar radiation data input, 122, 122f
Interior search algorithm (ISA), 251—252
Internal combustion engines (ICEs), 15
International Atomic Energy Agency (IAEA),
 43
International Energy Agency (IEA), 84—85
Internet of Things (IoT), 247

K

Karush-Kuhn-Tucker approach, 246, 250, 261

L

Landslide Susceptibility Zones (LSZ),
 203—204
Least squares (LS) technique, 157
Life cycle analysis (LCA), 314—320

Index **381**

Life cycle assessment (LCA)
  decarbonization, 101
  electricity and heat production systems, 98
  electric mix, 100
  environmental assessment tools, 99−100
  environmental loads, 100
  greenhouse gas (GHG) emissions, 98, 99f
  ISO 14040 and ISO 14044 standards, 100
  phases, 101−102
  sustainability, 98−99
Life cycle cost (LCC)
  circular economy, 105−106
  costs and benefits, 103−104
  decision-making/evaluation process, 102
  economic efficiency, 105, 105f
  elements, 104
  ISO 14040 standard, 102, 102f
  phases, 104, 104f
  sensitivity analysis, 105
  stages, 107, 107f
Linear programming (LP), 225−226,
    237−238
  adjusted optimal solution, 239
  optimal solution, 238, 239f−240f
Linear regression model, 152
LINGO optimization modeling software, 283
Long short-term memory (LSTM), 206

M

Machine learning (ML) methods, 151, 169,
    203, 279−280
Man-made disasters, 36, 42
Markov decision process (MDP), 256−257
Maximum power point tracking (MPPT), 173
  adaptive neuro-fuzzy inference system
    (ANFIS) architecture, 180, 181f
  algorithms, 177−178, 178f
  backpropagation algorithm, 180
  DC−DC boost converter, 179
  fuzzy logic controller (FLC), 178, 178f,
    179t, 180f
  trained networks, 179−180, 181t
Methane biocenosis, 131
Methane digestion process
  biochemical methods, 4, 11
  biogas production, 9
  biological and thermochemical
    gasification, 9
  biostimulants, 6−7
  chemical reagents, 6−7
  decomposition process, 8
  digester fermentation chamber, 5−6

ecosystems, 7−8
energy market, 8
liquid organic waste, 10
manure runoff, 3, 10
mechanical effect, 4−5, 11
methane-forming microorganisms, 7
microbiological methods, 4, 11
mixing methods, 6
organic biomass, 10−11
organic raw materials, 8−9
renewable energy sources, 8
temperature, 5
thermal and electromagnetic treatment, 4,
    11
volatile solids, 10
Microbial electrolysis cells (MEC), 20
MILP. *See* Mixed-integer linear programming
    (MILP)
Mixed-integer linear programming (MILP),
    246, 249
  four-layer biogas supply chain, 259
Monte Carlo method, 284
MPPT. *See* Maximum power point tracking
    (MPPT)
MSLiP, 283
Multiobjective/multilevel optimization
  adaptive cuckoo search, 247
  bilevel (BL) programming, 250−251
  bioenergy supply chains, 257−263
  biofuel supply chains, 257−263
  in civil engineering, 250
  control strategy, 246
  cuckoo search (CS), 247
  definition of, 245−246
  differential evolution (DE), 246
  economic load/emission dispatch (ELED),
    251−257
    applications in, 253−254, 255t, 258t
    IEEE30 and IEEE57 systems, 252
    Markov decision process (MDP),
      256−257
    nondominated sorting genetic algorithm
      (NSGA-II), 253
    swarm-based metaheuristics, 253
  genetic algorithms (GAs), 249
  industrial problems, 248
  Karush-Kuhn-Tucker conditions, 246
  in material and structural engineering,
    247−248
  mixed-integer linear programming (MILP),
    246, 249
  simulated annealing (SA), 250

382 Index

Multiobjective/multilevel optimization
(*Continued*)
  sustainable capacity planning and
      optimization, 263–267
    applications in, 265, 266t, 268t
    multiobjective particle swarm
        optimization (MOPSO), 264
    optimal energy mix, 264
  in sustainable energy economics, 250–251,
      251f
Multiobjective particle swarm optimization
    (MOPSO), 264
Multiple-criteria decision-making (MCDM),
    314

N

National Renewable Energy Laboratory
    (NREL), 113, 116–117
Natural disasters, 36–37, 39, 50
Network power absorption probability, 292
Nondominated sorting genetic algorithm II
    (NSGA-II), 247, 253
Nonparametric functions, 151–152

O

Occam's razor principle, 153
  multiple testing, 168–169
  optimal valid partitioning (OVP), 163
  parametric regression models
    conditional probability, 156
    least squares (LS) technique, 157
    null hypothesis, 154
    optimal function, 153–154
    permutation test technique, 154–157
    P-values, 154, 157, 159
    statistical significance, 153
  partitions families, 163–164, 164f
  piecewise linear models
    correlation coefficient, 160
    data set, 162
    parathyroid hormone (PTH), 159, 161f
    P-value, 162–163
    quadratic programming task, 159–160
    regression coefficients, 159–160, 162
    verification, 163, 163t
    vitamin D level, 159–160
    vitD concentration, 163
  statistical significance
    equal probability, 165
    explanatory variable, 165–166
    null hypothesis, 166–167

permutation test, 165
P-value, 167
vascular endothelial growth factor
    (VEGF), 167, 168f
Off-grid energy systems, 113
Optimal valid partitioning (OVP),
    163
Optimized designing spherical void
    structures, 3D domains, 331–332
  mathematical model, 333–334
    adjusted phi-function for, 334
    with balancing conditions, 334–336
    computational complexity, 334
    inequalities, 334–335
    material density, 334
    nonintersected polyhedrons, 334
    nonlinear programming problem, 335
    objective function, 333
    phi-functions approach, 333
  numerical experiments, 336–341
    radii and spherical objects, 338t–340t,
        342t–343t
    spherical object packing, 340f–341f,
        343f
  problem formulation, 332–333
    connected polyhedral convex
        components, 332
    containment conditions, 333
    disconnected domain, 332
    distance conditions, 333
    Euclidean distance, 333
    lower and upper bounds, 332
OVP. *See* Optimal valid partitioning (OVP)
Oxidative depolymerization, 68–69

P

Particle swarm optimization (PSO), 348–350
  advantages of, 350
  average time, computation for, 358–360,
      360t
  disadvantages of, 350
  fitness evaluation for, 358–360, 360t
  fitness value *vs.* number of runs for,
      356–358, 359f
  inertial weight function, 350
  iteration *vs.* fitness value, 356–358, 358f
  parameter settings for, 350t
  stages of, 350, 351f
Permutation test, 152
PHEVs. *See* Plug-in hybrid power vehicles
    (PHEVs)
Phi-functions modeling approach, 332–333

Index **383**

Piecewise linear models
  correlation coefficient, 160
  data set, 162
  parathyroid hormone (PTH), 159, 161f
  P-value, 162–163
  quadratic programming task, 159–160
  regression coefficients, 159–160, 162
  verification, 163, 163t
  vitamin D level, 159–160
  vitD concentration, 163
Plug-in hybrid power vehicles (PHEVs)
  effectiveness, 348
  implementation of, 347
  multiple battery loads, 347
  state of charge (SoC), 348
Proportional integral (PI) controller, 173–174
PSO. *See* Particle swarm optimization (PSO)

Q

Quesada Grossman (QG) algorithm, 289

R

Recuperative heat exchanger (RHE), 22–23
Recurrent neural networks (RNNs), 204, 218, 220f
Renewable resources (RERs), 173
Robust optimization, 281

S

SAMPL, 283
SCM. *See* Supply chain management (SCM)
Self-awareness, 280
Sequential quadratic programming (SQP), 249
Solar panels, Zimbabwe
  dummy point
    creation, 232, 233f
    identification, 232–234, 233f
    subtour, 232, 232f
  dummy schools, 232
    subtour eliminators, 234
  electricity supply, 223
  equipment allocation, 224
  traveling salesman problem (TSP), 223
    algorithm, 234–235
    assignment-based branch and bound methods, 225
    dartboard design, 241
    dummy node, 236
    dummy points identification, 236, 236f–237f

dynamic programming–based methods, 225
exhaustive enumeration method, 225
heuristics, 225
hospital layout, 241
linear programming (LP), 225–226, 237–239, 239f–240f
model, 224, 225f
network features, 226–231, 226f, 228f–230f
numerical illustration, 235–239, 235f
production process, 241
scheduling, 241–242
typewriter keyboard, 241
wiring problem, 240–241
Solid retention time (SRT), 138
Spanning tree, 230
  minimum spanning tree, 230
    algorithm, 231
Stackelberg game theory, 263
Statistical significance
  equal probability, 165
  explanatory variable, 165–166
  null hypothesis, 166–167
  permutation test, 165
  P-value, 167
  vascular endothelial growth factor (VEGF), 167, 168f
Stochastic program model, 281–283
  software for, 283
  structure of, 282
STOPGEN program, 283
Supply chain management (SCM), 301, 303
  frequencies of, 311, 312t
  organizational analysis, 311
Supply Chain Operations Reference (SCOR) model, 306
Supply chain (SC) sustainability, 302
Sustainable supply chain management (SSCM), 302–304
  category identification, 306
  descriptive analysis, 306
    journal distribution, 308, 309t–310t
    time distribution, reference papers, 308, 308f
  material collection, 305–307, 307f
    database sources, 305
    exclusion criteria, 306
    inclusion criteria, 306
    research questions, 305
    search terms, 305
  material evaluation, 306–307

384 Index

Sustainable supply chain management (SSCM) (*Continued*)
  methodology of, 305–307
  modeling dimension, 311–314
    analytic hierarchy process (AHP), 314
    frequencies of, 311, 313t
    input–output analysis (IOA), 314
    linear programming (LP), 314
    multiple-criteria decision-making (MCDM), 314
    sample paper, 314, 315t–319t
  research gaps and, 320–321
  sustainability dimension, 314–320, 320f
Sustainable supply chain network (SSCN), 302
Swarm-based intelligent strategies, 347–348
  accelerated particle swarm optimization (APSO), 350–352
    advantages of, 352
    disadvantages of, 352
    flow diagram of, 352, 353f
    general values for, 352
    parameter settings of, 352, 352t
  comparative analysis, 366–372
    convergence analysis, 367–368, 369t
    fitness value, 368, 370f
    robustness, 370–372, 371t
    time of computation, 369, 370f
  gravitational search algorithm (GSA), 352–354
    advantages of, 354
    disadvantages of, 354
    Euclidian distance, 352–354
    flow diagram of, 354, 355f
    gravitational force, 352–354
    parameter settings of, 355t
    random number, 354
  hybrid PSOGSA algorithm, 354–356
  particle swarm optimization (PSO), 349–350
    advantages of, 350
    disadvantages of, 350
    inertial weight function, 350
    parameter settings for, 350t
    stages of, 350, 351f
  problem formulation, 348–349
  study objectives, 348

T

Taguchi method, 247–248
Thermal power plants, 57
Traveling salesman problem (TSP), 223
  algorithm, 234–235
  assignment-based branch and bound methods, 225
  dartboard design, 241
  dummy node, 236
  dummy points identification, 236, 236f–237f
  dynamic programming–based methods, 225
  exhaustive enumeration method, 225
  heuristics, 225
  hospital layout, 241
  linear programming (LP), 225–226, 237–238
    adjusted optimal solution, 239
    optimal solution, 238, 239f–240f
  model, 224, 225f
  network features, 226–231
    justification, 227–229, 229f
    leaf, 230
    matrix coefficients, 227
    minimum spanning tree, 230
    minimum spanning tree algorithm, 231
    optimal tour, 231
    school/node routes, 226, 226f
    spanning tree, 230
    standard constraints, 226–227
    subtours, 227, 228f
    tour, 230
    tree, 230–231, 230f
    unimodular matrices, 227
  numerical illustration, 235–239, 235f
  production process, 241
  scheduling, 241–242
  typewriter keyboard, 241
  wiring problem, 240–241
Triple bottom line (TBL), 302–303

U

Uninterruptible power supply system (UPS)
  current density, 63
  daily power generation, 61
  efficiency of, 59, 59f
  electricity consumption, 58, 62, 62f
  energy distribution unit, 60
  fuel economy, 63
  maximum current, 61
  maximum power, 61
  power limiter, 60
  power redundancy, 61
  vs. power supply system, 62, 62f

Index **385**

schedule of
city with industrial enterprises, 57, 57f
electrical appliances, 60, 60f
fuel consumption, 58, 59f
rural town with private residential sector, 58, 58f
thermal power plants, 57
waste steam, 58

V

Vascular endothelial growth factor (VEGF), 167, 168f
Volatile fatty acids (VFAs), 68, 70f
Clostridium acetobutylicum, 74−76, 77f
glycerin, 76, 78f
physicochemical treatment, 73, 75f
pretreated sawdust, acidogenic fermentation, 74
straw biological acidogenesis products, 73−74, 75f
substrates conversion, 79
Vortex layer apparatus (VLA)
acoustic radiation, 144
biogas plants, 145−147
activated sludge, 146
efficiency, 146
electromagnetic devices, 145
electromagnetic mill, 146
energy consumption, 145
ferromagnetic elements, 145
liquid organic substrates, 146
specific energy consumption, 147
ultrahigh-frequency waves, 147
volatile solids, 147, 148f
capital expenditures, 143
economic effect, 142
electrochemical process, 144−145

elements, 139, 139f
Faraday's law, 144−145
ferromagnetic particles, 139, 140f
laboratory version, 139, 140f
magnetic fields, 142
magnetostriction, 144
material processing, 141
operation, 139
physicochemical properties, 144
process activation units, 142−143
wastewater treatment, 138

W

Wilcoxon signed rank tests, 251−252
Wind energy harvesting system (WEHS)
permanent magnet synchronous generator (PMSG), 174, 175f, 176−177, 177t
turbine model, 174−175, 176t
Wind speed, 182, 182f, 185f
active and reactive power, 187, 187f
adaptive neuro-fuzzy inference system (ANFIS) approach, 187, 189t
battery power, 183, 184f
boost converter power, 183, 184f
characteristics, 187, 188t
generator speed, 182−183, 183f, 186f
mechanical power, 182−183, 184f, 186f
peak phase voltage, 187, 187f
power coefficient, 182−183, 182f, 185f
speed controller, 190, 190t
tip speed ratio, 182−183, 183f, 186f
trial and error, 185
World Allergy Organization (WAO), 44
World Commission on Environment and Development (WCED), 301

Printed in the United States
by Baker & Taylor Publisher Services